Proceedings of the First International Conference on

Unsolved Problems of Noise

in Physics, Biology, Electronic Technology and Information Technology

Proceedings of the First International Conference on

Unsolved Problems of Noise

in Physics, Biology, Electronic Technology and Information Technology

Editors

Ch R Doering
University of Michigan, USA

LB Kiss
University of Uppsala, Sweden

MF Shlesinger
Office of Naval Research, USA

Sponsors: Office of Naval Research, European Office;
Office of Naval Research - USA;
U.S. Air Force European Office of Aerospace Research and Development;
Institute of Physics (IOP, UK);
IEEE and EDS (USA);
European Laboratory for Electronic Noise (ELEN, EU);
National Scientific Research Foundation (OTKA, Hungary);
OMFB (Hungary).

World Scientific
Singapore • New Jersey • London • Hong Kong

Published by

World Scientific Publishing Co. Pte. Ltd.
5 Toh Tuck Link, Singapore 596224
USA office: 27 Warren Street, Suite 401-402, Hackensack, NJ 07601
UK office: 57 Shelton Street, Covent Garden, London WC2H 9HE

British Library Cataloguing-in-Publication Data
A catalogue record for this book is available from the British Library.

UNSOLVED PROBLEMS OF NOISE IN PHYSICS, BIOLOGY, ELECTRONIC TECHNOLOGY AND INFORMATION TECHNOLOGY
Proceedings of the First International Conference (UPoN '96)

ISBN-13 978-981-02-3199-6
ISBN-10 981-02-3199-7

First International Conference on:

UNSOLVED PROBLEMS OF NOISE (UPoN'96)

INTERNATIONAL SCIENTIFIC COMMITTEE:

INTERNATIONAL ORGANISING COMMITTEE:

LOCAL ORGANISING COMMITTEE:
(mostly from Dept of Experimental Phys., JATE Univ., Hungary)

CONTENTS

NOISE IN NONLINEAR DYNAMICAL SYSTEMS

NOISE IN BIOLOGICAL SYSTEMS

NOISE IN HIGH-Tc SUPERCONDUCTORS

PROLOG

OPENING WORDS BY THE CHAIRMAN OF UPoN'96

LASZLO B. KISS
Uppsala University, Angstrom Lab, Material Science Department, Box 534, Uppsala, S-75121 Sweden

Why did we organise the conference on Unsolved Problems of Noise, **UPoN'96** ? Nowadays, the scientific community lacks forums where unsolved problems can be published. I have had the opportunity to visit many institutes of physics and technology in different countries and have often found that the most interesting experimental results have been laying on the table in the lab instead of being published. The scientists who discovered these results were very pleased to speak about and discuss these findings, however they did not want to publish them until a rational correspondence with some accepted theoretical models or views could be achieved. Without that, they felt it was hopeless (sometimes, even dangerous) to get the paper published. I have realised that they are right. Scientific refereeing practices and professional opinions do not really support the publication of strange or very unusual issues. Progressive ways of approaching scientific problems are often thought to be suspect or erroneous. Many recent discoveries heralded as revolutionary issues are relatively slight improvements of technical issues. Less significant scientific results obtain Nobel prices than that happened in the 1920-1960's years. Michelson and Morely of today would have had serious problems with the publication of their unbelievable experimental data. Therefore, the Einstein of today would not have had the chance to read these results and to develop the theory of relativity. Even if he could develop it somehow, his revolutionary theory would certainly be classified as a fiction and not a serious scientific result.

Inspired by these thoughts, the idea of the ***UPoN*** conference series is to provide a forum for discussing important unsolved problems of noise. This idea has been supported by many of today's scientific leaders as the international committees of the conference and the list of important sponsors indicate. We are very grateful also to the Department of Experimental Physics, JATE University, Szeged, Hungary for the essential help with the local organisation of UPoN'96. We hope that this meeting will be a kind of break-through in the objective and style of noise conferences as the logo of UPoN'96 shows.

The key-sentence of the UPoN'96 conference is a citation from the "Speech on the Mountain" of the Bible: "Blessed are they which do hunger and thirst after righteousness: for they shall be filled" (Mt 5:6). For the UPoN'96 Organisers the word "righteousness" has had multiple meanings:

1. The idea of UPoN conference is to point out the contradictions or imperfections of present scientific knowledge instead of hiding them.

2. Selection of invited speakers: anyone in the scientific community had the opportunity to submit a proposal for an invited talk.

3. Refereeing most of the proposals had been done in a double-blind way. Though, in some cases the reader could conjecture who was the author of the paper, in most of the cases, the Referee was uncertain about the identification of the author. Therefore, the name and affiliation of the authors had practically not influenced the refereeing process, dislike at some other scientific papers and conferences. Invited talk proposals (reviews of old unsolved problems) had been refereed by 3 members of the International Scientific Committee, while regular talk proposals (new unsolved problems) had been refereed by 2 members.

4. After the conference, the Editors have done a last thorough screening of the articles. This is partly the reason why some of the presented talks is not published in this book.

Before turning to the unsolved problems of noise presented in this book, I would like to take your attention to the probably most important unsolved problem of nowadays science, because this problem is also relevant to noise phenomena. We, noise researchers all know that a finite duration time-record of a noise does not have any meaning. Either, the record duration has to be infinite or we should have finite records from an infinite number of analogous physical systems to use the terms of nowadays science: distribution functions, noise spectra, etc. We are interested in the general properties, and we are unable to accurately predict details of single events. Scientists of classical physics believed that this is due to our limited knowledge and by solving the set of equations describing a complex system, we, in principle, could be able to predict even a single sequence of events accurately. Quantum physics has proved that it is not the case: the nature is fundamentally "noisy": the single event is basically unpredictable; the wavefuntion provides only a probability distribution for the elementary processes. The strange fact is that the experimentalist can record the whole sequence of elementary events, however, nowadays science is unable to deal with the accurate prediction of these elementary events, it is even unable outline a mathematical framework for that. It is like, we are studying an infinite *"Book"* entitled *"Nature"*, in which, the different letters correspond to different occurring values of physical quantities. The only thing what a scientist can do is to make/predict a statistics of the occurrence of different *"letters"* appearing in the book, including some cross-correlation effects. Mathematically, all these statistics are inherently based on some infinite ensembles of elements. On the contrary of that, in reality, experimentalists have always been facing ensembles with finite number of elements, a number determined by resolution of instruments and limited time for measurement. The experimentalist has to smooth, filter, *trash*, the vast amount of the particular information from the sequence of data and to compare the few remaining statistical results with theories for *inherently* infinite statistics. We cannot predict the particular sequence of letters on the next page what we have not seen yet in *"Book of Nature"*, we can talk only about *probabilities* of future events. It is a part of the basic problem: we do *NOT* understand that *"text"* in the book. Turning to noise processes, for example, a thermal noise of a given impedance is a timefunction which contains practically infinite amount of information, it looks like an encoded wideband signal with an encoding which provides the maximal efficiency of information transfer. However, nowadays science is unable to decode that information, it is even unable to formulate a proper initiative question about that, it can understand only just a few basic characteristics of this process, like amplitude distribution functions, power density spectra. The Copenhagen foundation of quantum physics, due to Heisenberg's uncertainty principles, even forbids us to ask such questions about the particular information carried by a given sequence of elementary events. Will we ever have the ability to decode what the thermal noise of a particular resistor is talking to us? If yes, that science will be *Another Science*, not the science which has been evolving since Gallilei/Newton. I believe that the way of science and mathematical approach which we have basically been following since Newton, which is a way based on finding generally valid laws and formulate equations of balance for *conserving physical quantities*, strictly does lead to the quantum physics of our century, which is a crown of this evolution on one hand, on the other hand, it is a sort of deadlock, especially, when we really want to read that *"Book of Nature"*.

I hope, by studying this book, the Reader will find interesting scientific problems to think about.

OPENING WORDS FROM THE INTERNATIONAL ORGANISING COMMITTEE

DEREK ABBOTT
University of Adelaide, EEE Department, Australia

What a marvellous idea to have a conference on "unsolved problems". Conferences have been traditionally about what we have come to know -now we have one about what we do not know. It is refreshing to have a new paradigm, do things differently and openly discuss vexing and unsolved problems.

Of all the disciplines, noise is probably the most challenging and only stubbornly releases its secrets to us. So let us all use this conference to have some fun and do things a little differently. Normally an audience cross-examines the speaker. Perhaps in a conference on unsolved problems this should be the other way around - the speaker now poses questions to the audience. Question time will not be a one-way affair, but will be a time of lively interaction or brainstorming. Let us remove our academic inhibitions and let our childish fascination of nature speak out. The British physicist William Henry Bragg (1862-1942) once said ``The important thing in science is not so much to obtain new facts as to discover new ways of thinking about them."

It is an honour to be addressing so many distinguished researchers in the field of noise. Noise is a very deep subject and it takes a special breed of person to explore it. Because of the all-pervading nature of noise, the researcher is forced to become an expert in many areas of physics and electrical engineering. The exploration of noise can involve studying device physics through to electronics, biology, earthquakes, thunderstorms, mechanics, quantum theory and gravitation to name a few. Noise appears everywhere.

It is appropriate that the birth of this conference is in 1996, which is the 70th anniversary of Johnson's famous experiment. Both Johnson and Nyquist were Swedish immigrants, working in the USA. They both attended Yale and completed their PhDs at the same time. Johnson shared his results with Nyquist, who only took a month to come up with his now well-known derivation. Although this result was foreshadowed by Einstein in 1905 and de Haas-Lorentz (the first woman in noise theory) in 1912, Johnson and Nyquist turned 1926 into a major milestone for electrical noise. Also exactly a hundred years earlier, in 1826, Brown carried out his microscope studies and opened the door for research into fluctuation phenomena.

The more we have studied fluctuations over the last 170 years, the more we have learned and yet we have discovered there is even more we do not know. Noise certainly keeps us humble.

The mathematician Jacob Bronowski (1908-1974) once wrote, "There is no absolute knowledge..... all information is imperfect. We have to treat it with humility."

The famous Danish philosopher Soren Kierkegaard (1813-1855) said "The paradoxical character of truth is in its objective uncertainty." It is fascinating that this was written 100 years before there was such a beast as a Copenhagenist! Danish philosophy found its way into Danish physics. In contrast, the Dutch philosopher Benedict de Spinoza (1632-1677) wrote, "Nothing in nature is random.... A thing only appears random through the incompleteness of our knowledge." He said that some 300 years before 'hidden variable' theories were debated.

FUNDAMENTAL PROBLEMS OF RANDOM PROCESSES

A BRIEF HISTORY OF RANDOM PROCESSES

MICHAEL F. SHLESINGER
Office of Naval Research
Physical Sciences Division
800 N. Quincy St.
Arlington VA 22217-5660, USA

This volume deals with unsolved problems of noise. Before we can understand and appreciate noise in natural and man-made systems it may be instructive to begin by exploring the beginnings of the field of mathematics known as random processes. In addition to being instructive it is also reassuring to see that previous generations also faced their own set of unsolved problems. Several of those who introduced new ideas became very well known for their efforts. A general set of references related to historical matters is given at the end.

Man's conscience deliberation with random processes is as least as old as the ancient game of throwing the bones . The bones in question, called astrogali, are the heel bones of animals including dogs and sheep. These bones fit nicely in the hand, and are the forerunners of dice. In fact, the French expression "jeu d'hasard" (games of hazard or chance) comes from the Arabic for dice "al-zar" . In reality, gambling (from dice to actuarial tables to risk-benefit analysis) has always been in the forefront of expanding the frontiers of probability theory. The bones in question have only four faces upon which they come to rest. The four-sided spinning dreidel is the most similar, in this respect, to throwing of the bones. In ancient times a mathematics of probability did not arise. Perhaps, this was because the early dice did not have equally likely outcomes and the occurrence of a rare event (landing on a less probable face) was ascribed to luck rather than sparking the development of the idea of the permanence of statistical ratios. The Greeks played with a roll of five dice while the Roman played with a roll of four. Each combination of outcomes was given a name, much the same as today we call a role of a pair of ones "snake eyes". For the Romans four ones were called "the dogs". They labelled the four faces as 4 (upper bone), 3 (opposite side of the upper bone), 1 (flat lateral side), and 6 (its opposite side). The Venus throw of 1,3,4,6 which represented beauty had one of everything. On the other hand, a mathematics of probability also did not arise from the ancient Chinese, *I Ching*, which has 64 equally likely outcomes. The *I Ching* was used as an oracle and perhaps has too many outcomes to provide the insight to inaugurate a new field of mathematics.

Despite some early writings on dice games, probability got its recognized start with a correspondence between Pascal and Fermat in 1654 which cleared up a misconception between the probability to win and the expected winnings. The problem was posed to Pascal by a French gambler, Chevalier de Mere. He found it was more likely to get a 6 with four throws of a die, than to get a pair of sixes with 24 throws of a pair of dice. Specifically, the expected number of sixes in four throws of a die is 4/6, which equals the expected number of pairs of sixes in 24 throws of a pair of dice. So why should the empirical probabilities differ? The probability to throw at least one six in four throws is $1 - (5/6)^4 \cong 0.5177$, and the probability to get at least one pair of six in 24 throws of a pair of dice is $1 - (35/36)^{24} \cong 0.4914$. Equal expectations do not imply equal probabilities. Having easily settled this matter, Pascal and Fermat wrestled with other problems such as, how to split up the ante in games of chance which are not completed. With more than two players considered, these problems would tax many modern practitioners. For example, if A, B, and C play a game of chance and A needs 1 more win, and B and C both need two more wins, then if the game is not continued show that the ante should be split as A:B:C is to 17:5:5.

Early confusion also existed between the difference between combinations and permutations. A gambler brought the following problem to Galileo's attention. Throw 3 dice and you are more likely to get their sum adding to 10, before a role gives a sum of 9. Yet, there are six combinations which give 10 [6,2,2], [5,2,3], [4,2,4], [6,3,1], [4,3,3], and [5,4,1], and six combinations which give rise to a 9. So why is getting a 10 first a better bet? As Galileo correctly pointed out there are 6X6X6=216 possible permutations of three throws of a die. For example the [6,2,2] combination can occur 3 ways as [6,2,2], [2,6,2], and [2,2,6]. Adding up all the permutations gives the probability to get a 6 as 27/216, while the probability to get a 9 is only 25/216. In any event, these early concepts of probability were perplexing to early gambling practitioners.

Lest we moderns feel too superior, let me relate a recent question of probability which caused contestants in a TV game show to scream and many times make wrong decisions. The game had three curtains with a prize hidden behind only one of the curtains. The contestant picks a curtain, say #3, which is not yet opened. Of the two remaining curtains, at least one (and maybe both) harboured no prize. An unpicked curtain with no prize behind it is opened, say #2. The contestant is then asked if he would like to switch the curtain he picked, #3, with the other unopened curtain #1. In effect, the contestant is being offered curtains #1 and #2, and these have a probability of 2/3 of leading to the prize. Curtain #3 has probability of 1/3 of being right, which is one-half of the probability if the switch is made. So the switch should always be made. If you still need convincing, try the game yourself.

The Pascal-Fermat letters stimulated Huygens to write a treatise on probability theory. This, in turn, peaked the interest of Jacob Bernoulli who turned his attention to questions of combinations and permutations, and games of chance.

His works were published posthumously, in 1713, under the title *Ars Conjectandi* (Art of Conjecture). Jacob's brother John, also an eminent mathematician disapproved of probability and effectively held up publication of his brother's book. When it eventually appeared, John called it "a monster which bears my brothers name". What was in this monster book? Jacob introduced the Bernoulli process where a player has a probability p to win in each trial of a game, and probability $q = 1 - p$ to lose. We would today equate Jacob Bernoulli's analysis to a random walk on a lattice by keeping track of a player's status along a one dimensional axis, with success equated to a jump of one unit to the right, and failure a jump of one unit the left. The probability, $p_n(k)$, of k successes in n trials is given by $[n!/((n-k)!k!]\, p^k q^{n-k}$. In the limit of large n, and for p=q, DeMoivre, showed, in 1756, that the probability to get $2h$ more successes than failures (or *vice versa*) in n Bernoulli trials is $(2pn)^{-1/2} \exp(-h^2/2n)$. This can be seen by expanding the factorials in $p_n(k)$ with $k = n/2 \pm h$. This probability function is today called a Gaussian because Gauss showed, in 1809, that DeMoivre's result holds in a more general situation of being the probability limit distribution when adding up identically distributed random variables with finite second moments. An intimate connection with these types of random walk processes and Gaussian probabilities was latter made to the phenomenon of diffusion called Brownian motion. For example, in Bernoulli's game (process) consider two players each starting with R/2 coins and with equal probability to win or lose a coin in each trial. The mean number of trials, $\langle N(R) \rangle$, before one player loses his fortune can be calculated to be $R(R+1)/6$, and for large R, $\langle N(R) \rangle \approx R^2$. In Brownian motion, the mean square distance a particle moves after N steps satisfies $\langle R^2(N) \rangle \approx N$. In effect, Einstein moved the averaging to the space variable, instead of Bernoulli's time variable N. Bernoulli calculated what we would call today a mean first passage time, i.e., the number of games played until one player's stake is exhausted. His was the "*first*" first passage time calculation.

Poisson, in the 1830's, found another possible limit for the Bernoulli process; let $p \to 0$ and $n \to$ infinity such that $np \to 1$, a constant. Then $p_n(k) = n(n-1) \cdots (n-k+1)\, p^k (1-p)^{n-k}/k! \approx (np)^k e^{-np}\, e^{pk}/k! \bullet 1^k e^{-1}/k!$. This is called a Poisson probability distribution. Poisson wrote a book on probability, but devoted to social problems, such as the probability of a jury reaching the correct verdict. The first reported use of the Poisson distribution came from Germany, at the late date of 1894, where it was used to model the number of soldiers kicked to death by their horse. As this was a rare event the Poisson distribution provided a good fit.

Over in London, Bernoulli's contemporary De Moivre, wrote in 1718, the first edition of his famous treatise on probability, *The Doctrine of Chances*, which provided explicit odds for popular card games, and more importantly introduced new concepts, including the use of generating functions, Sterling's formula for log (n!), the derivation of the formula $(\cos J + i \sin J)^n = \cos nJ + i \sin nJ$, and proving that the Gaussian is the many trial limit of a Bernoulli process (in the 1756 3rd edition). A problem which could be solved by generating functions can be found in Crystal's book on Algebra " By French Law an illegitimate child receives one third of the proportion of the inheritance that he would have received had he been legitimate. If there be p legitimate and n illegitimate children show that the portion of the inheritance due to a legitimate child is $1/p - n/[3p(p+1)] + n(n-1)/[3^2 p(p+1)(p+2)] - \cdots + n!/[3^n p(p+1) \cdots (p+n)]$. Even back then it was not unimportant to have an application for your work. Being a French Huguenot, DeMoivre was never able to secure an academic position in England. He managed by setting himself up in Slaughter's Coffee House where, for a fee, advice could be garnered. De Moivre's other probability book, *Annuities Upon Lives*, inaugurated actuarial science. De Moivre also predicted his own death. At about the age of 80 he noticed that he was sleeping more. He then kept a log of his hours and extrapolated that in seven years he would be sleeping 24 hours per day. His prediction was correct.

The Gaussian (also called the normal) is the limit probability distribution for a *sum* of identically distributed random variables with finite second moments. This was shown by Gauss in 1809, and is called today the Central Limit theorem of probability. Actually, this result was obtained in 1808 by Robert Adrain, an American of Irish descent. Adrain published in an obscure American journal called The Analyst, which only lasted for one volume. Nevertheless, Adrain was one of the several scientific and mathematical leaders who flourished during the Jefferson presidency. Adrain proved his result, (that adding up many identical random variable with finite moments, leads to the Gaussian probability distribution) in the context of the study of errors in measurement. The Gaussian is also called the Law of Errors. Galton, the English statistician was so taken with the law that he wrote "I know of scarcely anything so apt to impress the imagination as the wonderful form of cosmic order expressed by the Law of Frequency of Error. The law would have been personified by the Greeks and deified, if they had know of it. ··· The larger the mob and the greater the apparent anarchy, the more perfect is its sway." This last sentence refers to the better convergence to a Gaussian the more terms in the sum of random variables.

It was not until 1879 that MacAlister introduced the distribution for a *product* of random variables called the log normal distribution, as it is a Gaussian in the logarithm of the variable. Schockley, the inventor of the transistor, studied the productivity of scientists by investigating the number of papers published by individuals. He found this to obey the lognormal distribution. He argued that several factors were necessary to successfully publish, including getting ideas, having technical expertise, the ability to fight with referees, etc. He further argued that a product of these random factors were related to one's productivity. And a product of random variables would be governed by a lognormal distribution. It was said that he gave pay raises to his staff based on the logarithm of the number of papers published, and noted this with a Marxist joke, --- Let each be rewarded according to the logarithm of his abilities.

Laplace in his 1815 treatise *Theorie analytique des probabilities* continued the tradition of introducing new tools in applied mathematics through probability. Before Laplace, most work in probability focused on the discrete where all the possible outcomes of a trial could be enumerated. Card and dice games are good examples of this genre. Laplace brought the integral and continuum mathematics to the fore of probability theory with questions like; if an urn contained an infinite number of black and white tickets and one chooses A tickets of which p turn out to be white and q are black, then what is the probability that if an additional B tickets are chosen that m are white and n are black. Laplace's solution was an integral from 0 to 1 with an integrand of the form $x^{p+m}(1-x)^{q+n}$ led the way for calculating a wide variety of integrals, many of which are well known today.

Probability was still fringe area of mathematics with a somewhat disreputable reputation, until the work of Kolmogorov and Khintchine, in the 1930's, place probability on a firm mathematical basis. This foundation was necessary because probability ran into trouble early on. In the early 1700's, Nicolas Bernoulli discovered a troubling game of chance. The game was to throw a coin until a head resulted. With probability 1/2 this occurs on the first throw, and a win of one coin would result. If N tails proceeded a head throw, this event of probability $1/2^{N+1}$ would yield a win of 2^N coins. One wins an order of magnitude more, but with an order of magnitude less probability. The expected winning from a round of this game would be the probability of winning times the win amount, summed over all possible outcomes. Well, this would be $1/2 \times 1 + 1/4 \times 2 + \cdots + 1/2^{N+1} \times 2^N + \cdots =$ infinity. This game is called the St. Petersburg Paradox, because Daniel Bernoulli wrote about it in the commentary of the St. Petersburg Academy in 1722. The paradox arises when trying to determine a fair ante to play this game. The banker wants the player to ante up an infinite number of coins because this is the bank's expected loss. The player only wins a single coin half the time, two coins one-quarter of the time, etc. Thus a player would not like to place a large ante because most of the times his winnings are small. Today we would recognize this as an example which produces a normalized probability distribution, but one which has an infinite first moment. The paradox is trying to find a characteristic size from a distribution which does not possess one. In a distribution with an infinite mean, sampling produces values of all size, but in such a proportion that no single size is dominant or characteristic.

Statistics could also led to controversy, then and now. In the 1700's small pox was epidemic. It was discovered in Turkey that a primitive inoculation made from cowpox could create immunity to smallpox. Jenner, in England, later correlated that milkmaids in England were pretty (they didn't get smallpox and its facial scars) because they contracted cowpox which protected them from smallpox. Jenner developed a safe vaccine. In any event, earlier on inoculations began and statistics were collected. Seven percent of an untreated population died from smallpox. For an inoculated population .5% died from the inoculation. What to do, to inoculate or not? Daniel Bernoulli lobbied vigorously in favour of inoculation, while D'Alembert took the opposite case. D'Alembert said if left alone you have a 93% chance to live. Get an inoculation and 1 out 200 would die from the inoculation. The probability was great that the death would be from the 93% that would have lived. Either choice has its risks for the individual, but for the population at large inoculation was clearly optimal. These argument reached Benjamin Franklin in America, where at his urging George Washington had the Continential Army inoculated at the time of the Revolutionary War. The army escaped decimation by smallpox while some town were scourged. If the army was taken ill the outcome of the war might have been entirely different.

Another paradox was discovered by Betrand. Take a circle of unit radius and randomly draw a chord. What is the probability that the chord length is larger than the side of an equilateral triangle inscribed inside the circle? Draw a random chord. Now draw an inscribed equilateral triangle with a vertex at one end of the chord. If the chord intersects the side of the triangle opposite the vertex then it is longer than the side of the triangle. This occurs with probability 1/3. Now inscribe a circle inside of the triangle. Its area is 1/4 of the original circle. If a chord has its midpoint inside this inner circle its length will be longer than the side of the triangle. Picking a point at random inside the large circle will give a probability of 1/4 that it is also in the small circle, thus giving that as the probability that a random chord is longer than the side of an inscribed equilateral triangle. So is the probability 1/3 or 1/4? This is really a question of how one measures success, as a one dimensional section of arc, or as a two dimensional area. One must describe the measurement process one uses to obtain the correct probability.

In addition to these paradoxes, the connection of probability to gambling gave it an unsavoury reputation. The Edict of Louis IX said "They shall abstain ... from dice and chess, from fornication and frequenting taverns. Gaming houses and the manufacture of dice are prohibited." Kendall the English mathematician said "During the Dark Ages gambling was prevalent throughout Europe. Efforts ... of Church and State to control the evils ... were ineffective. Nothing is more indicative of the persistence of gambling than the continual attempts made to prevent it. Montroll and the author wrote on this same subject saying, "Since travelling was onerous (and expensive) and eating hunting and wrenching generally did not fill the 17th century gentleman's day, two possibilities remained ... praying and gambling; many preferred the latter. But despite, opposition to probability as a real mathematics, due to a number of seeming paradoxes, and prohibitions from church and state, due to its gambling connections, the ideas set in motion by the above mentioned books would spread and be carried on by new generations.

Poisson's book found its way to Russia where Chebyshev read it and did his 1846 Master's thesis on probability, and also started a school in this area. His star students were Markov and Lyapunov whose Markov chains and Lyapunov exponents are of great importance in modern work. The year 1905 saw the name random walk used for the first time. Pearson, the English mathematician wrote to Nature on "The Problem of the Random Walker" He posed the following question. "A man starts from a point zero and walks L yards in a straight line: he then turns through any angle whatever an walks another L yards in a second straight line. He repeats the process n times. I require the probability that after these n stretches he is at a distance between r and r+dr from his starting point 0." Lord Rayleigh wrote back that this was a problem which he had already solved in the context of adding up waves of equal amplitude, but random phases. Today this is called Pearson's random walk.

The early 1900's saw the introduction of Brownian motion by Einstein and Bachelier, and reaction kinetics by Smoluchowski. Langevin, in 1908, studied equations of motion with an additive random force. Jumping over potential barriers in noisy systems was investigated by Pontryagin, Andronov, and Witt in Russia in 1933 and Kramers in Holland in 1940. These works initiated the field of noise-induced transitions. Schottky in 1922 studied the phenomena of shot noise in diodes. Specifically, if electrons arrive at a detector, with a rate ν and cause a decaying time dependent current i(t), then the total measured current is $I(t) = \sum_{j=1}^{\infty} i(t - T_j)$, where T_j represents the past arrival time of the jth electron.

This shot noise has an average value of $\nu \int_0^{\infty} i(t)dt$ and variance $\nu \int_0^{\infty} i^2(t)dt$.

This is also known as Campbell's theorem to mathematicians and it was introduced in 1909. Chandrasekar's review of random processes in 1943 including random walks, Brownian motion, escape over barriers, and stellar dynamics dominated the field for many years. Rice reviewed advances in analyzing noise in electrical devices. The analysis of telephone switching systems, by Fry at Bell Telephone in the 1920's, and by Erlang in Denmark in the 1940's produced advances in the statistical dynamics of congestion of networks. Fisher and Tippett in 1928 introduced extreme value theory to find the largest or the smallest value from a set of random variables. Statistical physics focused on noise, for its own intrinsic properties, and introduced the Master equation, entropy and fluctuation-dissipation theorems. Lévy, in the 1920's and 1930's, introduced scale invariant random walks and distributions for random variables with infinite moments, which were an important precursor for the field of fractals. Montroll beginning in the 1940's opened the field of random walks to physicists with a very clear and easy to use Green's function-generating function approach. With these advances an tools science was ready to pursue the question of noise in natural systems and man-made devices. The remainder of this volume will detail some of these exciting advances and point out still unsolved problems.

REFERENCES

1. A. DeMoivre, *The Doctrine of Chances, Or a Method of Calculating the Probabilities of Events of Play*, (Chelsea Publishers 1718).
2. I. Todhunter, *History of the Theory of Probability*, (Chelsea Publishers 1865).
3. F. N. David, *Games, Gods, and Gambling*, (Hafner Publishers 1962).
4. S. Chandrasekar, *Rev. Mod. Phys.* **15**, 1 (1943)
5. M. G. Kendall, *Biometrica, 43*, 1 (1956).
6. N. Rabinovitch, *Probability and Statistical Inference in Ancient Medieval Jewish Literature*, (U. Toronto Press 1973).
7. E. W. Montroll and M. F. Shlesinger, *The Wonderful World of Random Walks*, in *Studies in Statistical Mechanics* **11**, (North-Holland Publishers 1984).
8. Gábor J. Székely, *Paradoxes in Probability Theory and Mathematical Statistics*, (D. Reidel Pub. Co., Dordrecht, 1986)
9. F. Moss and P.V.E. McClintock, *Noise in Nonlinear Dynamical Systems*, **1-3** (Cambridge University Press. New York 1990)

THE FOKKER-PLANCK BOUNDARY LAYER, RECURRENCE TIMES, AND WANG & UHLENBECK'S UNSOLVED PROBLEMS

CHARLES R. DOERING

Department of Mathematics, University of Michigan, Ann Arbor, MI 48109-1109 USA

The Fokker-Planck boundary layer is essential to understanding quantites ranging from first passage times, to recurrence times, to the distribution of maximum excursions for a Brownian particle with inertia. These problems were described by Wang and Uhlenbeck in 1945 in their list of unsolved problems in the theory of Brownian motion, and in this paper we discuss how they are all manifestations of the phase space boundary conditions. We review the progress that has been made during the last half of the twentieth century, and we describe related unsolved problems which remain in applications of diffusion theory to problems with concentration boundary conditions, such as transport across biological channels.

1 Introduction

Following the seminal 1930 paper "On the Theory of Brownian Motion" by Ornstein and Uhlenbeck [1] and the next decade's progress by (among others) Kramers [2] and Chandrasekhar [3], Wang and Uhlenbeck produced an up-to-date review of the field in 1945 entitled "On the Theory of Brownian Motion II" [4]. In the final section of that paper, Wang and Uhlenbeck identified five particular unsolved problems within the theory of stochastic processes. Specifically, they listed

(a) The Approach to the Barometric Distribution,
(b) First Passage Time Problems,
(c) The Recurrence Time Problem,
(d) The Distribution of the Average Value, and
(e) The Distribution of the Absolute Maximum of a Random Function $y(t)$ in a Given Time Interval T.

Problems (a), (b), (c) and (e) share the feature that they are rooted in questions of boundary conditions in phase space. Problem (d) is a very general question in probablility theory, and we will not consider it here.

Unsolved problem (a) illustrates the boundary condition issue in a clear and basic physical setting. Consider a particle of mass m diffusing in an isothermal bath under the influence of gravity in the half-space $z > 0$ with an impermeable reflecting floor at $z = 0$. Let p denote the momentum, and $P(z,p,t)$ be the time-dependent phase space probability distribution satisfying the Klein-Kramers-Fokker-Planck equation

$$\frac{\partial P(z,p,t)}{\partial t} = -\frac{p}{m}\frac{\partial P}{\partial z} + mg\frac{\partial P}{\partial p} + \gamma\frac{\partial}{\partial p}\left(pP + mk_B T\frac{\partial P}{\partial p}\right), \tag{1}$$

where g is the acceleration of gravity, γ is the friction coefficient, and $k_B T$ is Boltzmann's constant times the temperature. Equation (1) must be supplemented with initial and boundary conditions appropriate for the physical situation under consideration. Wang and Uhlenbeck raised the question of the solution starting from definite position and momentum values,

$$P(z,p,0) = \delta(z - z_0)\,\delta(p - p_0). \tag{2}$$

The other boundary conditions are natural, i.e., $P \to 0$ for $z \to \infty$ and $p \to \pm\infty$, but Wang and Uhlenbeck did not sound completely convinced about the appropriate boundary condition for the reflecting surface. They said, "We feel sure that this means the condition:

$$P(0,p,t) = P(0,-p,t)." \tag{3}$$

This reflecting boundary condition supports the equilibrium (barometric) distribution

$$P_{eq}(z,p) = g\sqrt{\frac{m}{2\pi(k_B T)^3}}\,\exp\left(-\frac{p^2}{2mk_B T}\right)\exp\left(-\frac{mgz}{k_B T}\right). \tag{4}$$

The problem posed by Wang and Uhlenbeck, then, was to determine the fully time-dependent exact analytic solution of (1) in order to study the approach to the barometric distribution. It's the reflecting boundary condition (3) that complicates the problem; in the absence of the floor where (3) is replaced by natural boundary conditions ($P \to 0$ as $z \to -\infty$), the exact time-dependent solution is straightforward to write down. In that case the phase space process is jointly gaussian in z and p and the distribution is completely determined by the time dependent means and variances which are easily computed from (1).

Obvious attempts of applying the method of images do not work for $g \neq 0$, and I don't know if Wang and Uhlenbeck's problem (a) has actually been solved as of 1996. I do not believe it is a particularly pressing issue, but it is interesting because it shows that (i) even very simple physical systems may resist exact solution and, perhaps more significantly, (ii) nontrivial phase space boundary conditions are possible for reasonable physical problems, and such boundary conditions may introduce new challenges.

Unsolved problems (b), (c) and (e) are very closely related for phase space diffusion processes, and these are what we will focus on in this paper. Consider unsolved problem (b) in its simplest setting: a diffusing particle is placed in motion at time $t = 0$ in the half-space $z > 0$ and we ask for the statistical properties of the random time when the particle first achieves $z = 0$. This is the classical "first-passage" or "exit time" problem, and the boundary at $z = 0$ is called an absorbing boundary. For overdamped motion the analogous problem is exactly soluble. For the particle with inertia, however, it remains to this day to be solved in full generality by exact analytic methods. The problem may be precisely stated as follows: the time-dependent phase space probability distribution satisfies

$$\frac{\partial P(z,p,t)}{\partial t} = -\frac{p}{m}\frac{\partial P}{\partial z} + \gamma\frac{\partial}{\partial p}\left(pP + mk_BT\frac{\partial P}{\partial p}\right) \tag{5}$$

with inital conditions corresponding to definite position and momentum values,

$$P(z,p,0) = \delta(z - z_0)\,\delta(p - p_0). \tag{6}$$

The boundary conditions are again natural ($P \to 0$) for $z \to \infty$ and $p \to \pm\infty$, but as in the previous problem, Wang and Uhlenbeck expressed some discomfort with their proposal for the appropriate boundary condition at the absorbing surface. They said, "We feel sure that this means the condition:

$$P(0,p,t) = 0 \;\; \forall\, p > 0." \tag{7}$$

No condition is given for $p < 0$ at $z = 0$. That must be determined *a posteriori* from the solution. Although this "half-space" condition may seem odd, it is physically correct and mathematically well posed. It says that no particles are allowed to enter the region $z > 0$ from below, so when the particle crosses this boundary it never returns. The total probability remaining in $z > 0$ at time t is the probability that the particle has not reached $z = 0$ by time t, from which all of the statistics of the first-passage time may be computed. Equation (6) is parabolic in p, but it has a hyperbolic character in z with the associated speed being p/m. So while "initial" data is appropriately specified at $z = 0$ for $p > 0$, giving information that may propagate into the equation's domain, the solution for $p < 0$ propagates from within the domain and may not be specified independently.

Wang and Uhlenbeck's unsolved problems (c) and (e) are closely related to (b). The recurrence time is the random time that it takes a particle to return to a point from which it starts. In the overdamped, high friction limit this time is precisely zero, but in full phase space it may be cast as a nontrivial exit-time problem as in problem (b). Likewise, the problem in (e) may be rewritten as a first-passage time excercise, as was already recognized by Wang and Uhlenbeck.

The fundamental problem (b) was solved analytically in its simplest form only in the 1985 [5], while the general exit time problem remains unsolved in full exact detail to this day. In the remainer of this paper we describe the progress that has been achieved and the continuing relevance of this problem to present day applications. Modern applications of these ideas include the problem of ion diffusion and transport through channels in biological membranes and important unsolved problems of stochastic motion with concentration boundary conditions where both the exit time and the recurrence time problems find a natural expression.

2 Solved and unsolved problems for Brownian motion

Consider the first-passage time problem (exit time problem) Wang and Uhlenbeck's problem (b). In the overdamped limit (inertia-less limit, or high friction limit) of Brownian motion, the probability density of the particle's position, $\rho(x,t)$, obeys the simple diffusion equation

$$\frac{\partial \rho}{\partial t} = D\frac{\partial^2 \rho}{\partial x^2} \tag{8}$$

where the diffusion coefficient is

$$D = \frac{k_B T}{m\gamma}. \tag{9}$$

For the exit time problem out of an interval $[0,L]$ starting from $x_0 \in (0,L)$, the initial condition is

$$\rho(x,0) = \delta(x - x_0) \tag{10}$$

and the "absorbing" boundary conditions are

$$\rho(0,t) = 0 = \rho(-L,t). \tag{11}$$

Let t_{ex} be the random time at which the particle reaches the boundary. Then the probability distribution of t_{ex} can be computed from the solution of Eqs. (9–11) via

$$\text{Prob}\{t_{ex} > t\} = \int_0^L dx\, \rho(x,t)\,. \tag{12}$$

The mean exit time is particularly straightforward to compute:

$$\langle t_{ex} \rangle = \frac{x_0(L-x_0)}{2D}\,. \tag{13}$$

Now consider the analgous problem for a particle with inertia. The time-dependent phase space probability distribution $\rho(x,p,t)$ satisfies

$$\frac{\partial \rho(x,p,t)}{\partial t} = -\frac{p}{m}\frac{\partial P}{\partial x} + \gamma \frac{\partial}{\partial p}\left(p\rho + mk_BT\frac{\partial \rho}{\partial p}\right) \tag{14}$$

with inital conditions corresponding to definite position and momentum values,

$$\rho(x,p,0) = \delta(x - x_0)\,\delta(p - p_0). \tag{15}$$

The boundary conditions are again natural ($P \to 0$) for $p \to \pm\infty$, but they are complicated at the absorbing boundaries. At $x = 0$ and L the physical condition is that no particles are entering the interval:

$$\rho(0,p,t) = 0 \text{ for } p > 0 \text{ and } \rho(L,p,t) = 0 \text{ for } p < 0. \tag{16}$$

These boundary conditions say nothing about the distribution for exiting particles; that information must be determined from the solution. If Eqs. (14–16) could be solved, then the probability distribution of t_{ex} could be computed according to

$$\text{Prob}\{t_{ex} > t\} = \int_{-\infty}^{\infty} dp \int_0^L dx\ \rho(x,p,t). \tag{17}$$

Consider the recurrence time problem, Wang and Uhlenbeck's problem (c). Here the question is of the random time t_{rec} between the particle successively visiting a certain position, say $x = 0$. The relevant initial-bounday value problem is similar to the exit time problem, only in this case the particle starts at the absorbing boundary. The problem to be solved is the first passage time problem from the initial position back to the initial position. For purposes of explanation consider

$$\frac{\partial \rho}{\partial t} = D\frac{\partial^2 \rho}{\partial x^2} \tag{18}$$

with the initial condition just above the boundary at 0,

$$\rho(x,0) = \delta(x-\varepsilon), \tag{19}$$

along with the absorbing boundary condition

$$\rho(0,t) = 0, \tag{20}$$

and the natural condition ($\rho \to 0$) for $x \to \infty$. The recurrence time distribution is computed from

$$\text{Prob}\{t_{rec} > t\} = \lim_{\varepsilon\to 0}\int_0^\infty dx\ \rho(x,t). \tag{21}$$

Eqs. (18–20) are easily solved by the method of images, and it is easy to see that the recurrence time is precisely zero, with probability one, in the limit $\varepsilon \to 0$ when the particle starts at the boundary. The can also be correctly deduced from the mean exit time formula (13) which shows that if $x_0 = 0$ then $\langle t_{ex}\rangle = 0$, which implies $t_{ex} = 0$ almost surely, uniformly in the length L.

If the particle has inertia, however, then "starting at the absorbing boundary" is a more involved condition. For along with the initial position of the particle we must also specify the initial momentum. In this case the initial-bounday value problem is

$$\frac{\partial \rho(x,p,t)}{\partial t} = -\frac{p}{m}\frac{\partial P}{\partial x} + \gamma\frac{\partial}{\partial p}\left(p\rho + mk_BT\frac{\partial \rho}{\partial p}\right), \tag{22}$$

with

$$\rho(x,p,0) = \delta(x)\,\delta(p-p_0) \tag{23}$$

where for definiteness we'll take $p_0 > 0$. Natural boundary conditions ($\rho \to 0$) are appropriate for $p \to \pm\infty$, and at $x = 0$ the physical condition is that no particles are passing from the region $x < 0$ to the region $x > 0$:

$$\rho(0,p,t) = 0 \quad \text{for } p > 0 \text{ and } t > 0. \tag{24}$$

This boundary condition say nothing about the distribution of particles crossing from $x > 0$ back to the region $x < 0$; as with the exit time problem, that distribution must be determined from the solution. If Eqs. (22–24) could be solved, then the probability distribution of t_{ex} could be computed according to

$$\text{Prob}\{t_{rec} > t\} = \int_{-\infty}^{\infty} dp \int_{0}^{\infty} dx \; \rho(x,p,t). \tag{25}$$

The question of the distribution of the absolute maximum of a diffusion process in a given time interval, Wang and Uhlenbeck's problem (e), is equivalent to the first-passage time problem. To see this, let $M(T,x_0)$ be the maximum absolute value of the stochastic process X_t representing the particle's position during the time interval $[0,T]$ when the particle starts from x_0. That is, $M(T,x_0)$ is the random variable defined by

$$M(T,x_0) = \sup_{0 \le t \le T} |X_t|. \tag{26}$$

For a given value m, if $x_0 \in [-m,m]$ then $M(T,x_0) < m$ is the event that X_t does not reach $\pm m$ by time T, i.e., that $t_{ex} > T$ where t_{ex} is the exit time from $[-m,m]$:

$$\text{Prob}\{M(T,x_0) < m\} = \text{Prob}\{t_{ex} > T\}. \tag{27}$$

Of course for $m \le |x_0|$, $\text{Prob}\{M(T,x_0) < m\} = 0$. Hence the solution of the exit time problem, for both cases of the simple overdamped limit and the full phase space problem, immediately yields the statistics for the maximum value.

Although each of these problems may be fully analyzed for the case of simple diffusion describing overdamped motion, none of them has been solved in general for the Klein-Kramers-Fokker-Planck equation with the phase space boundary conditions. What has been solved [5,6] is the Klein-Kramers-Fokker-Planck equation with boundary conditions appropriate for a single absorbing boundary at, say, $x = 0$:

$$\frac{\partial \rho(x,p,t)}{\partial t} = -\frac{p}{m}\frac{\partial P}{\partial x} + \gamma\frac{\partial}{\partial p}\left(p\rho + mk_BT\frac{\partial \rho}{\partial p}\right) \tag{28}$$

for $x > 0$ and $-\infty < p < \infty$ with boundary condition

$$\rho(0,p,t) = 0 \text{ for } p > 0. \tag{29}$$

For example, in order to sustain a steady state with these boundary conditions, a steady flux $J = -|J|$ of probablility from the right must be maintained. The marginal density for particle positions near the absorbing boundary is then

$$\rho(x) = \int_{-\infty}^{\infty} dp\, \rho(x,p) = \frac{|J|}{D}(x + x_M) + O(e^{-x/\lambda}), \tag{30}$$

where x_M is the "Milne extrapolation length" for the Klein-Kramers-Fokker-Planck equation,

$$x_M = |\zeta(\tfrac{1}{2})|\lambda = (1.4603...)\times\lambda, \tag{31}$$

λ is the mean free length,

$$\lambda = \frac{1}{\gamma}\sqrt{\frac{k_BT}{m}}, \tag{32}$$

and $\zeta(z)$ is the Reimann zeta function. In the overdamped limit $\lambda \to 0$, and the density is just what is expected from the simple diffusion equation, i.e.,

$$\rho(x) = \frac{|J|}{D}x. \tag{33}$$

The effect of a finite mean free length, in comparison with the elementary diffusion equation for this problem, is for a finite particle density to exist at the absorbing boundary. The "extrapolation length" is so named because the linearly extrapolated density vanishes a distance x_M beyond the physical boundary. The boundary layer near the absorbing boundary has the thickness on the order of the mean free length and is characterized by corrections to the simple-diffusion linear behavior of the density, corrections which decay exponentially away from the boundary. These results are in excellent agreement with detailed numerical studies of the Klein-Kramers-Fokker-Planck boundary layer problem [7].

Solving the exit time problem from an interval $[0,L]$ starting from $x_0 \in [0,L]$ involves two absorbing boundaries, and this has not yet been computed exactly.

However, if the mean free length is small compared to all other length scales in the problem, i.e., if the particle motion is in the small Knudsen number regime with

$$\frac{\lambda}{L} << 1, \quad \frac{\lambda}{x_0} << 1, \quad \text{and} \quad \frac{\lambda}{|L - x_0|} << 1, \tag{34}$$

then the boundary layers from the two ends of the interval interfere only weakly with each other and with an initial layer and a matched asymptotic analysis may be performed. For the case of initial momentum $p_0 = 0$, the mean exit time starting from $x_0 = L/2$, the generalization of the result $\langle t_{ex}\rangle = L^2/(8D)$ from Eq. (13), is

$$\langle t_{ex}\rangle = \frac{(L+x_M)^2}{8D} + \frac{\kappa}{4\gamma} + O(e^{-L/\lambda}) \tag{35}$$

where $\kappa = .22749...$ is computed from the asymptotic analysis of the Klein-Kramers-Fokker-Planck equation. The accuracy of the asymptotic results have been confirmed in numerical simulations [8].

3 Other unsolved problems

Consider the problem of diffusion of particles through a channel connecting two reservoirs in which fixed concentrations are maintained. An important application of this kind of set-up is the modeling and analysis of drift and diffusion across biological channels [10], although in practice those problems may be significantly more involved [11,12] than the basic problem posed in this section. Some quantities of interest in such systems are:

(i) the magnitude of the steady state current through the channel,
(ii) the statistics of the current fluctuations in the steady state,
(iii) the residence time in the channel,
(iv) conditional probablities.

These are all relevant and observable quantities in experiments and direct (Monte Carlo or molecular dynamics) simulations. Only (i) can be computed in simple diffusion theory. The others require some more sophisticated modeling and analysis.

In the high friction, overdamped Brownian motion picture, the problem for diffusion along a channel of length L connecting reservoirs at the left and right with concentrations C_{in} and C_{out} is formulated as a boundary value problem on the interval $[0,L]$:

$$\frac{\partial \rho}{\partial t} = D\frac{\partial^2 \rho}{\partial x^2} \tag{36}$$

with the boundary conditions

$$\rho(0,t) = C_{in}, \quad \rho(L,t) = C_{out}, \tag{37}$$

and initial density in the channel,

$$\rho(x,0) = \rho_0(x), \quad \text{for } 0 < x < L. \tag{38}$$

The steady state solution is elementary:

$$\rho(x) = C_{in} - (C_{in} - C_{out})\frac{x}{L} , \tag{39}$$

yielding the expected expression for the steady state current,

$$J = -D\frac{\partial \rho}{\partial x} = D\frac{(C_{in} - C_{out})}{L} . \tag{40}$$

For channels long compared to the mean free path of a particle, this magnitude for the current is expected to be quantitatively correct even for particles with inertia.

But the other quantities (ii)–(iv) cannot be fully studied in the high friction limit. Fluctuations in the current cannot be computed, because to do so we must have some information on velocity fluctuations and variations of the time-of-flight of particles through, and/or residence times of particles in, the channel. In simple Brownian motion the particle motions are not differentiable and velocites are undefined, so these quantites either make no sense or take on trivial values. Likewise, conditional probabilities, such as the probability of a particle exiting at one end given entrance into the channel at the other, are not really meaningful in this limit: the probability of "transmission" before "reflection" is zero for Brownian motion, which paradoxically seems to imply that no particle flux is possible! Effectively this is just what the formula in Eq. (40) tells us because the diffusion coefficient, $D = k_B T/m\gamma$, is inversely proportion to the friction coefficient and, strictly speaking, $J \to 0$ when $\gamma \to \infty$ and all other parameters are held fixed. Items (ii)–(iv) are simply too much for elementary diffusion theory to explain.

For some quantities we may even be led to qualitatively wrong conclusions by the overdamped analysis. For example, as discussed in the last section and repeated in the paragraph above, the fact that the recurrence time for Brownian motion is zero implies that a particle that starts at the boundary (the entrance to the channel) leaves immediately. However, if one considers proper physical units and takes particle inertia into account from the beginning, we can conclude that a particle entering a long channel at one end spends a nonvanishing amount of time in the channel, proportional to the length of the channel, before exiting at one end or the other *even in the high friction limit*. The argument goes as follows: a particle with inertia entering the channel at $x = 0$ does so with a momentum on the scale of the thermal momentum

$$p_{th} = \sqrt{mk_BT} \ , \tag{41}$$

and travels a time proportional to the momentum relaxation time $\tau = \gamma^{-1}$ before equilibrating again. Hence a particle is effectively inserted into the channel a distance proportional to the mean free length,

$$x_0 \sim \frac{p_{th}}{m} \tau = \lambda. \tag{42}$$

But from Eq. (13), the mean exit time out one end of the channel or the other is

$$\langle t_{ex} \rangle = \frac{x_0(L-x_0)}{2D} \sim \frac{\lambda L}{D} \ , \tag{43}$$

so an estimate for the residence time in the channnel is

$$\langle t_{ex} \rangle \sim \frac{\lambda L}{D} = \sqrt{\frac{mL^2}{k_BT}} \tag{44}$$

uniformly in the friction coefficient γ. Of course the particle really starts out in the Klein-Kramers-Fokker-Planck boundary layer and so Eq. (43) is not really precise, but the argument is compelling. The point is that in the high friction limit, even if the particle starts nearly at the boundary, the timescale of the motion is slowed down so much that it ends up taking a finite time to exit again.

These observations suggest that it will be essential to take particle inertia into account to full treat the channel diffusion problem. The situation is even more complicated, though, because one is then faced with the problem of defining the concentration boundary conditions in the phase space picture, and this is not an easy task.

Consider the problem of a channel of length L extending from $x = 0$ to $x = L$ with reservoirs extending from $x = -\infty$ to $x = 0$ on the left, and from $x = L$ to $x = +\infty$ on the right. It is tempting to say that the reserviors consist of particles in thermal equilibrium at the stated concentrations, but this is not a well posed problem in phase space. At the boundaries between the channel and the reservoirs we are *not* allowed to specify the distributions of particle momenta of particles exiting the channel; those distributions must be computed from the solution to the full problem. There will be boundary layers, similar to those described in section 2, on both sides of the boundaries, in the channel and in the reservoirs.

The best one might hope for, then, is to impose thermal equilibrium distributions at the given concentrations asymptotically as $x \to \pm\infty$. But even in one dimension this obscures where the boundary between the reservoir and the channel is (in the absence of energy barriers at the borders of the channel), and we might guess that this will also bring up the issue of the existence of a steady state at all. That is, for the one-dimensional problem in the simple diffusion picture, if we impose asymptotic boundary conditions $\rho \to C_{out}$ and $\rho \to C_{in}$ as $x \to \pm\infty$ respectively, then we observe that the time asymptotic solution is $\rho = (C_{in}+C_{out})/2$ with *no* steady state particle flux (even for finite γ). The ever-increasing "depletion zone" in the reservoirs render this steady state trivial. These considerations suggest that we must consider a more realistic geometry to describe the problem.

If the simple one-dimensional version of the problem appears to breakdown, then it is natural to attempt a two-dimensional formulation. It makes sense that there must be some geometric structure to define the border between reservoir and channel, and in two-dimensions this may be achieved by considering the left hand reservior as extending throughout the half-space ($-\infty < x < 0, -\infty < y < \infty$), the channel of width w as the region ($0 < x < L, 0 < y < w$), and the right reservoir as the half-space ($L < x < \infty, -\infty < y < \infty$). In what is perhaps the simplest formulation of the problem, reflecting boundary conditions may be imposed on all "rigid" boundaries defining the geometry of the domain, and the appropriate equilibrium distributions may be imposed far from the channel openings. However, this formulation in two-dimensions presents the same problem of the constantly growing depletion zone as the one-dimensional case. This may be inferred by considering the high friction limit due to the logarithmic behavior of the Green function for the Laplacian in two-dimensions (the same problem arises with an absorbing disk in two-dimensions, i.e., the classical Smolukowski problem, where just as in one-dimension, the depletion zone around an absorber grows forever). The three dimensional version of the problem holds more promise, because we expect the depletion zone to saturate and reach a steady state in that case. Hence it appears that the full three dimensional version of the problem must be considered! In applications to biological channels there may be additional modeling complications due to the channel opening and closing stochastically, surface diffusion, and channel migration [12]. All these effects can be expected to complicate the analysis further, but it remains an unsolved problem to uncover the behavior of the simple system described here.

4 Conclusion

Although the twentieth century has witnessed tremendous progress in the theory of stochastic processes and Brownian motion, fundamental unsolved problems remain as challenges for the twenty first century. As new experimental methods are developed, and as new applications arise, we can expect the theory of diffusion to continue to play a central role in our understanding of many fundamental physical, chemical and biological processes. In particular the problem of diffusion and current fluctuations through a channel connecting reservoirs at fixed particle concentrations presents an important and basic unsolved problem.

Acknowledgements

I am grateful to my many collaborators and friends who have patiently taught me much over the years. I am especially grateful to Zeev Schuss helpful discussions on matters pertaining to this paper. The nice argument about finite channel residence times in the overdamped limit comes from Ref. [10]. The original research reported here was supported in part by the US National Science Foundation and the US Department of Energy.

References

1. G. E. Uhlenbeck and L. S. Ornstein, Phys. Rev. **36**, 823 (1930).
2. H. A. Kramers, Physica **7**, 284 (1940).
3. S. Chandrasekhar, Rev. Mod. Phys. **15**, 1 (1943).
4. M. C. Wang and G. E. Uhlenbeck, Rev. Mod. Phys. **17**, 323 (1945).
6. M. A. Burschka and U. M. Titulaer, J. Stat. Phys. **25**, 569 (1981).
7. T. W. Marshall and E. J. Watson, J. Phys. A **18**, 3531 (1985).
8. C. R. Doering, P. S. Hagan and C. D. Levermore, Phys. Rev. Lett. **59**, 2129 (1987); P. S. Hagan, C. R. Doering and C. D. Levermore, J. Stat. Phys. **54**, 1321 (1989); P. S. Hagan, C. R. Doering and C. D. Levermore, SIAM J. Appl. Math. **49**, 1480 (1989).
9. R. F. Fox, J. Stat. Phys. **54**, 1353 (1989).
10. R. S. Eisenberg, M. M. Klosek and Z. Schuss, J. Chem. Phys. **102**, 1767 (1995).
11. An excellent example of the sophistication of current experimental methods and the complexity of real systems is: S. M. Bezrukov, I. Vodyanoy and V. A. Pargesian, Nature **370**, 279 (1994).
12. S. M. Bezrukov, UPoN proceedings contribution, contained in this volume.

LOW-FREQUENCY CONDUCTANCE NOISES

ELECTRICAL NOISE IN SEMICONDUCTORS: THE ROLE OF PHONONS

B. K. JONES
School of Physics and Chemistry, Lancaster University, Lancaster LA1 4YB, UK
B.JONES@LANCASTER.AC.UK

The excess (1/f) noise in conductors is a resistance fluctuation. There is some support for this noise in semiconductors to be linked with phonons, although no specific mechanism has been proposed. The possibility of phonons causing the noise in other systems has also been suggested, although other mechanisms seem more appropriate in many other materials. The evidence for a possible phonon mechanism is reviewed and experiments to aid our understanding are proposed.

1 Introduction

Excess electrical noise, that is the noise which is in addition to the well understood thermal and shot noise, with a $1/f$ spectral density occurs in many systems. For many of these systems there are good physical models that are generally accepted. These are usually based on processes that are specific to that system but general principles can be stated. The basic fluctuator consists of a two state system with a characteristic time constant. Each fluctuator produces a spectral density with a Lorentzian spectrum with some characteristic time. If the time constants of the members of an ensemble of fluctuators varies exponentially with some parameter, such as energy or distance, and there is a uniform distribution of fluctuators in this variable then a $1/f$ spectrum results. Examples are thermally activated processes and tunnelling processes for the above two example parameters. In all these cases the fluctuator is based on some type of defect, such as a mobile atom moving between interstitial sites or the change in occupancy of the discrete energy level or trap produced by a chemical or structural impurity.

In bulk semiconductors and a few other systems there is still no generally accepted detailed mechanism for $1/f$ noise. In some cases authors assume a defect model as above but in other cases the analysis is often performed assuming that at least some of the noise is due to mobility fluctuations produced by interaction with phonons. No specific mechanism or fundamental fluctuator for this has been proposed. The exception is the universal quantum $1/f$ noise model which is predicted to be so all pervading that it produces noise in all systems, however, its predicted magnitude is much smaller than the observed noise so that we will not discuss it here.

Here we will review the possible role of phonons in the electrical noise process of bulk semiconductors and other systems where it has been invoked[1-8]

2 Electrical $1/f$ noise

It is generally accepted that the excess noise is a resistance fluctuation. The basic equations used in the analysis of $1/f$ noise are

$$\frac{S_R}{R^2} \equiv \frac{\overline{r^2}}{R^2 \Delta f} = \alpha_{meas} \frac{1}{Nf} \tag{1}$$

$$\frac{S_R}{R^2} \equiv \frac{\overline{r^2}}{R^2 \Delta f} = \alpha_{latt} \left(\frac{\mu}{\mu_{latt}}\right)^2 \frac{1}{Nf} \tag{2}$$

$$\frac{S_R}{R^2} \equiv \frac{\overline{r_n^2}}{R^2 \Delta f} = \alpha \frac{1}{Vf} \tag{3}$$

where the normalised spectral density is inversely proportional to the frequency and N , the total number of carriers in the sample. Let us first consider Eq.1 which is simply one way of relating the noise to the sample properties. The intensity parameter α_{meas} was originally considered a constant for all systems but now it is taken as a measure of the strength of the noise in any one system. The implication of this equation is that each carrier is subject independently to fluctuations. There is no general justification for this. Since α_{meas} is no longer a universal constant, N can be replaced by nV the carrier density and volume. Then a new strength parameter α (equal to α_{meas}/n in the above but not necessarily dependent on n) can be made as in Eq.3. The noise then varies as the $(\text{volume})^{-1}$ which would be appropriate if there were a uniform distribution of independent noise sources throughout the volume. In bulk semiconductors it is not easy to separate these two possibilities experimentally. In systems such as the silicon MOS transistor it is clear that the noise varies as $(\text{surface area})^{-1}$ and the number of traps in the surface of the oxide above the inversion layer so that a different formulation using area rather than volume is more appropriate.

The conductance depends on the product, μn, of the mobility and the carrier number density. There has been considerable discussion about which of these quantities fluctuates. Whereas n can change by trapping, which immobilises the carrier with no change to the electric field pattern, the meaning of mobility fluctuations is less clear since mobility is a quantity which represents the average drift velocity of all the charges. Note that in some semiconductors the carriers are in valleys which have very different properties and are very anisotropic. Can we distinguish between a carrier immobilised at a trap for a time and one scattered to a very low mobility state on the same or another valley?

Let us look at two systems with well established mechanisms. In metals there is no carrier trapping and the resistance fluctuations are due to mobility or scattering fluctuations caused by moving defects rather than by phonons. In MOS channels the noise is caused by the trapping of carriers and the trapped carrier can produce local scattering by its potential so that number and mobility fluctuations are correlated.

To determine whether the mobility or number fluctuates requires special experiments on well characterised samples. As well as the noise in the resistance, the noise should be measured in other quantities such as the magnetoresistance, the Hall coefficient or thermoelectric effect which weight the variables differently. The interpretation is not easy because of the energy and direction dependence of the carrier properties. One effect that is very noticeable in silicon is the piezoresistance in an extrinsic sample which is exceptionally large because of the change of some of the fixed number of carriers from high to low mobility valleys. This could be a good experimental system. More definitive experiments are still needed and we will discuss these later.

The assumption that phonons are responsible for the $1/f$ noise means that only the phonon (lattice) scattering contributes to the resistance fluctuations. The scattering due to static lattice defects is assumed to be time independent. Matthiessen's Rule says that in bulk, low defect concentration, samples the carrier scattering ($\approx$ inverse mobility) from different causes is additive. This results in the extra reduction factor, $(\mu/\mu_{latt})^2$ in Eq.2. In this case the strength parameter, α_{latt} should be a more fundamental quantity than α_{meas}. It should be noted that there are departures from the Rule, notably in samples with small lateral dimensions where there is boundary scattering. Eq.2 could be easily verified with a careful set of experiments using different amounts of inert impurities which do not create electrical traps or affect the surface condition of the sample.

There is a basic problem with Eq.2 since it states that the noise decreases if defects are added in order to decrease the mobility (μ_{latt} remains constant). This is in opposition to the well established observation that samples with almost any type of defect show large noise[3–9]. To account for this Eq.2 has now been interpreted that α_{latt} is not just due to lattice noise but can change its magnitude significantly if the lattice contains defects[10,11] It is not yet clear which defects are to be included into the factor α_{latt} and which into $(\mu/\mu_{latt})^{2,3,4,12}$. This is most unsatisfactory since now the raw data has been thoroughly manipulated, with consequent error propagation, while the simplicity of the original equation (Eq.1) is lost. At this point some of the common assumptions made in the literature about mobility fluctuations should be pointed out. Sometimes Eq.1 is referred to as appropriate to mobility fluctuations. In fact it is only an empirical equation that infers no mechanism, and is often not even an appropriate equation since α_{meas} is not found to be a constant of the system when the number of carriers is changed. Approaching this from the other direction, if one is going to use a mobility fluctuation assumption then Eq.1 or 2 has to be used because there is no model with a predictive equation. Because of the extra evidence that has been accumulated Eq.2 should probably be used.

3 Phonon properties

Phonons are the natural quantised lattice vibrations of a solid. In the low frequency limit they are sound waves. The energy range of the normal modes is about 10^{13} Hz (10^{-21}J, 41meV or 480K). Because of the unit cell shape and the form of the dispersion curve the spectrum of the allowed states (density of states) has considerable structure. In semiconductors there can be acoustic and optical modes. The dispersion curves of the different modes are very different. There is considerable anisotropy in the phonon properties including the phase and group velocity. The interaction of the phonons with electrons is also very different for the different modes.

The occupation of the allowed states, to give the actual mode density, is governed by Bose-Einstein statistics so that the occupation probability of any one state is slowly varying with temperature in the normal experimental range. We would thus expect to see any contribution to the electron noise due to the phonons to be very variable between different semiconductors, especially polar and non-polar, and only slowly temperature dependent near room temperature. That is there should be a major difference in the noise between silicon and gallium arsenide and no thermal spectroscopy is likely. The anisotropy is not likely to be observed since resistance effects are integrals over all states. However some models do suggest effects due to individual sections of the phonon phase space.

4 Phonon noise

If we assume that the electrical noise is due to phonons then the fluctuation may be in the number of active phonons or in the interaction cross-section. We have the same dilemma as in the earlier number or mobility fluctuation argument but with even less experimental basis. It is not easy to think of a mechanism for slow fluctuations in the interaction cross-section. The phonon density is given by the density of states and the occupation probability. The latter is well defined and has known statistical fluctuations. In the experimental region (near the Debye temperature) all the states have significant occupancy so that for a measurable effect there must be a significant fluctuation in the occupation of some *integrated* number of states. This can be compared with the success of the Debye spectrum which is used in the theory of specific heats which is a gross approximation to the real density of states because the integrated quantity is used. It should be remembered that there are energy and momentum conservation rule requirements to be satisfied. Another problem is in the presence of the factor $1/N$ in Eq.1. This implies that each carrier fluctuates slowly independently. Each carrier has only a short life between scatterings which bring it back to a completely different electron state. The suggestion is that there are many independent phonon density fluctuations so that at any time each electron state is subjected to a slowly varying scattering and hence the ensemble mobility fluctuation varies as $1/N$.[10,13]. This is thought to be unreasonable[7] since there needs to be continuity of that scattering channel. In fact the resolution of this problem may not be difficult since there is very little evidence for a $1/N$ dependence in semiconductors and many experiments indicate other variations[14-18]

Let us now look at the spectrum, and in particular, the presence of significant low frequency intensity. As with the basic discussion of low- frequency resistance fluctuations it is not easy to think of a possible phenomenon for phonons which has a very wide range of time constants extending down to very low frequencies. The phonons mainly concerned with electron scattering are near 1THz where the sound transit time and damping time are both much shorter than the observed noise periods. The phonon mean free path is about 5nm with a relaxation time of a picosecond. Although low frequency phonons do have long decay times, they have low density and weak interaction with electrons and low frequency normal modes do not exist in small samples.

In equilibrium, a phonon density fluctuation corresponds to a temperature fluctuation. The amplitude and time constants of these depend on thermal diffusion through the sample and to its heat sink. These were once considered as a cause of $1/f$ noise in a different context, but have been discounted in all systems except those with very large temperature coefficient of resistance. In order to produce a $1/f$ spectrum the fluctuations at different phonon energies are often considered as independent and with different characteristic times.

The slow processes which cause resistance noise in other systems are two level systems which change state by thermal activation or quantum mechanical tunnelling causing number (trapping and detrapping) or scattering noise. The phonon system does have similar defects. A two level or mobile lattice defect, such as an interstitial or a mobile atom, will cause a local phonon mode and a scattering. These will both move with the atom but not cause changes in the number of phonons or the scattering strength because of the translational invariance of the scattering. However, there is the possibility of interference between two such scatterers. More important, there exist two level systems which have resonant absorption and emission, rather like a phonon trap, which will remove specific phonon modes for a time. Although the total number of phonons in an equilibrium system is constant there could be relative occupation fluctuations between modes. Thus a fluctuating longitudinal to transverse mode conversion could cause large resistance fluctuations if the scattering were very different between these two modes.

5 Existing models

Models take various forms. Mathematical methods are needed to produce the basic structure of the spectrum and the magnitude of the electron phonon system interaction. Often treated separately is the physical mechanism for the phonon fluctuations and the details of the interaction.

Sato[19] and Musha[20] showed how phonon number fluctuations, due to temperature fluctuations, could produce volume resistance fluctuations, but they both assumed a $1/f$ spectrum. An attempt at generating a $1/f$ spectrum from a distribution of relaxation times has been made[21] but this had no physical basis and included many assumptions.

The most complete model of noise caused by phonon fluctuations has been by Jindal and van der Ziel[22–24] which has been refuted by Yevick et al[25]. In this it was pointed out that there is a large isotope scattering effect in silicon (but not all other semiconductors). The scattering of the acoustic phonons by these lattice atoms of different mass leads to fluctuations in the phonon numbers of different modes which have reasonable time constants. However the phonon mean free paths which result are unreasonable. Although this could be rationalised if there was no surface scattering, the values are much longer than observed.

The basic problem with such models has been illustrated by Weissman[26]. In order to obtain a range of time constants, to produce the $1/f$ spectrum (and also to be physically realistic), it is necessary to assume that the scattering (mobility) varies throughout the electron distribution. Normally the direction dependence is ignored and only an energy dependence is assumed. Thus the mobility at each electron energy is assumed to fluctuate independently because it interacts with one of a set of individually fluctuating phonons. This has been shown to produce large and varied departures from the simple theory because in a real material there are different energy dependencies of the different quantities and processes involved. This contrasts with the general features of the noise which are shown experimentally by many semiconductors.

6 Phonon fluctuations

If the basic source of the noise is fluctuations in the phonon density then a direct observation of this would be useful. The observed $1/f$ noise in different variables observed in quartz crystals[27], triglycine sulphate[28] and optical fibres[29] can be interpreted as appearing in the loss. This parallels nicely the observed resistance fluctuations in conductors. Direct observations have been made of the phonon density fluctuations by Musha et. al. on quartz and water[30,31]. Observations were made of the Brillouin scattering. These measurements are very difficult because great stability is needed to observe frequencies below 10^{-3} Hz and, since noises add, there are many possible sources of $1/f$ noise in experimental systems. The noise was found to scale with the experimental volume which could be interpreted as the number of independent modes. In the calculations there seems to be an error by a factor of $(kT/h\omega)$, the phonon state occupancy, which is about 200. This reduces the fractional phonon number fluctuations per mode. Since T is included this suggests a possible more direct role of temperature fluctuations. More experiments, using different methods, by independent workers are needed.

7 Phonon noise spectroscopy

A series of reports by Mihaila have shown peaks in the excess noise intensity at temperatures corresponding to the energies of peaks in the density of states of specific phonon modes. The systems reported include Cu, Ag and Si thin films[32] and discontinuous Pt films[33]. He has also reported noise peaks in silicon bipolar transistors as the base current is varied[34–36]. These results are surprising. The density of states has sharp peaks, the occupation probability at, and near, room temperature is slowly varying with phonon energy and temperature. Thus the number of excited states also has peaks. In order to perform spectroscopy there must be a sharp energy selection process. In electron systems this is accomplished by the sharp change in the occupation at the Fermi energy. In a bulk semiconductor there will be phonon interactions over a wide range of energies and angles. As the temperature rises one would not expect to see a sudden change in the interaction and noise at the density of states peak since the occupation of that level only changes slowly with temperature.

For the thin metal films a possible mechanism is phonon assisted tunnelling across the insulating gaps between the metal islands. It is not clear how this could give a sharp change since the gaps will vary in width and the voltage drops across the gaps will vary a lot in such an irregular system. In the bipolar transistors the excess noise arises from the non- ideal, or generation current in the base-emitter junction and this usually occurs at the surface where the depletion layer reaches the surface. Again one can imagine some way that phonons may assist the charge flow (for instance in a tunnelling component) but it is not clear why it is the whole base-emitter voltage that is the relevant quantity.

Since the effects seen are very small, it is necessary for these experiments to be repeated by independent workers, preferably with an automated system which can show unbiased consistency between different samples, experiments and a decrease or increase in the experimental variable. There are also problems in the interpretation which need to be investigated. Are bulk or surface phonons involved? Why is it the metal and not the insulator phonon that is involved? Why are peaks, dips and inflection points seen in the different cases and what are the selection rules which produce large effects? One interesting experiment, which does not seem to have been done, might be to investigate the noise in tunnel junctions under conditions where the current can be seen to be phonon assisted[37–41]. In practice, the noise in junctions is normally dominated by fluctuations caused by trapping and detrapping of traps in the insulator.

8 Possible experiments

In assessing the experimental evidence one must be aware of the difficulty of performing reliable measurements because almost any deficiency in the apparatus, experimental method or quality of the specimen is known to produce $1/f$ noise. This can be illustrated in many cases, but naturally few results have been published by careful experimenters who have eliminated that source of noise. This is especially true of impurities and defects in the samples which are known to cause excess noise of the generation- recombination (GR) or $1/f$ variety in many systems. There are well known cases of bulk defects and surface states in semiconductors and dirt in liquids.

This point is important because mobility fluctuations due to phonon scattering should be an intrinsic property of the system and should produce the same effects for all samples and therefore there should be no sample-to-sample variation. This type of variation is a common feature of noise experimental results, especially when the samples are from different sources and made in different ways.

Other features that should be verified in each experiment is whether the noise is a volume or surface effect and in fact obeys Eq.1. This is rarely performed on each experimental system since the work involved is considerable. The determination for semiconductors is particularly important since there are known surface and boundary noise effects.

Weissman[26] has shown that rather than just studying the magnitude and perhaps the temperature dependence of the resistance fluctuations, considerable information may be obtained from Hall and thermoelectric voltage noise, the higher order statistical fluctuations of these signals and the correlations between them.

Semiconductors are difficult materials for the study of noise since they contain many forms of defect which are known to cause noise. It is very well established that there is a noise contribution from the surface, both from adsorbed chemical effects, interface states at the surface-oxide layer and in the band bending region near the surface. The source of the noise from interface states is familiar in silicon MOST, but there are similar effects with other materials. A band bending at a surface or interface is always present due to oxide charges or pinning by surface or interface states. Deep levels are known to produce GR noise and the bent band or the continuous distribution of surface states will ensure that some states are at the Fermi level and therefore are active noise generators. This type of noise is well established and must appear with a broad spectrum since there will be a spread of time-constants in these cases. Such boundary effects have been well investigated in silicon[14,15] and GaAs[16,17,42]. Number fluctuations in the bulk are less likely to generate $1/f$ noise because suitable levels will not coincide with the Fermi level. The experimental results should therefore always be compared with the theory for surface number fluctuations as well as bulk mobility fluctuations and bulk number fluctuations. Surface effects have been shown to account for the contact noise results originally ascribed to mobility fluctuations[43].

Because of these edge effects the extensive studies on simple bulk or epitaxial layers is suspect. There is probably a surface noise but it cannot be investigated without the variables provided by a more complex geometry. Since interfacial noise has been found to be larger than depletion layer edges, the ideal sample is a conductor completely encased in depletion regions. A bias on the junction gives another experimental variable. A suitable sample might be a high quality n-type layer between two p-type layers with a pattern of gate electrodes to deplete the channel to form a Kelvin and Hall geometry, as is done for 2DEG quantum contact samples.

Some of the specific experiments which have been used to demonstrate the existence of mobility fluctuations can be explained by trapping effects. Excess noise in a diode, or bipolar transistor, exists in the generation current, I', so that $S_i \approx I'^2$. The diffusion current varies as $I \approx \exp(eV/kT)$ and the generation current varies as $I' \approx \exp(eV/nkT)$ with n $\approx$ 2 so that if $I > I'$, $S_i \approx I$[44].

The persistent current produced in AlGaAs by exciting the DX centre has been used to change the number of carriers, N[45]. Unfortunately the system is not simple and the creation of these carriers also creates the same number of charged centres. Simple, but reasonable assumptions could then produce a number fluctuation with the properties observed.

The reduction in the noise as the drift velocity saturates at high fields[46] needs further investigation and deeper interpretation since the results could be interpreted that α varies as μ^2, where μ is here the effective mobility, and this would agree well with the μ^2_{meas} variation often found[10].

9 Conclusions

There is good evidence in semiconductors, as in other systems, that the excess noise increases as the defect density increases. Also there is a well established mechanism for excess noise in semiconductors in the generation-recombination noise due to the trapping/detrapping process. This is Lorentzian for a single time constant and will be $1/f$ for a suitable distribution of time constants. Although there is still no complete model for such a noise mechanism it seems likely to be the basis of the process. There seems little need or justification for a mobility fluctuation model based on phonon fluctuations.

Acknowledgements

I would like to thank A Kozorezov and J K Wigmore for valuable discussions.

References

1. N Lukianchikova, *Noise Research in Semiconductor Physics*, (1997) to be published.
2. M B Weissman, Rev. Mod. Phys. **60** (1988) 537-71.
3. L K J Vandamme, IEEE Trans. Electron Devices **41** (1994) 2176- 87.
4. F N Hooge, Noise in Physical Systems (1991) 7-14.
5. F N Hooge, Noise in Physical Systems (1995) 8-13.

6. F N Hooge, T G M Klienpenning and L K J Vandamme, Rep. Prog. in Phys. **44** (1981) 479-32.
7. Sh M Kogan, Sov. Phys. Usp. **28** (1985) 170-95.
8. V Palenskis, Lith. Phys. J. **30** (1990) 107-52.
9. B K Jones, Advances in Electron. And Electron Phys. **87** (1994) 201-57.
10. F N Hooge, IEEE Trans ED41 (1994) 1926-35.
11. L Ren and F N Hooge, Physica B176 (1992) 209-12.
12. F N Hooge, Noise in Physical Systems and $1/f$ Noise (1995) 8-13.
13. F N Hooge, Physica B162 (1990) 344-52.
14. B K Jones, J. Phys. D. **14** (1981) 471-90.
15. K Kandiah, IEEE Trans. Electron Devices **41** (1994) 2006-15.
16. P A Folkes, J Appl. Phys. **68** (1990) 6279-88.
17. M A Abdala, M A Iqbal and B K Jones, Solid-State Electron. **39** (1996) 287-95.
18. S Hashiguchi, N Aoki and H Ohkubo, Solis-State Electron. **29** (1986) 745-9.
19. H Sato, J Phys Soc Japan 49 (1980) 2087-8.
20. T Musha, Phys Rev B26 (1982) 1042-3.
21. B Dierickx and E Simoen, IEEE Trans ED38 (1991) 1913-7.
22. R P Jindal and A Van der Ziel, Noise in Physical Systems and $1/f$ Noise (1981) 173-7.
23. R P Jindal and A Van der Ziel, J Appl Phys 52 (1981) 2884-8.
24. R P Jindal and A Van der Ziel, Appl Phys Lett 38 (1981) 290-1.
25. D Yervick et al, J Appl Phys 59 2983-5.
26. M B Weissman, Physica 100B (1980) 157-62.
27. M Planat and J J Gagnepain. Noise in Physical Systems (1986) 323-6.
28. T Musha, A Nakajima and H Akabane, Jap. J. Appl. Phys. **27** (1988) L311-3.
29. A J van Kemenade, P Herve and L K J Vandamme, Electon. Lett. **30** (1994) 1338-9.
30. T Musha et al, Phys Rev Lett 64 (1990) 2394-7.
31. T Musha and G Borbely, Jap J Appl Phys 31 (1992) L370-1.
32. M Mihaila, Physics Lett, **107A** (1985) 465-7.
33. M Mihaila, Noise in Physical Systems and $1/f$ Noise (1991) 17-22.
34. M Mihaila, Noise in Physical Systems and $1/f$ Noise (1987) 343-6.
35. M Mihaila, Physics Lett 104A (1984) 157-8.
36. M Mihaila, Noise in Physical Systems (1995) 433-5.
37. A M Speakman and C J Adkins, J. Phys. Condens. Matter 4 (1992) 8053-72.
38. M Pepper, J Phys. C: Solid St. Phys., **13** (1980) L709-19.

39. T Carruthers, Appl. Phys. Lett., **18** (1971) 35-7.
40. B Koslowski, C Baur, R Moller and K Dransfeld, Suface Science **280** (1993) 106-14.
41. K Maeda, S Sugita, H Uota, S Uchida, M Hinomura and Y Mera, J. Vac. Sci. Technol. **B12** (1994) 2140-3.
42. M A Abdala and B K Jones, Solid-State Electron. **35** (1992) 1713- 9.
43. R D Black, P J Restle and M B Weissman, Phys. Rev. **B28** (1983) 1935-43.
44. D Pogany and J A Chroboczek, Microelectron. Eng. **28** (1995) 83- 6.
45. F Hofman and R J J Zijlstra, Solid-State Comm 72 (1989) 1163-6.
46. G Bosman et al, Physics Lett 80A (1980) 57-8.

SOME UNSOLVED PROBLEMS IN 1/f CONDUCTANCE NOISE

M. B. WEISSMAN
Department of Physics, University of Illinois at Urbana-Champaign
1110 West Green Street, Urbana, Illinois 61801-3080

Abstract: A selective survey of 1/f resistance noise in a variety of materials shows that there are diverse detailed mechanisms. The mathematics accounting for the 1/f spectral form turns out to arise in a trivial fashion from generic properties of disorder. However, specific puzzles about the noise origens arise for many dissimilar materials, e.g. polycrystalline Bi films, spin-density waves in Cr, spin-glasses, amorphous metals, and even crystalline Cu with defects. Two themes emerge for future study- collective dynamics versus local effects in disordered systems, and the connection between local effects and measured properties in materials with strong local conductivity inhomogeneity.

The origins of 1/f noise in conducting materials provide one of the most notorious problems in condensed matter physics. In this paper I shall try to sort out which aspects of the problem are solved and which remain unsolved. The general theme will be that the spectral form is so easy to produce by a variety of mechanisms, that it provides little guidance in unraveling the particular question of the noise origin in a given material.

We shall present an argument for the ubiquity of 1/f noise that makes no reference to non-equilibrium dynamics. This approach will come as something of a surprise to those who seek an explanation for 1/f noise in general properties of driven dynamical systems[1].

The essential experimental evidence for the irrelevance of dynamical systems has been described previously[2]. In most resistors, the apparent fluctuations in R are independent of whether R is measured using a dc current, a pulse train, or ac currents, and also independent of the current amplitude for small currents. In some cases, it has been possible to use fluctuations in the Johnson noise to show that the 1/f resistance noise is present in the absence of any applied current. We may add another obvious point: dynamical systems which exhibit scaling can be constructed with all sorts of scaling exponents, with no special preference for f_{-1}[3]. The very fact that the noise exponent in condensed matter clusters around -1 should strongly suggest that algebraic dynamical scaling laws are not the key. Of course, when noise with very different exponents (e.g. Barkhausen noise)[4] turns up, the argument may be turned around to imply that non-equilibrium explanations are likely suspects (as is obviously true anyway for Barkhausen noise).

The essential argument for why condensed matter is full of 1/f noise is as follows[5,6]. The rates of the low- frequency noise processes are typically in the neighborhood of 1 Hz to 100 kHz. Such frequencies are very low compared with the typical microscopic frequencies (e.g. Arrhenius attempt rates) in condensed matter, which are usually greater than 10^{12} Hz. Thus there is some big factor, ordinarily an exponential, giving the ratio of the relaxation time to the microscopic times. These exponential factors can either be the Arrhenius factors of thermally activated processes or the exponentials of tunneling processes[2]. In either case, if the parameters in the exponents (barrier heights, effective masses, tunneling distances) have some distribution of values, due to disorder, a distribution of characteristic rates will be found.

The way in which such distributions add up to give 1/f noise is illustrated in Figure 1, which shows spectra from a collection of small resistors on a silicon wafer. Although each spectrum is distinct (and not 1/f) because it comes from a small number of traps, each with a particular characteristic time, the average is close to 1/f.

Will those distributions give 1/f noise? The distribution of characteristic rates or frequencies, $r(f_c)$, is simply related to the distribution of exponential parameters, say activation energies E_A, by $\rho(f_c)\, df_c = \rho\,(E_A)\, dE_A$. Therefore $\rho(f_c) = \rho\,(E_A)\, dE_A/df_c$. Because E_A depends logarithmically on f_c, dE_A/df_c = constant/f_c. Therefore $\rho\,(f_c)$ = (a function of log $(f_c))/f_c$. So long as the distribution $\rho(EA)$ is algebraic, the corrections to the 1/f form will be logarithmic in f_c. Thus, in the presence of disorder one expects low frequency noise to be of the 1/f form with log corrections, although if one looks in regions corresponding to exponential or Gaussian tails of the distribution, other power laws can be found.

Since explanations of the above type involve quasi-equilibrium noise, it has been awkward that examples of conductance noise demonstrably obeying the fluctuation- dissipation theorem have not been available. The main excuse has been that the variety of properties at local fluctuating sites precludes finding a suitable parameter to use in the dissipation experiment. Recently, the 1/f noise in giant-magnetoresistive materials has provided a nice example in which the magnetic field is a suitable perturbation parameter. In the large regime in which the fluctuating magnetization is the biggest noise source, comparison of the 1/f noise and the out-of-phase response of R to ac magnetic fields can be made via the fluctuation dissipation relation with no adjustable parameters. The agreement is very good, despite the fact that both the noise and the out-of-phase response show large hysteresis as the magnetic field is changed. (See Figure 2.) Thus one does not need true equilibrium, only quasi-equilibrium on the relevant time scale, for an equilibrium picture to work[7]. The unsolved

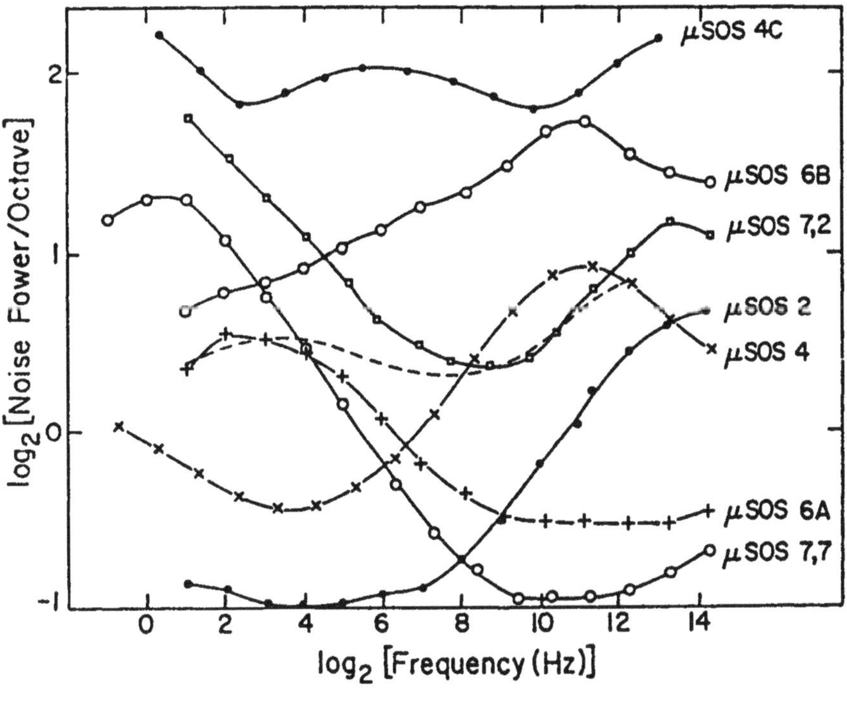

Figure 1: Spectral densities (integrated over octaves, so that 1/f noise gives a horizontal line) are shown for a series of nominally identical mm-scale resistors from a silicon-on-saphhire wafer. The dotted line is the average, much closer to 1/f [43]. The vertical scale has been expanded more than a factor of four compared to the horizontal scale, to magnify the remaining deviations.

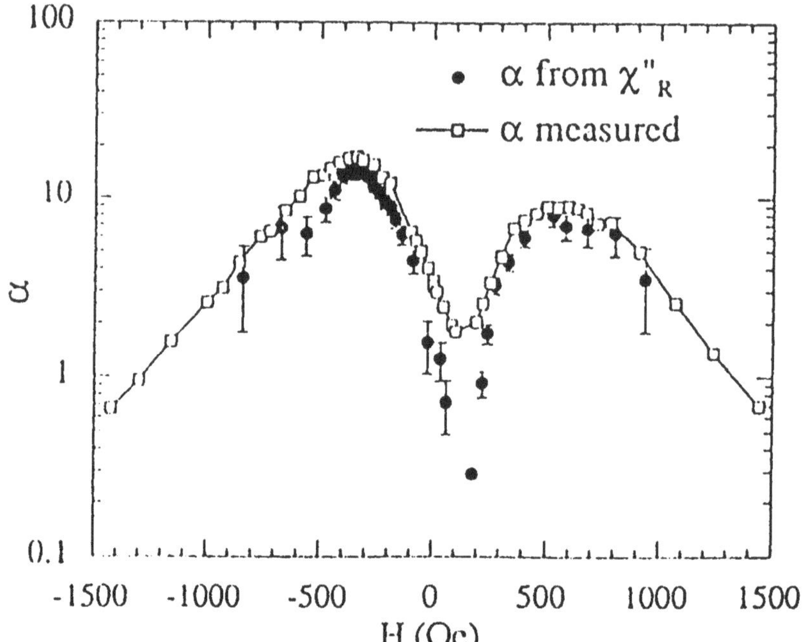

Figure 2: A comparison between the standard 1/f noise parameter a (defined as the mean-square fractional resistance fluctuatiions over a factor of e in frequency, multiplied by the number of atoms in the sample) and the out-of-phase ac magnetoresistivity of a Co-Cu giant magnetoresistive multilayer. The normalization of the out-of-phase response gives a predicted a based on the fluctuation-dissipation relation, with no adjustable parameters[7].

problem of the origin of 1/f noise then splits into many little unsolved problems: what things are actually present in different materials to give the spread of relaxation rates? Within this collection of problems, new broad themes emerge: does the 1/f noise come from a simple sum over many separate sites, each in a fixed, disordered environment, or does it result from collective modes of some complicated, interacting physics? We shall see that this broad question has no general answer, but only specific answers for individual systems.

Although there are few cases in which a first-principles explanation of observed 1/f noise is available, we will discuss a few of the many cases in which the detailed ingredients of the noise have at least been characterized. (A better-known case[8] is provided by the noise in oxide layers on Si, for which technology has driven more detailed materials-oriented studies.) Because 1/f noise is measurable in very small samples, it provides a useful tool for probing slow dynamics in disordered systems. In small samples, even though many models can give the same average spectrum, the statistical details of the noise can be extremely sensitive to the form of the model.

A few examples will illustrate the main points. Antiferromagnetic Cr, crystalline Cu, polycrystalline Bi, spin-glasses, amorphous metals at low temperature, and hydrogenated amorphous Si (a-Si:H) provide good illustrations of what is known and what remains to be found out.

Good crystals of Cr provide one of the better examples of why universal theories of 1/f noise don't work. Above the Neel temperature, some 1/f noise of unknown origin is found. Since the magnitude of this noise grows as the crystal quality is lowered[9,10], it is a safe guess that in some way or other it comes from defect motions, like typical 1/f noise in metals[2,5,6]. On lowering the temperature a few degrees, the noise power increases almost three orders of magnitude, without changing its 1/f form[9–11]. (See Figure 3.) Measurements of the tensor symmetry of the noise[9] and of the very large individual fluctuations of which it consists indicate[10] that in this regime thermal switching of the q-vector of domains of the spin-density-wave order between allowed crystal orientations provides the main noise source in this regime. (See Figure 4.) Just why the q-order breaks up into domains isn't understood, although it must have something to do with a combination of disorder and strain constraints. If the temperature is then lowered another few tens of K, the noise spectral density does not change dramatically. However, a check of the tensor symmetry properties, and of the individual domain noise in small samples, shows that the q-vector fluctuations have frozen out, and that fluctuations of domains of the polarization planes of the spin-density wave now dominate, just as in the acoustic attenuation[12]. The formation of these domains is more puzzling than the q-domains, because even in free single crystals neutron scattering

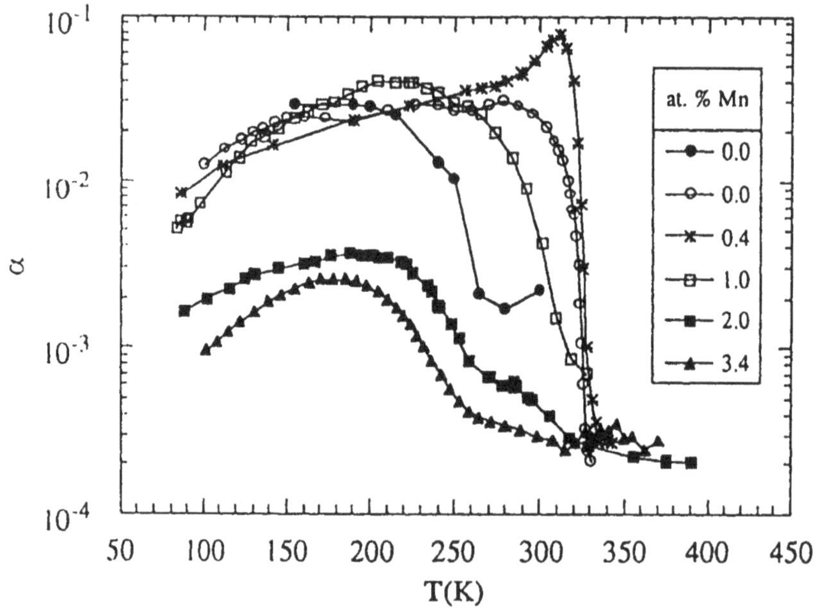

Figure 3: The noise parameter a vs T for several Cr alloy films, all epitaxial except the lower-quality film represented by filled circles. The dependence of the noise onset tracks the transition into the incommensurate spin-density wave phase, with the paramagnetic phase and the commensurate phase relatively quiet[14]. Noise in the paramagnetic phase increases with increasing defect scattering[11].

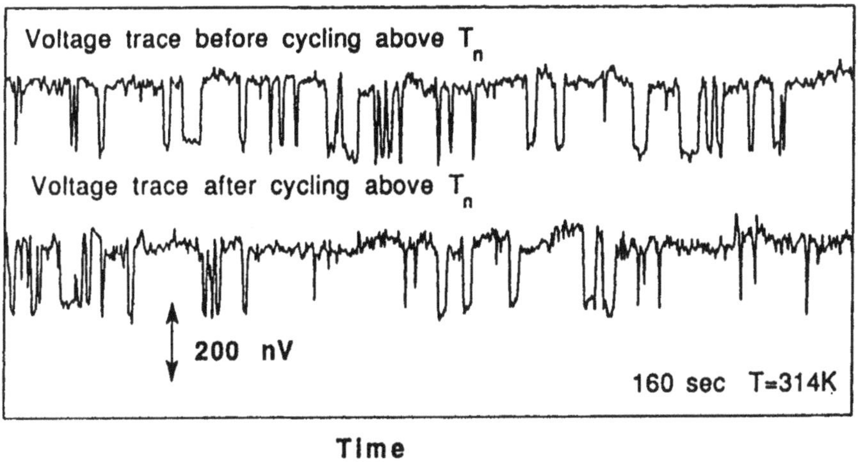

Figure 4: The resistance switching in a Cr epitaxial film is shown just below the Neel temperature. These large switches, reproducible after cycling to the paramagnetic phase, come from q-domains of the spin-density wave[10].

shows that such domains spontaneously form even after the polarization has been aligned with a field, once the field is removed[13]. The thermodynamics of this spontaneous break-up don't make sense unless some residual material disorder is invoked. Experiments on doped material show that the incommensurability of the SDW and the lattice plays a key role[14], (see Figure 3 again.) but no theoretical understanding of how disorder and incommensurability determine the domain dynamics is available. One of the most intriguing results arises in the 1/f noise of a nominally simple system- single crystal Cu nanobridges[15–17]. It had long been suspected that defect motions were responsible for most 1/f noise in metals. The experiments of Ralls et al showed that this was indeed the case, with a variety of different types of defects involved. These include pointlike defects which can diffuse through the sample as well as larger defects. The mystery arises in the temperature dependence of the noise statistics. Below about 150 K, the noise could be decomposed into the sum of several random-telegraph-like components. At higher T, even though the average spectrum was essentially unchanged, these components "melted" into some pattern which continually changed over time[15–17]. How is this glass-like melting-freezing to be understood? What are the stable modes in the low-T frozen state? What are the connections between interacting defects and spin-glass Hamiltonians? Do ordinary metals have some sort of defect glass state, and if so what detrmines the glass melting temperature?

Polycrystalline Bi films provide a different sort of puzzle. In Bi, there is a large, reproducible peak in the 1/f noise near room temperature, so long as the films are sufficiently annealed to give large crystalline domains, and so long as the surface-to-volume ratio is small[2]. The noise has distinctive tensor symmetry (nearly traceless) corrsponding to rotations in a high-symmetry environment[18]. The dependences of the noise amplitude on sample properties, the narrow distribution of frequencies, and the symmetry all point toward the same conclusion: the noise comes from the good crystalline regions, quite unlike typical defect noise in metals. Although the phenomenon has been known for many years, no one knows the explanation. Does Bi have some special defect, whose concentration is highly reproducible? Is there some more exotic source, related to weak symmetry-breaking effects such as found in Cr?

In spin-glasses, both the magnetization itself and (in conductors) the conductance are found to show 1/f noise[19-21]. The famous universal conductance fluctuation effect[22,23] provides the main mechanism by which the conductance noise comes from the same slow spin fluctuations which give the magnetization noise. The magnetization noise fits the fluctuation-dissipation theorem, even when the spin-glass is not in long-term thermal equilibrium[19,20]. However, spin-glass theories remain highly speculative, so that the basic physical picture of the spin fluctuations is not established[24-26]. At two limits of the possible descriptions are the droplet picture[27] and a hierarchical picture, related to the Parisi solution of an infinite-range model[28]. The simplest droplet picture envisions the low frequency dynamics as coming from distinct compact spin droplets which are thermally allowed to turn over, and for which a natural range of sizes translates to the distribution of characteristic rates. Hierarchical descriptions assume that on some length scale the sets of spins which flip are highly diffuse and mutually overlapping, with some special statistical features describing those overlaps. The set of metastable spin configurations form an ultrametric space, where the metric is given by the fraction of the spins that are different. This metric also corresponds roughly to the log of the time required to go from one configuration to another. Various properties of the dynamics on such hierarchical spaces have been worked out[29].

The noise in these dissimilar descriptions is expected to be of a 1/f form for the usual reason. It would be misleading to say that the origins of the spin noise are known since one cannot even distinguish between such qualitatively distinct pictures.

Our recent experiments on small samples of the spin-glasses CuMn[30,31] and AuFe [32] showed that no universal description will work to describe the spin fluctuations even in conducting spin-glasses with oscillatory spin-spin interactions. In CuMn, nothing like persistent two-state droplets showed up; instead we found complicated multi-state fluctuators. (See Figure 5.) In AuFe, however, droplet-like fluctuators, switching repeatedly between two spin states even as the temperature or magnetic field was changed, were a major part of the noise. (See Figure 6.) Especially in CuMn, the scaling of higher moments of the noise disagreed with droplet pictures and fit expectations for a hierarchy, indicating that the fluctuators must be diffuse. A very convenient measure of htese non-Gaussian effects is provided by the "second spectrum", $S_2(f_2,f)$ the spectrum of the fluctuations at frequency f_2 in noise power measured in a band around frequency f. (See Figure 7.) An actual physical picture of what sort of spin collections fluctuate coherently, what sort of spin states they fluctuate between, how they interact with each other, etc. remains undetermined for both materials. A particular open question concerns the role of the local anisotropy seen by the spins (very different in magnitude for these two materials) in determining the global noise statistics. Genuinely amorphous materials have long been known to have low temperature excitations with a broad distribution of slow relaxation rates[33]. The microstructure of these excitations, and the reason for an approximate universality of their (properly normalized) density in a variety of materials are not known[34]. This same phenomenon shows up as low-temperature 1/f noise in amorphous conductors, allowing experiments to be made on samples so small as to isolate a few of the fluctuating sites[35]. In several materials such as amorphous co-sputtered Si-Au, these sites are, as had been hypothesized, approximately two-state or few- state systems[35]. (See Figure 8.) Their kinetics are governed either by thermal activation or by tunneling, depending on T; either way of course gives 1/f for the average spectrum. Their interactions, determined from the slow fluctuations in the spectral shape, appear stronger than would be expected for strain interactions among randomly located defects- most sites show clear signs of interacting enough with other sites to exhibit slow fluctuations in their spectral density. (See Figure 9.) The detailed pattern of sites changes after the sample is heated up to temperatures (e.g. 60 K) far too low to show gross annealing. Do the interactions among sites leave many different ways in which the collective system can freeze? What determines what floppy sites remain at low-T? What are the connections between the "universal" properties of low-temperature excitations in glasses and the analagous effects at higher T is defctive Cu? Because noise experiments can view individual sites as well as average properties, they have a better chance of answering such questions than do standard thermal

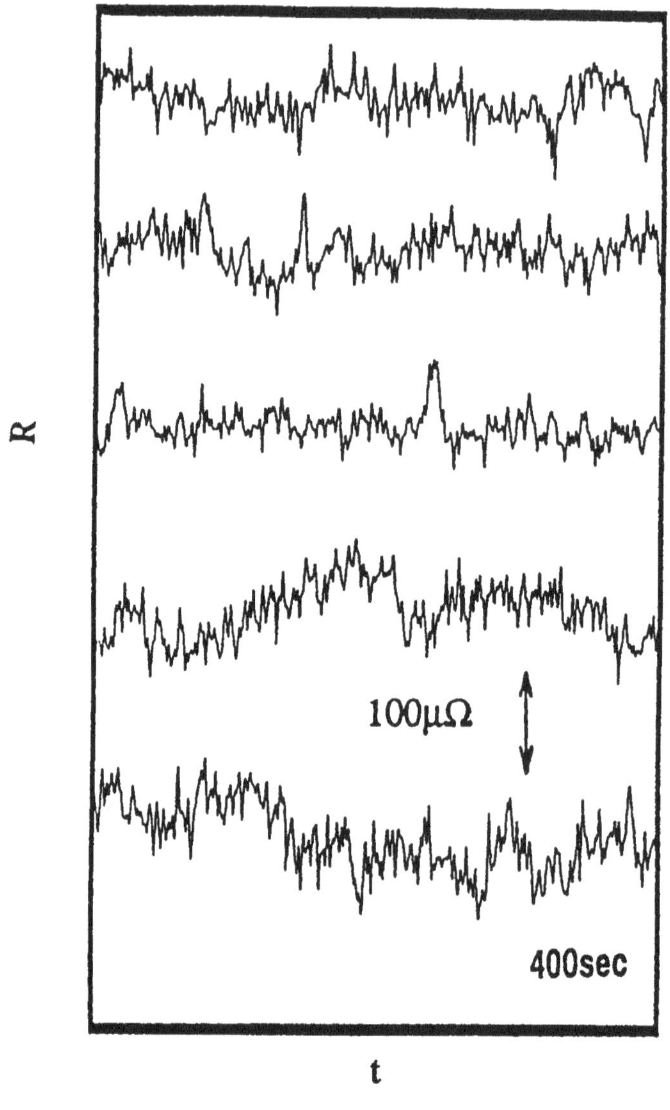

Figure 5: An example of $\delta R(t)$ for a small (under 10^8 atoms) CuMn sample in the spin-glass phase. Although the noise is non-Gaussian, discrete random telegraph events are hard to find and not persistent[44].

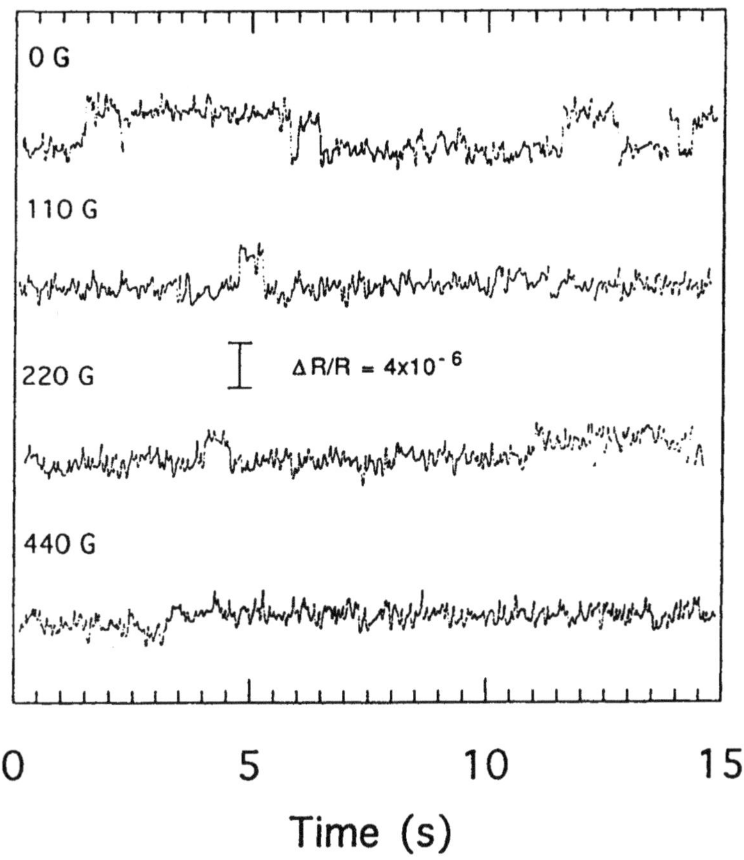

Figure 6: An example $\delta R(t)$ for a small (under 10^8 atoms) AuFe sample in the spin-glass phase, at 4.7 K. The noise is dominated by discrete random telegraphs, whose sensitivity to magnetic field is similar to that expected for spin-glass droplets[32].

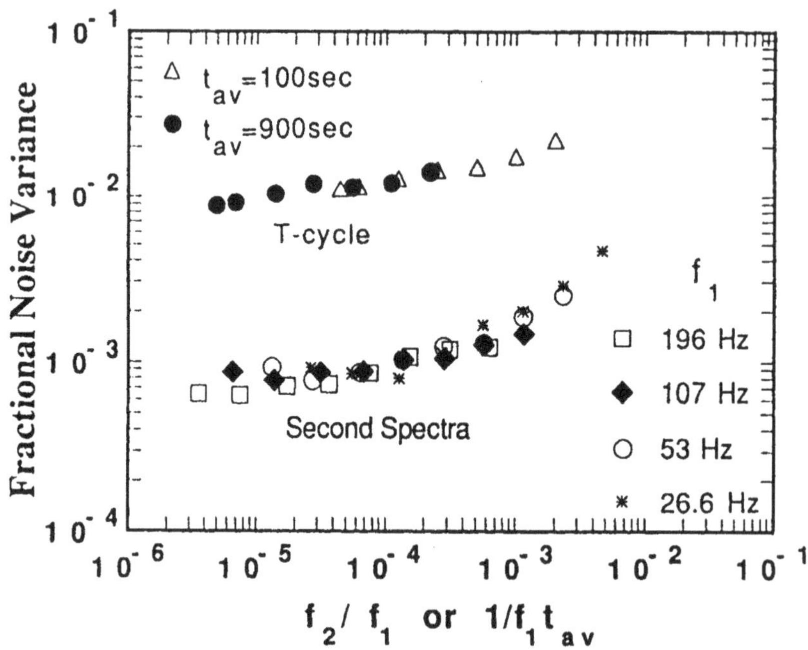

Figure 7: The upper curves give the fractional variance of the noise power for the small CuMn sample on repeated thermal cycles; the lower curves give non-Gaussian second spectra $f_2S_2(f_2,f_1)$ in units of fractional variance. f_1 is the ordinary first-spectral frequency. The scaling properties of these plots (independence of f_1 for fixed f_2/f_1) are consistent with hierarchical pictures but not droplet pictures. The magnitude and shape of the second spectrum gives a great deal of other information on the dynamical pattern[30].

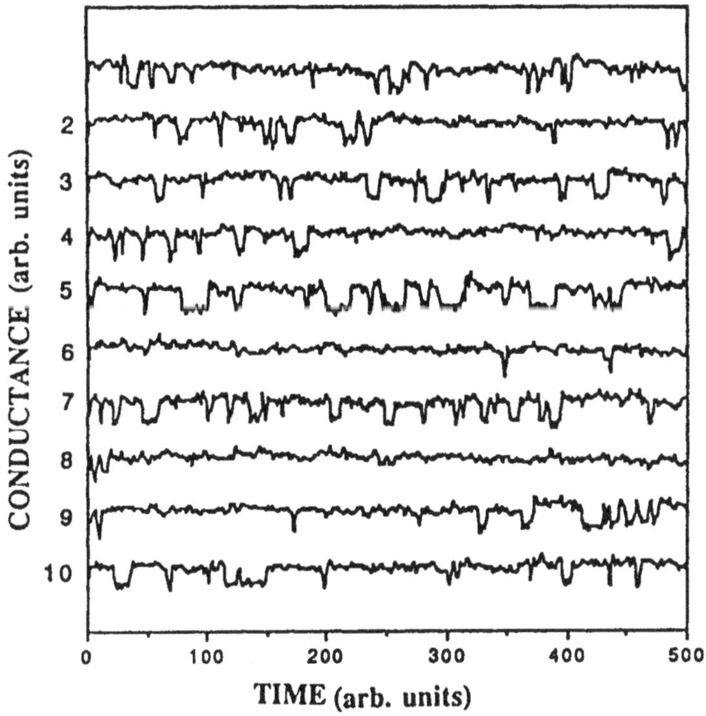

Figure 8: A typical $\delta R(t)$ for a small (under 10^8 atoms) sample of amorphous cosputtered C-Cu at 4.3 K. Each sweep represents 500 data samples with a sampling time of 1.3 ms. The steps are from fractional resistance changes of about 0.3 percent. The random-telegraph signal shows that a two- state picture is approximately right[35].

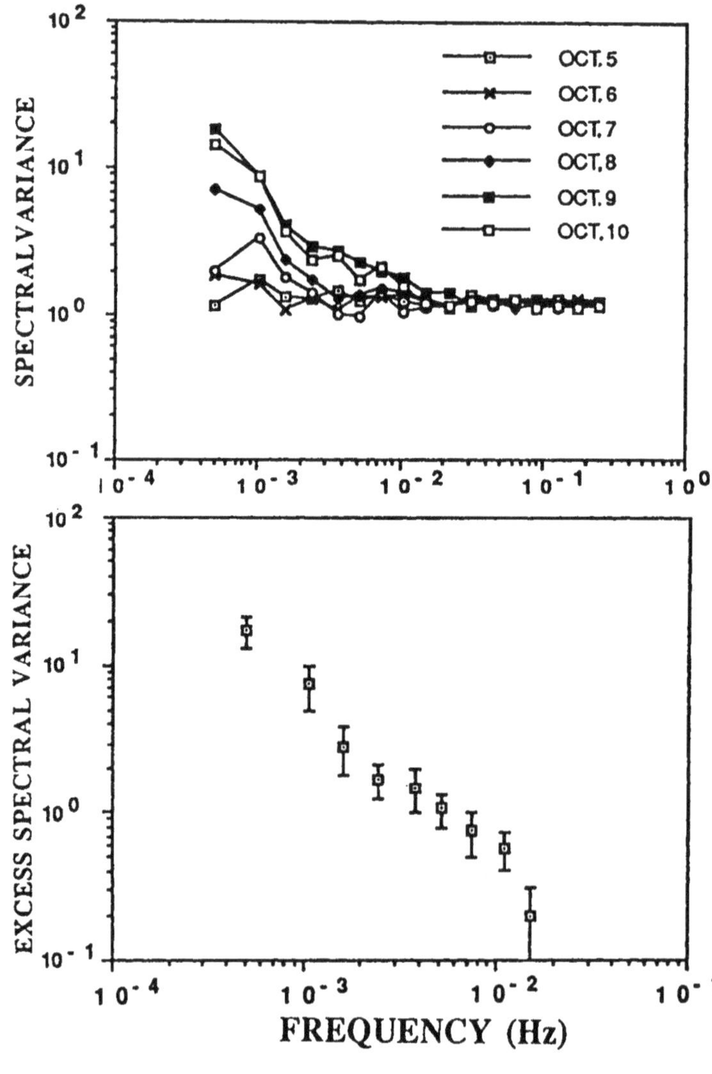

Figure 9: a) Second spectra $S_2(f_2, f)$ (normalized so that Gaussian noise would give 1) are shown for a C-Cu sample at 5.3 K. The octave number refers to which first-spectrum octave is used, e.g. octave 9 here is from 106 Hz to 213 Hz. Part b shows the data from octave 9 with the (mainly Gaussian) white component subtracted. The low-f_2 contributions show that the nominal two-state system must in fact couple to other slow degrees of freedom[35].

measurements.

The 1/f noise in these amorphous metals does not go away at room-temperature. In fact, the noise becomes non- Gaussian even in samples for which it is Gaussian between 100 K -200 K[35]. The question again comes up: what sort of collective motions are involved? How general is this phenomenon in amorphous conductors?

Hydrogenated amorphous silicon (a-Si:H) has been found[36–38] to have huge 1/f noise that is very non-Gaussian even in large samples, e.g. 10^6 mm^3. Big conductance steps were often found. The noise is, in detail, non-linear in the applied field, suggesting that motions of charged H+ are involved, since uncharged objects would not be sensitive to the applied field. The non-Gaussianity requires highly inhomogeneous current paths and/or big blobs of hydrogen moving collectively, which would no doubt be accompanied by highly inhomogeneous current anyway. Interesting fluctuations in the detailed spectral shape (random telegraphs come and go) are found[39], and these turn out also to show up in simulations of noise in percolating networks. (The reason is obvious: a path on which one object is fluctuating can be shut down by a sloer fluctuation of another object on the same path, so long as the local fluctuations are large.)

Several questions come up right away: If the local conductances among neighboring sites are sufficiently spread out, will highly inhomogeneous percolation like conductance result even if the sample is overall homogeneous? What actually are the local objects which fluctuate- can they be single H+ near the high-current links, or are there really H blobs? Here, solving the 1/f noise questions would also mean understanding the conduction mechanism better.

A related problem has come up in a quite distinct material. In partially annealed, still somewhat disordered, large samples of a manganate "colossal magnetoresistance" material, giant conductance steps appear[40]. Here, no large H blobs can be invoked. However, there must still be large local inhomogeneities in the conductivity, related to the magnetoresistance itself. In fact, the non-Gaussian effects can be turned off by applying a sufficiently large magnetic field at low temperature, aligning most of the magnetic domains and making the conductivity more homogeneous.

Is this non-Gaussianity also due to inhomogeneous current paths, and should it be expected as a general feature in variable-range hopping problems, since these are known[41] to give percolation-like paths? Work on noise in such highly disordered conductors has begun[42], but most of the detailed predictions needed to understand non- Gaussian effects have not yet been made. Simulations of noise in variable-range hopping are harder than the equivalent simulations in discrete percolation problems, but might turn out to have wider physical relevance.

We have seen that a general theme (collective vs. individual dynamics) emerges when searching for the origins of 1/f noises. Also, in strongly disordered conductors (i.e. with big ranges of local conductivity), the relation between the local fluctuations and the measured macroscopic fluctuations has not been fully worked out. However, these themes are not very relevant to the question of "Why 1/f?", which has a very simple answer: $d\ln(f)/df = 1/f$.

Acknowledgments

This work was supported by the NSF DMR 96-23478 and, through the facilities of the Materials Research Lab, by NSF DMR 89-20538.

References

1. P. Bak, C. Tang, and K. Wiesenfeld, *Phys. Rev. Lett.* **59**, 381 (1987).
2. M. B. Weissman, *Rev. Mod. Phys.* **60**, 537 (1988).
3. J. Kertesz and L. B. Kiss, *J.Phys. A* **23**, L433. (1990).
4. G. Montalenti, Z. Angew. *Phys.* **28**, 295 (1970).
5. P. Dutta and P. M. Horn, *Rev. Mod. Phys.* **53**, 497 (1981).
6. Sh. M. Kogan, *Usp. Fiz. Nauk.* **145**, 285 (1985). [*Sov. Phys. Solid State* **24**, 1921 (1982)].
7. H. T. Hardner, M. B. Weissman, M. B. Salamon, and S. S. P. Parkin, *Phys. Rev. B* **48**, 16156 (1993).
8. D. M. Fleetwood, P. S. Winokur, R. A. Reber, T. L. Meisenheimer, J. R. Schwank, M. R. Shaneyfelt, and L. C. Riewe, *J. Appl. Phys.* **73**, 5058 (1993).
9. N. E. Israeloff, M. B. Weissman, G. A. Garfunkel, D. J. VanHarlingen, J. H. Scofield, and A. J. Lucero, *Phys. Rev. Lett.* **60**, 152 (1988).
10. R. Michel, N. E. Israeloff, M. B. Weissman, J. Dura, and C. P. Flynn, *Phys. Rev. B* **44**, 7413 (1991).
11. J. H. Scofield, J. V. Mantese, and W. W. Webb, *Phys. Rev. B* **34**, 723 (1986).

12. W. C. Muir, E. Fawcett, and J. M. Perz, *J. Magn. Magn. Mater.* **69**, 113 (1987).

13. E. Fawcett, *Rev. Mod. Phys.* **60**, 209 (1988).

14. R. P. Michel, M. B. Weissman, K. Ritley, J. C. Huang, and C. P. Flynn, *Phys. Rev. B* **47**, 3442 (1993).

15. K. S. Ralls and R. A. Burhman, *Phys. Rev. Lett.* **60**, 2434 (1988).

16. K. S. Ralls, D. C. Ralph, and R. A. Buhrman, *Phys. Rev. B* **40**, 11561 (1989).

17. K. S. Ralls and R. A. Burhman, *Phys. Rev. B* **44**, 5800 (1991).

18. R. D. Black, P. J. Restle, and M. B. Weissman, *Phys. Rev. Lett.* **51**, 1476 (1983).

19. P. Refrigier, M. Ocio, J. Hammann, and E. Vincent, *J. Appl. Phys.* **63**, 4343 (1988).

20. W. Reim, R. H. Koch, A. P. Malozemoff, M. B. Ketchen, and H. Maletta, *Phys. Rev. Lett.* **57**, 905 (1986).

21. N. E. Israeloff, M. B. Weissman, G. J. Nieuwenhuys, and J. Kosiorowska, *Phys. Rev. Lett.* **63**, 794 (1989).

22. S. Feng, A. J. Bray, P. A. Lee, and M. A. Moore, *Phys. Rev. B* **36**, 5624 (1987).

23. B. L. Al'tshuler and B. Z. Spivak, *Pis'ma Zh. Eksp. Teor. Fiz.* **42**, 363 (1985). *JETP Lett.* **42**, 447 (1985).

24. K. Binder and A. P. Young, *Rev. Mod. Phys.* **58**, 801 (1986).

25. K. H. Fischer and J. A. Hertz, *Spin Glasses.* (Cambridge, Cambridge, 1991).

26. V. S. Dotsenko, M. V. Feigel'man, and L. B. Ioffe, *Spin Glasses and Related Phenomena.* (Harwood, Glasgow, U.K., 1990).

27. D. S. Fisher and D. A. Huse, *Phys. Rev. B* **38**, 386 (1988).

28. M. Mezard, G. Parisi, and M. A. Virasoro, Eds., *Spin Glass Theory and Beyond* (World Scientific, Singapore, 1987).

29. A. T. Ogielski and D. L. Stein, *Phys. Rev. Lett.* **55**, 1634 (1985).

30. M. B. Weissman, N. E. Israeloff, and G. B. Alers, *J. Magn. Magn. Mater.* **114**, 87 (1992).

31. M. B. Weissman, *Rev. Mod. Phys.* **65**, 829 (1993).

32. K. A. Meyer and M. B. Weissman, *Phys. Rev. B* **51**, 8221 (1995).

33. W. A. Phillips, *Rep. Prog. Phys.* **50**, 1657 (1987).

34. C. Yu and A. J. Leggett, *Comments Cond. Matter Phys.* **14**, 231 (1988).

35. G. A. Garfunkel, G. B. Alers, and M. B. Weissman, *Phys. Rev. B* **41**, 4901 (1990).

36. C. E. Parman, N. E. Israeloff, and J. Kakalios, *Phys. Rev. B* **44**, 8391 (1991).

37. C. E. Parman and J. Kakalios, *Phys. Rev. Lett.* **67**, 2529 (1991).

38. C. E. Parman, N. E. Israeloff, and J. Kakalios, *Phys. Rev. Lett.* **69**, 1097 (1992).

39. L. M. Lust and J. Kakalios, *Phys. Rev. Lett.* **75**, 2192 (1995).

40. H. T. Hardner, M. B. Weissman, M. Jaime, R. E. Treece, P. C. Dorsey, J. S. Horwitz, and D. B. Chrisey, *J. Appl. Phys.* (in press).

41. V. Ambegaokar, B. I. Halperin, and J. S. Langer, *Phys. Rev. B* **4**, 2612 (1971).

42. Sh. M. Kogan and B. I. *Shklovskii, Fiz. Tekh. Poluprovodn.* **15**, 1049 (1981). *Sov. Phys. Semicond.* **15**, 605 (1981).

43. P. J. Restle, R. J. Hamilton, W. M. B., and M. S. Love, *Phys. Rev. B* **31**, 2254 (1985).

44. N. E. Israeloff, G. B. Alers, and M. B. Weissman, *Phys. Rev. B* **44**, 12613 (1991).

DEFECTS AND THE Δn–$\Delta \mu$ CONTROVERSY IN SEMICONDUCTOR 1/f NOISE

F.N. HOOGE
Department of Electrical Engineering
Eindhoven University of Technology
5600 MB Eindhoven, The Netherlands

We shall discuss the role defects play in the generation of 1/f noise in homogeneous semiconductors. Three models of 1/f noise are reviewed.

1 Stating the problem

Damaging the crystal lattice of semiconductors strongly enhances the 1/f noise. This is a well established experimental fact. The experimental fact leads to the following problem: Is all 1/f noise caused by defects? Or do defects create additional 1/f noise on top of the "normal" 1/f noise, which is always found, even in perfect crystals?

I shall discuss the problem from the perspective of three noise groups, Eindhoven, Kiev and St. Petersburg. I am well aware that at this conference it is forbidden to present solutions. Nevertheless, as long as I am discussing one model I shall try to prove its correctness. Although each of the models, on its own, looks correct, the three solutions exclude and contradict each other. I can demonstrate the correctness of each model, but I cannot reconcile the three models. That is the real problem.

I shall not discuss each model in its historical development but as we understand it now. I shall give the most decisive experimental evidence that we have after all these years of research. Finally, I shall mention experimental facts that seriously are at variance with the model.

It is convenient to express the conductance noise of homogeneous samples as the quantity α, defined by the relation

$$\frac{S_R}{R^2} = \frac{\alpha}{fN} \tag{1}$$

where N is the total number of free carriers [1]. Relation (1) makes it easy to compare the experimentally determined noise from different samples, and to compare the predictions of a theoretical model with experimental results. By using (1) one does not say anything about the origin of the noise. In particular, one has not confessed to which faith one adheres: mobility fluctuations or number fluctuations.

I cannot fully discuss the problem of mobility versus number noise in all its aspects. I shall restrict myself to the influence of defects on noise. I shall use the experimentally well established influence as an argument in the discussion on the nature of the 1/f noise. This will prove a very efficient way of generating unsolved problems.

2 The Eindhoven model

The Eindhoven model is the simplest one regarding the role of defects. The defects play no role at all. It explains the "normal" 1/f noise, which is found even in the best crystals [1,2]. There is one hypothesis to start with.

Hyp: The number of phonons, ϕ, in a mode of the lattice vibrations fluctuates with a 1/f spectrum.

1 When particles are scattered by a mode the scattering cross section fluctuates with a 1/f spectrum. This holds true for electrons (mobility noise) and for photons (Musha's optical experiments [3]).

2 For each mechanism that scatters electrons, we define α_i, analogously to (1)

$$\frac{S_{\mu_i}}{\mu_i^2} = \frac{\alpha_i}{fN} \tag{2}$$

where S_{μ_i} / μ_i^2 is proportional to $S_{\phi_i} / \overline{\phi}_i^{\,2}$, which is proportional to $\exp(\theta_i/T)$.

3. When there are several scattering mechanisms, Matthiessen's rule

$$\frac{1}{\mu_\Sigma} = \sum_i \frac{1}{\mu_i} \tag{3}$$

leads to

$$\alpha_\Sigma = \sum_i \left(\frac{\mu_\Sigma}{\mu_i}\right)^2 \alpha_i \tag{4}$$

where the index Σ indicates the result of all processes, like in the total mobility μ_Σ and in the overall α_Σ, as defined in (1).

α_i may be zero; for example for impurity- or for surface scattering where no modi are involved. α_{latt} can be further decomposed in terms of acoustic deformation (ad) and polar optical (po) vibrations.

$$\alpha_{latt} = \left(\frac{\mu_\Sigma}{\mu_{ad}}\right)^2 \alpha_{ad} + \left(\frac{\mu_\Sigma}{\mu_{po}}\right)^2 \alpha_{po} \tag{5}$$

4. Values of α_{latt} are usually found experimentally in the range 10^{-4} to 10^{-3}.

This straightforward presentation, starting from one single hypothesis, differs from the way in which our ideas developed over the years. Some people may find the historical development confusing, but the final result is simple.

Relation (4) predicts that α_Σ is proportional to μ_Σ^2 for a series of samples with different doping levels and, therefore, with different contributions of impurity scattering. There is ample experimental evidence for the correctness of relation (4) [1,2]. As an example, Fig. 1 shows log α versus log μ for InP at 77 K, recently published by Chen and Leys [4]. In the case of number fluctuations the experimental points should follow a horizontal line.

Since the source of the electrical 1/f noise is the 1/f noise in ϕ, the number of phonons per mode, there seems to be no role for defects to play. Nevertheless, we are well aware of the experimental fact that defects increase the noise considerably. Not only do we accept the experimental evidence from other groups, our own work on irradiated semiconductors confirms that defects increase the 1/f noise [5,6]. Therefore we suggest that defects create additional 1/f noise on top of the "normal" mobility noise expressed by α_{latt}.

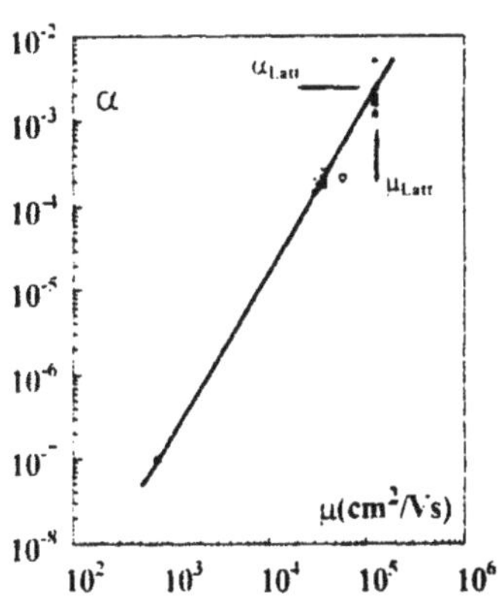

Fig. 1. α_Σ versus μ_Σ at 77 K InP samples with different doping. Measurements by Chen and Leys[4].

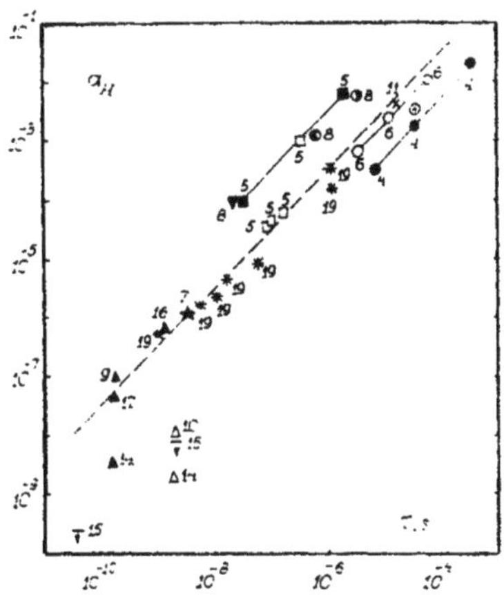

Fig. 2. α versus τ,, after Lukyanchikova[7]. Resistors and devices of different materials.

The 1/f fluctuations in the energy of a lattice mode could be the result of a special type of coupling between the modes. This special coupling is the foundation for purely mechanical models and computer simulations of lattice vibrations. Defects may play a role in the coupling of modes that are assumed to be independent harmonic oscillators in a perfect crystal only.

3 The Kiev model

Lukyanchikova collected α-values from her own work and from literature [7,8]. She considered resistors and simple devices, where a characteristic time τ for generation-recombination processes could be found. This provided ample support for the empirical relation she proposed

$$\alpha = \beta\tau \tag{6}$$

with $\beta = 300\ s^{-1}$. The g-r processes could be either in the bulk or at the surface. In the latter situation τ is the time the carriers need to diffuse to the surface.
See Fig. 2. Relation (6) should not be applied to very thin resistors of good semiconductor material. In such samples the surface only acts as a scatterer, not as a recombination centre[9]. If we leave out such samples from Lukyanchikova's survey, no harm is done to the basic idea that g-r processes strongly influence the 1/f noise. The processes that connect the two types of noise are unknown. Only some vague suggestions have been made. On might think of slow processes in the lattice or of slowly diffusing defects.

What are the consequences for the present discussion on the role of defects ? Introduction of defects enhances the g-r processes. Either the defects themselves are the relevant g-r centres; their τ then appears in relation (6). Or they influence the τ by influencing the kinetics of the g-r centres already present.

Anyway, the defects determine the α value. This cannot be a straightforward relation, since naïve reasoning leads to: more defects give more transitions with lower τ and hence lower α according to relation (6). This is not in agreement with general experience where more defects give rise to more noise. A recent publication[10] on poly-silicon confirms that there is a relation between α and τ. However, the experimental relation is not in agreement with relation (6).
It is $\alpha \propto \tau^{-n}$ with $0{\cdot}5 < n < 2{\cdot}5$.

4 The St. Petersburg model

The St. Petersburg group, D'yakonova, Levinshtein and Rumyantsev studied 1/f noise mainly in GaAs. They summarized their work in an extensive review paper[11]. The 1/f noise is explained as g-r noise from states in an exponential tail connected to the conduction band. See Fig. 3. The exponential shape is required for obtaining a 1/f spectrum. The tail states are due to defects. It is not theoretically explained how the defects create a tail that is exponential. But abundant experimental evidence is presented for the existence of the tail states:

1. The noise does not depend monotonically on illumination intensity.
2. The tail explains persistant photoconductivity with a long relaxation time.
3. The tail relates the noise quantitatively to damage.

One could oppose that since the defects are essential, the model would predict that there is no 1/f noise in perfect semiconductors. The normal value of α, about 10^{-4}, requires 10^{13} states in the tail. The authors consider this to be a low concentration of defects that will always be present.

According to the St. Petersburg model, 1/f noise is a fluctuation in the number of free carriers due to g-r noise from states created by defects. This model cannot be reconciled with the idea of mobility noise. It has been demonstrated experimentally that 1/f noise in fairly good material is mobility noise with $\alpha_{latt} \sim 10^{-4}$. The same α-values have been found in the best materials we can grow.

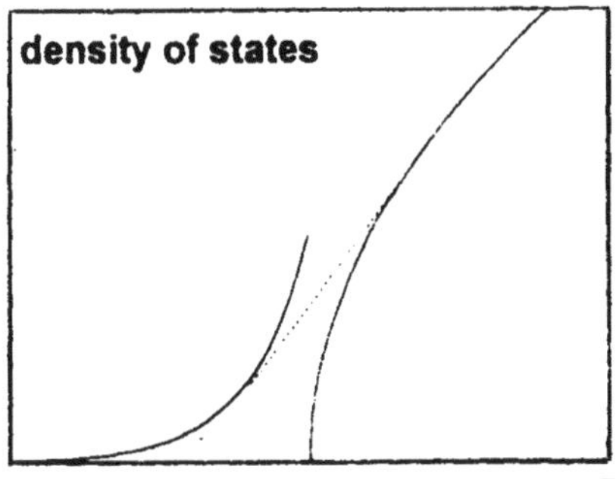

Fig. 3. Conduction band with exponential tail according to the St. Petersburg model[11].

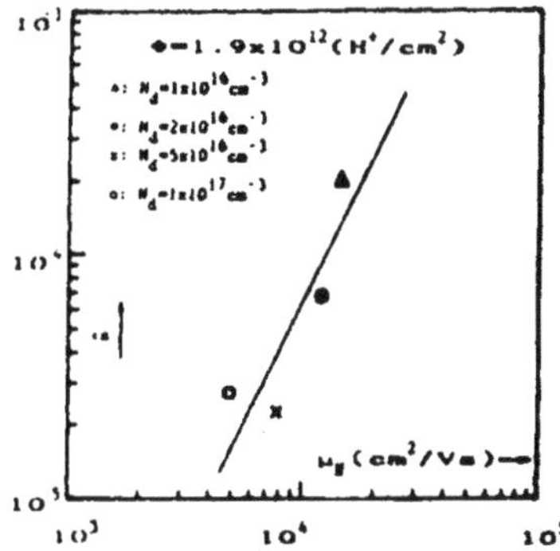

Fig. 4. α versus μ, measured at 77 K. GaAs samples with different original doping, irradiated with the same proton dose[15].

5 Discussion

How strongly the three models conflict can best be illustrated by the following table.

Group	Δn ? $\Delta\mu$?	Model	Role of defects
Eindhoven	$\Delta\mu$	fluctuation of cross-sections of lattice modes	No direct role. (coupling of modes ?)
Kiev	Δn (?)	related to normal g-r noise $\alpha = \beta\tau$	?
St. Petersburg	Δn	g-r noise from states in tail of conduction band	defects create tail

As promised, I shall not come up with a solution. But I think I am permitted to make two suggestions that may help to find a solution.

1. The 1/f noise in disordered metal layers is adequately described by models like the local interference model (LI), universal conductance fluctuation model (UCF), and two-level systems (TLS) [12,14]. Perhaps, a theoretician should tell us how many defects are required to make such models applicable to perfect epitaxial layers of semiconductors.
2. Experimentalists should investigate whether the 1/f noise induced by defects is of the Δn or $\Delta\mu$ type.

There already is an example of such an experiment. Lin Ren [15] of the Eindhoven group induced 1/f noise by proton irradiation of GaAs. The induced noise was proportional to the radiation dose. He investigated a series of GaAs samples with different original impurity contents, and hence different contributions of impurity scattering. All samples were irradiated with the same proton dose.
Fig. 4 shows log α versus log μ, plotted analogously to Fig. 1. Lin Ren's induced 1/f noise looks very much like mobility noise. What about Ukrainian and Russian 1/f noise ?

References

1. F.N. Hooge, T.G.M. Kleinpenning and L.K.J. Vandamme, *Reps. Prog. Phys.* **44**, 479 (1981).
2. F.N. Hooge, *IEEE Trans. El. Dev.* **41**, 1926 (1994)
3. T. Musha, G. Borbély and M. Shoji, *Phys. Rev. Lett.* **64**, 2394 (1990)
4. X.Y. Chen and M.R. Leys, *Solid State Electronics* (in press) (1996)
5. L. Ren, *J. Appl. Phys.* **74**, 4534 (1993)
6. X.Y. Chen and V. Aninkevicius, Proc. 7th Vilnius Conference on Fluctuation Phenomena in Physical Systems. Palanga (1994) p. 260.
7. N.B. Lukyanchikova. *Phys. Lett.* **A180**, 285 (1993).
8. N.B. Lukyanchikova, Proc. 7th Vilnius Conference on Fluctuation Phenomena in Physical Systems. Palanga (1994) p. 105.
9. F.N. Hooge, Proc. 7th Vilnius Conference on Fluctuation Phenomena in Physical Systems. Palanga (1994) p. 61.
10. N.N. Tkachenko, *JETP Lett.* **59**, 276 (1994).
11. N.V. D'yakonova, M.E. Levinshtein and S.L. Rumyantsev, *Sov. Phys. Semicond.* **25**, 1241 (1991).
12. M. Kogan and K.E. Nagaev. *Solid State Comm.* **49**, 387 (1984).
13. J. Pelz and J. Clarke. *Phys. Rev. B***36**, 4479 (1987).
14. N. Giordano, *Rev. Solid State Sc.* **3**, 27 (1989).
15. L. Ren and F.N. Hooge. Noise in Physical Systems and 1/f Fluctuations. St. Louis 1993. Editors P.H. Handel and A.L. Chung. *AIP Conference Proceedings* **285**, 65 (1993).

ON 1/f NOISE AND FREQUENCY INDEPENDENT LOSS TANGENT

T.G.M. KLEINPENNING
Department of Electrical Engineering,
Eindhoven University of Technology,
Eindhoven, The Netherlands

A new model has been proposed, which relates the 1/f noise in semiconductors to the frequency independent loss tangent in dielectrics. It is demonstrated that this model can explain the Hooge relation. The model supports the opinion that 1/f noise is caused by mobility fluctuations.

1 Introduction

The open-circuit voltage noise S_V of a capacitor with a dielectric is given by

$$S_V = 4kT\,\mathrm{Re}(Z) = \frac{4kTR}{1+\omega^2R^2C^2} = \frac{4kT\varepsilon''/(\varepsilon'\omega C)}{1+(\varepsilon''/\varepsilon')^2} \approx \frac{4kT\tan\delta}{\omega C}. \tag{1}$$

Here ε' and ε" are the real and imaginary part of the dielectric constant $\varepsilon_r = \varepsilon' - j\varepsilon''$. The loss tangent is defined by $\tan\delta = \varepsilon''/\varepsilon'$, the capacitance is $C = \varepsilon'\varepsilon_0 A/L$, and the resistance $R = L/\omega\varepsilon''\varepsilon_0 A$ with L the length and A the cross-section of the capacitor. In many dielectrics the loss tangent is usually found to be almost frequency-independent and much smaller than one[1], hence according to Eq. (1) the voltage noise S_V is inversely proportional to the frequency, just like 1/f noise. Now we can ask ourselves whether there is a relation between 1/f noise and loss tangent in materials. In this paper we try to relate 1/f noise and constant loss tangent. More specified, we try to relate Hooge's parameter α with the loss tangent δ.

2 Outline of the model

The electric charge density $\rho(\vec{r})$ of an atom, built up of nuclear and electronic charge, can fluctuate. On the condition that the total charge is constant we have

$$\Delta\rho(\vec{r},t) = \rho(\vec{r},t) - \langle\rho(\vec{r},t)\rangle \quad \text{and} \quad \int \Delta\rho(\vec{r},t)\,d\vec{r} = 0. \tag{2}$$

These fluctuations lead to a fluctuating electric dipole. If we put such an atom in between two parallel metal electrodes, then we observe voltage fluctuations across the electrodes. On the other hand we find the cross-section for electron scattering by such an atom to fluctuate.

A dielectric in between two metal electrodes shows voltage fluctuations according to Eq. (1). A free electron, moving in such a dielectric, will be scattered by the electric field fluctuations due to $\Delta\rho(\vec{r},t)$. These fluctuations lead to fluctuations in the electron mobility with spectral density S_μ. Now the question arises: "What is the relation between S_μ on the one side and $S_V = 4kTRe(Z)$ on the other side ?".

3 Theoretical approach

3.1. Atomic scattering of electrons

The scattering of electrons by an atom is represented by the scattering amplitude[2]

$$f(\theta) = f(\vec{k},\vec{k}') = \frac{-2\pi m}{h^2}\int e^{-i\vec{k}'.\vec{r}}\phi(\vec{r})e^{i\vec{k}.\vec{r}}d\vec{r} = \frac{2}{a_o qK^2}\int \rho(\vec{r})e^{i\vec{K}.\vec{r}}d\vec{r}\,, \tag{3}$$

where m is the electron mass, h is Planck's constant, $\phi(\vec{r})$ the electrostatic potential of the atom, $\vec{k}$ and $\vec{k}$ the wave vectors of the incident and scattered electron, $\vec{K} = \vec{k} - \vec{k}'$, and $a_o = h^2\varepsilon_o/\pi q^2 m$ the Bohr radius, with q the elementary charge and ε_o the permittivity of vaccum. Taking the nucleus at $\vec{r} = 0$, we obtain

$$f(\theta) = \frac{2Z}{a_o K^2}\left[1 + \frac{1}{Zq}\int \rho^e(\vec{r})e^{i\vec{K}.\vec{r}}d\vec{r}\right] = \frac{2Z}{a_o K^2}\left[1 - F(K)\right]\,, \tag{4}$$

where Z is the atomic number, θ the scattering angle, $\rho^e(\vec{r})$ the electronic charge density, and F(K) the atomic scattering factor with F(0) = 1. For the nuclear charge density the integral at the r.h.s. of Eq. (3) leads to Zq. Fluctuations $\Delta\rho^e(\vec{r})$ lead to fluctuations in f(θ) and thus to fluctuations in the differential cross-section σ(θ) = f(θ)f*(θ). The asterisk denotes the complex conjugate. We have

$$\Delta\sigma(\theta) = f(\theta)\Delta f^*(\theta) + f^*(\theta)\Delta f(\theta)\,, \tag{5}$$

with

$$\Delta f(\theta) = (2/a_o qK^2)\int \Delta\rho^e(\vec{r})e^{i\vec{K}.\vec{r}}d\vec{r}\,. \tag{6}$$

Assuming a sphere symmetrical charge distribution, then we have[2]

$$F(K) = (-4\pi/ZqK)\int_0^\infty r\rho^e(r)\sin(Kr)dr\,. \tag{7}$$

With the help of Eqs. (4,5,6) we find the fluctuations Δσ(θ) to be

$$\Delta\sigma(\theta) = \frac{8Z[1-F(K)]}{qa_o^2K^4}\int \Delta\rho^e(\vec{r})\cos(\vec{K}.\vec{r})d\vec{r}\,. \tag{8}$$

At room temperature the wave vector k of thermal electrons is of the order of $2\pi mv_{th}/h \approx 10^9\ m^{-1}$. The atomic radius a is of the order of 10^{-10} m. Therefore we make the approximations

$$Kr << 1,\ \ \sin(Kr) \approx Kr - (Kr)^3/6,\ \ \text{and}\ \ \cos(\vec{K}.\vec{r}) \approx 1 - (\vec{K}.\vec{r})^2/2\,. \tag{9}$$

With Eqs. (7,9) and the relation $\int 4\pi r^2\rho^e(r)dr = -\,Zq$ we find

$$1 - F(K) \approx (-K^2/6qZ)\int 4\pi r^4\rho^e(r)dr \approx K^2a^2/6\,. \tag{10}$$

Combining Eqs. (8-10) and using the relation $\int \Delta\rho^e(\vec{r})d\vec{r} = 0$ we obtain

$$\Delta\sigma(\theta) \approx \frac{-2Za^2}{3qa_o^2}\cdot\int \Delta\rho^e(\vec{r})\left[\frac{\vec{K}.\vec{r}}{K}\right]^2 d\vec{r}\,. \tag{11}$$

The fluctuations in the total cross-section are given by

$$\Delta\sigma = \int_0^{2\pi}\int_0^{\pi} \Delta\sigma(\theta)\sin\theta[1-\cos\theta]d\theta d\varphi\ . \tag{12}$$

3.2 Mobility fluctuations and cross-section fluctuations

Consider a dielectric where the electronic charge density around each atom fluctuates. The fluctuations at different atoms are assumed to be uncorrelated. A free electron moving in this dielectric will have a scattering cross-section per atom σ, a free path λ and a mobility μ. Fluctuations $\Delta\sigma$ lead to fluctuations $\Delta\lambda$ and $\Delta\mu$. During the free path the electron passes $p = \lambda/2a$ atoms. A relative fluctuation $\Delta\sigma/\sigma$ of an atom leads to a relative fluctuation in the free path $\Delta\lambda/\lambda = -(\Delta\sigma/\sigma)/p$. Since the fluctuations $\Delta\sigma$ of different atoms are uncorrelated, the relative spectral noise density in the free path and in the mobility for a single electron is found to be p times the contribution of one atom

$$S_\lambda/\lambda^2 = S_\mu/\mu^2 = p\cdot(S_\sigma/\sigma^2)/p^2 = 2n^2a\lambda S_\sigma, \tag{13}$$

where $\lambda = 1/(n\sigma)$ with n is the density of atoms, $n \approx 1/(2a)^3$.

3.3 Loss tangent and electronic charge density fluctuations

We apply two electrodes to the dielectric. There are no free electrons. Each atom has a dipole $\vec{p}$

$$\vec{p} = \int \rho^e(\vec{r})\vec{r}\,d\vec{r} \text{ and } \Delta p = \int \Delta\rho^e(\vec{r})\vec{r}\,d\vec{r}. \quad (14)$$

The fluctuations $\Delta\vec{p}$ lead to voltage fluctuations across the electrodes

$$\Delta V = \Delta p_x/(\varepsilon_r\varepsilon_o A) \text{ and } S_V = (\varepsilon_r\varepsilon_o A)^{-2} S_{p_x}, \quad (15)$$

where Δp_x is the component of $\Delta\vec{p}$ perpendicular to the electrodes. If the dipoles of the $N = nAL$ atoms fluctuate independently and if the orientation of the dipoles is random, then we obtain with $\langle\Delta p_x^2\rangle = \langle\Delta\vec{p}^2\rangle/3$ and $C = \varepsilon_r\varepsilon_o A/L$

$$S_V = N(\varepsilon_r\varepsilon_o A)^{-2} S_{p_x} = [n/(3\varepsilon_r\varepsilon_o C)]S_p. \quad (16)$$

From Eq. (14) it follows

$$S_p = \iint S_{\rho}e(\vec{r},\vec{r}')\vec{r}.\vec{r}'\,d\vec{r}\,d\vec{r}', \quad (17)$$

with $S_{\rho}e(\vec{r},\vec{r}')$ the cross-correlation spectral noise density of the fluctuations $\Delta\rho^e(\vec{r})$. Combining Eqs. (1,16) yields

$$S_p = 6\varepsilon_r\varepsilon_o kT \tan\delta/(\pi n f). \quad (18)$$

From Eqs. (17,18) we observe that dielectrics with constant $\tan\delta$ yield $S_{\rho}e(\vec{r},\vec{r}') \sim 1/f$. Consequently we can expect a $1/f$ spectrum for S_σ (see Eqs. (11,12)).

3.4 Relation between Hooge's parameter α and $\tan\delta$

Putting $S_\mu/\mu^2 = \alpha/f$ and using Eqs. (13,18), we obtain a relation between α and $\tan\delta$

$$\gamma = \alpha/\tan\delta = \frac{12}{\pi} na\lambda\varepsilon_r\varepsilon_o kT\, S_\sigma \Big/ S_p. \quad (19)$$

Using Eqs. (11,12,17) and taking $n = 1/(2a)^3$ the factor γ becomes

$$\gamma = \frac{2Z^2a^2\lambda\varepsilon_r\varepsilon_o kT}{3\pi\, q^2 a_o^4}\, r_o^2, \quad (20)$$

where the quantity r_o^2 is given by

$$r_o^2 = \frac{\iiiint\!\!\iint S_{\rho^e}(\vec{r},\vec{r}')\left(\frac{\vec{K}\cdot\vec{r}}{K}\right)^2\left(\frac{\vec{K}\cdot\vec{r}'}{K}\right)^2 g(\theta,\theta')d\vec{r}\,d\vec{r}'d\theta d\theta' d\varphi d\varphi'}{\iint S_{\rho^e}(\vec{r},\vec{r}')\,\vec{r}\cdot\vec{r}'\,d\vec{r}\,d\vec{r}'}, \quad (21)$$

with $g(\theta,\theta') = \sin\theta\sin\theta'(1-\cos\theta)(1-\cos\theta')$.

To evaluate r_o^2, we have to know how the fluctuations $\Delta\rho^e(\vec{r})$ go on. There are several possibilities.

According to Eq. (11) a transfer of electronic charge from the spot $\vec{r}$ to the spot $-\vec{r}$ leads to $\Delta\sigma(\theta) = 0$ and thus to $\gamma = 0$. Consequently, an electric dipole, which jumps between two opposite directions, gives no fluctuations in the cross-section σ.

If the fluctuations $\Delta\rho^e(\vec{r})$ are at random, then we have approximately $S_{\rho^e}(\vec{r},\vec{r}') = H(\vec{r})\delta(\vec{r}-\vec{r}')$. In this case the quantity r_o^2 becomes

$$r_o^2 = \frac{\iiiint\!\!\int H(\vec{r})(\vec{K}\cdot\vec{r}/K)^4 g(\theta,\theta')d\theta d\theta' d\varphi d\varphi' d\vec{r}}{\int H(\vec{r})\vec{r}^{\,2}\,d\vec{r}}. \quad (22)$$

Since the fluctuations occur at the edge of the atom $r \approx a$, and

$$\int_0^{2\pi}\int_0^{\pi} \sin\theta(1-\cos\theta)d\theta d\varphi = 4\pi, \qquad \left\langle(\vec{K}\cdot\vec{r})^4\right\rangle \approx \left\langle K^4 r^4\cos^4\beta\right\rangle \approx (3/8)K^4r^4,$$

with β the angle between $\vec{K}$ and $\vec{r}$, we expect r_o^2 to be of the order of $6\pi^2a^2$. Now we find

$$\alpha = \gamma\tan\delta \approx [4\pi Z^2\lambda\ \ \varepsilon_r\varepsilon_o kT(a/a_o)^4/q^2]\tan\delta. \quad (23)$$

Applying Eq. (3) to electrons in the conduction band of a semiconductor, we have to replace m by m* the effective mass, and a_o by $a_o\varepsilon_r m/m^*$.

For fluctuations $\Delta\rho^e(\vec{r},t)$, which are not uncorrelated with respect to the position $\vec{r}$ and which are not caused by dipoles jumping between opposite directions, we expect to have γ values of the order of magnitude as given by Eq. (23).

3.5 Numerical values for $\alpha = \gamma \tan\delta$

For Si, both n- and p-type, the value of γ at room temperature is of the order of

$$\gamma = 4\pi Z^2 \lambda \varepsilon_r \varepsilon_o kT (m^* a/m\varepsilon_r a_o)^4/q^2 \approx 0.1 \ .$$

Here we have used $\lambda \approx 4 \times 10^{-8}$ m and $m^*a/ma_o \approx 1$. For Ge we obtain $\gamma \approx 0.05$. For pure dielectrics the value of $\tan\delta$ is often found to be in the range of 10^{-5} to 10^{-3}. So α values for Si and Ge can be expected to be in the range of 10^{-6} to 10^{-4}. Usually the free path λ is inversely proportional to the temperature, in that case γ is independent of T.

4 Factor 1/N

For a single electron moving in a dielectric we obtain $S_\mu/\mu^2 = \alpha/f$ with $\alpha = \gamma\tan\delta$. What happens if N electrons move criss-cross through the dielectric ?

For a fluctuation $\Delta\rho(\vec{r}_1) = -\Delta\rho(\vec{r}_2)$ and $\left|\vec{r}_1\right| = \left|\vec{r}_2\right|$, Eq. (12) leads to $\Delta\sigma \sim \sin^2\beta_1 - \sin^2\beta_2$. Here $\beta_{1,2}$ is the angle between $\vec{k}$ and $\vec{r}_{1,2}$. For $\beta_2 = \beta_1$ and $\beta_2 = \pi - \beta_1$, we have $\Delta\sigma = 0$, The fluctuation $\Delta\sigma$ depends on the direction of $\vec{k}$, the average over all k-directions is $<\Delta\sigma>_{\vec{k}} = 0$. Since $<\vec{k}_i \cdot \vec{k}_j> = 0$ for $i \neq j$ and $i,j = 1$ to N we have $<\Delta\sigma_i \cdot \Delta\sigma_j> = 0$. As a result the contributions of the individual electrons to the fluctuations are uncorrelated. For the relative resistance noise we obtain[3]

$$S_R/R^2 = S_{\mu^*}/\mu^{*2} \approx \gamma \tan\delta/fN \approx \alpha/fN \ ,$$

where μ^* is the average mobility of the N electrons, i.e. $\mu^* = (1/N) \sum_{i=1}^{N} \mu_i$.

It is obvious that the model is compatible with the mobility fluctuation model, not with the McWhorter model for number fluctuations. The model predicts values for α of the right order of magnitude.

References

1. A.K. Jonscher, *Nature* **267**, 673 (1977).
2. J.L. Powell and B. Crasemann, *Quantum Mechanics*, (Addison-Wesley Publ. Company, London, England, 1962).
3. F.N. Hooge, *Physica* **114B**, 391 (1982).

NOISE IN THIN METAL FILMS AFTER LOW-TEMPERATURE ELECTRON IRRADIATION

K. ARMBRUSTER,* E. OCHS, A. SEEGER, A. STACH,† H. STOLL
Max-Planck-Institut für Metallforschung, Institut für Physik, Heisenbergstr. 1, D-70569 Stuttgart, Germany

Electrical resistance fluctuations in thin polycrystalline films (Al, Au) are investigated by means of a highly sensitive noise spectrometer. The analysis of noise in the temperature range between 10 K and 400 K according to the model of thermally activated processes proposed by Dutta, Dimon, and Horn indicates that defects are a source of $1/f$ noise. Other sources of resistance fluctuations observed are macroscopic temperature fluctuations and the relaxation of mechanical stress in the samples. The introduction of atomic defects by low-temperature electron irradiation increases the noise intensity drastically. The annealing behaviour of this noise component supports the view that atomic defects may contribute significantly to $1/f$ noise in metals even at rather low temperatures. In Au, an additional noise component, proportional to $1/f^2$, showed prominent maxima during the annealing procedure which might be related to the creation of metastable defects.

1 Introduction

The detailed investigation of $1/f$ noise in metals is a rather challenging experimental task owing to the smallness of the typical resistance fluctuations involved. Only by the development of high-resolution noise-measurement setups [1,2] did the investigation of noise in micron-scaled metal films at low temperatures become possible.

There is increasing evidence that defects may contribute to a large extent to $1/f$ noise in metals. Low-temperature irradiation with electrons or ions, which introduces atomic defects into the metal films, increases the $1/f$ noise drastically [3,4]. Further evidence for the influence of defects on noise in metals is provided by mesoscopic samples in which two-level resistance fluctuations have been observed [5,6] that are presumably caused by changes in the scattering cross-section of conduction electrons as defects move between two metastable states. The resulting noise spectra are Lorentzian with the knee frequency given by the characteristic time of the transition between these states.

The question still under debate is to what extent other effects contribute to noise observed in metal films. The present paper presents results on noise of Al and Au films before and after low-temperature electron irradiation and discusses them with respect to the individual sources of electrical noise.

---

*née: Dagge

†Present address: Compuserve, Munich, Germany

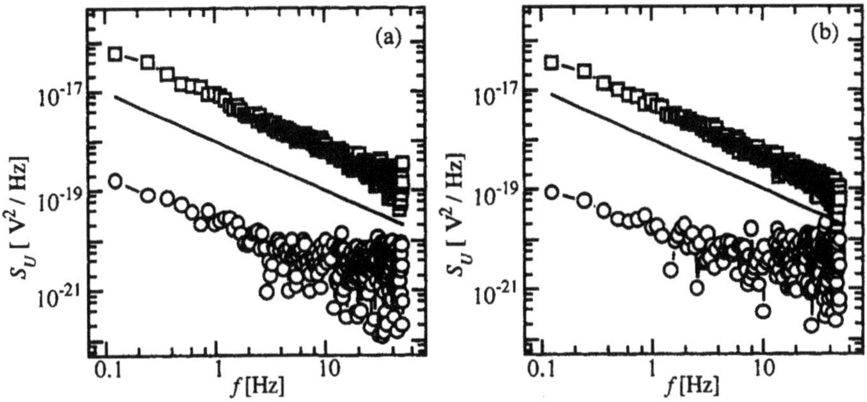

Figure 1: Spectral density of noise power $S(f)$ at 10 K (circles) and 300 K (rectangles) of a thin Al (a) and Au (b) film. The solid lines correspond to $S(f) \propto 1/f$.

2 Experiments

Polycrystalline samples (300 μm×5μm) were produced by electron-beam evaporation with thicknesses of 100 nm (Al) or 110 nm (Au)[‡]. The grain size distribution was rather wide with an average size of about 100 nm. The residual resistance ratios, $R_{300\,K}/R_{10\,K}$= 3.13 for Al and 6.8 for Au, are typical for metal films of this thickness.

The samples were irradiated below 10 K in a high-voltage electron microscope with 3.7×10^{23} e^-/m^2 at 1 MeV. In subsequent annealing experiments the recovery of both electrical resistance and noise was observed as a function of the annealing temperature T_a. Measurement of the low noise intensities occurring at low temperatures was made possible by application of a special phase-sensitive correlation technique [2] which averages out thermal noise and external interference. Thus the measurement technique allows us to rule out fluctuations of the output signal other than resistance fluctuations of the sample. During the noise measurements temperature fluctuations were kept below ±2 mK.

3 Results and Discussion

In the whole temperature range investigated the noise intensities of all samples, irrespective of whether they were undamaged or irradiated, were proportional to $1/f^m$ with an exponent m close to 1 (Fig. 1). In contrast to other measurements on metal films,[9] we found no increase of the noise power at low temperatures. Our experience is that in order to avoid any contributions to

‡sample preparation: A. H. Verbruggen, DIMES, TU Delft, The Netherlands

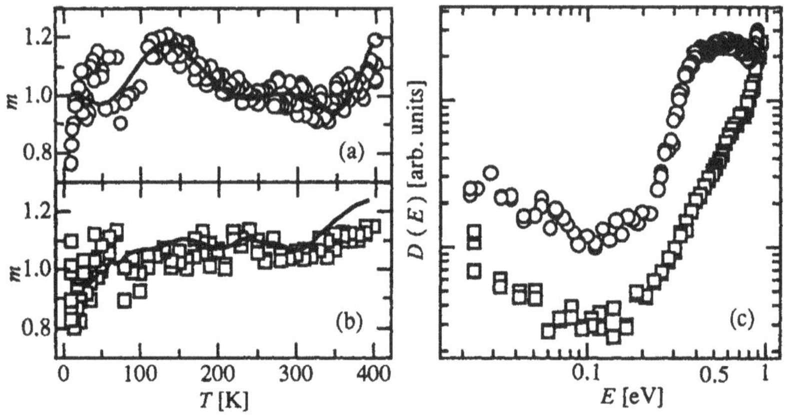

Figure 2: Analysis of $1/f$ noise according to the model of Dutta, Dimon, and Horn.[7,8] Measured frequency exponent m of Al (a, circles) and Au (b, rectangles) as a function of temperature. The solid lines give the exponent m derived from the above model. (c) Distribution of activation energies $D(E)$ of Al (circles) and Au (rectangles) calculated from the temperature dependence of the $1/f$ noise at 1 Hz.

the low-temperature noise spectrum due to macroscopic temperature fluctuations it is necessary to stabilise the temperature to about ± 2 mK, since the influence of temperature fluctuations increases drastically below about 40 K.

3.1 Unirradiated Specimens

The exponents m derived according to the model of Dutta, Dimon, and Horn[7,8] from $S(T)$ were in good agreement with the frequency dependence of the noise power (Fig. 2a and b). In the case of Al, the analysis of the temperature dependence of the spectral density of noise power, $S(T)$, within the framework of Dutta, Dimon, and Horn gave a maximum in the distribution of activation energies $D(E)$ at about 0.65 eV (Fig. 2c) and an attempt frequency ν_0 of the resistance fluctuations of 10^{12} Hz. The maximum of the activation energy at 0.65 eV is in agreement with the typical activation energy for grain boundary diffusion[10] and the attempt frequency of the resistance fluctuations ($\nu_0 = 10^{12}$ Hz) is compatible with the motion of atomic defects. The activation energy for grain boundary diffusion in Au is about 1.0 eV[11] and thus outside the range of $E < 0.9$ eV (corresponding to $T = 400$ K) accessible in the present experiment. These results support the view that the diffusion of defects can give rise to resistance fluctuations in metal films.

In some of the unirradiated gold samples, at low temperatures a $1/f^2$ component was found in addition to the $1/f$ noise (Fig. 3a). Its occurrence is correlated with sudden large fluctuations in the voltage across the sample

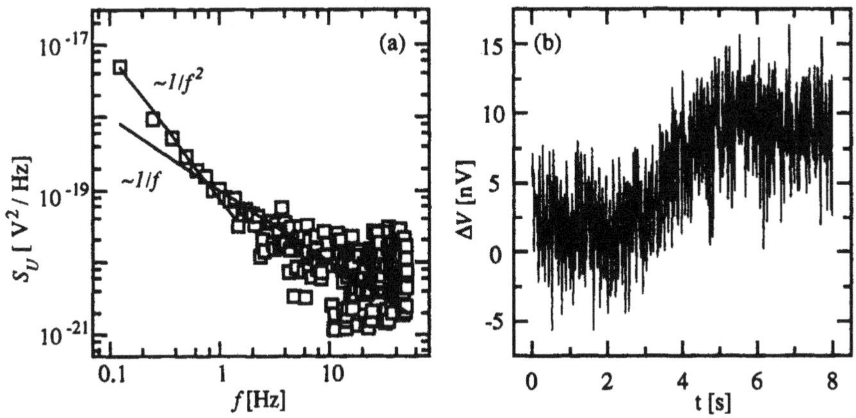

Figure 3: (a) Spectral density of noise power of an Au film at 90 K. The solid lines, which are proportional to $1/f$ and $1/f^2$, are given for comparison. (b) Typical voltage fluctuations $\Delta V(t)$ across the sample as a function of time.

(Fig. 3b) on the time scale of a few seconds. This suggests that the detected $1/f^2$ spectrum is probably the high-frequency part of a Lorentzian. Characteristically, the samples showing the $1/f^2$ component had a poor adhesion to the substrate and partial detachment of these films took place during the thermal cycling of the measurement procedure. An explanation for the $1/f^2$ noise could thus be that resistance fluctuations occur due to fluctuations of the thermal conductivity to the substrate, which would cause temperature fluctuations of the sample or fluctuations in the mechanical stress of the film. Such a connection between resistance fluctuations and mechanical stress was already described by Fleetwood and Giordano.[12]

3.2 Irradiated Specimens

For both metals, Al and Au, the typical increase of the residual resistance due to the irradiation with 3.6×10^{23} e^-/m^2 was about 20 % whereas the noise increased about one order of magnitude. In the case of Al the typical energy deposited by the electrons is sufficient for the creation of vacancies and interstitials in the bulk material. By contrast, the threshold for the creation of vacancy–interstitial pairs in bulk Au at 10 K is higher than the energy deposited by 1 MeV electrons. The damage occurs in this case at imperfections, e.g. grain boundaries or surfaces, and is known as subthreshold damage.[13]

In subsequent isochronal annealing experiments the noise and the resistance at fixed measuring temperature T_m were determined as a function of the annealing temperature T_a. The spectra of the irradiated Al specimens were proportional to $1/f$ during the whole annealing procedure. The irradiation-

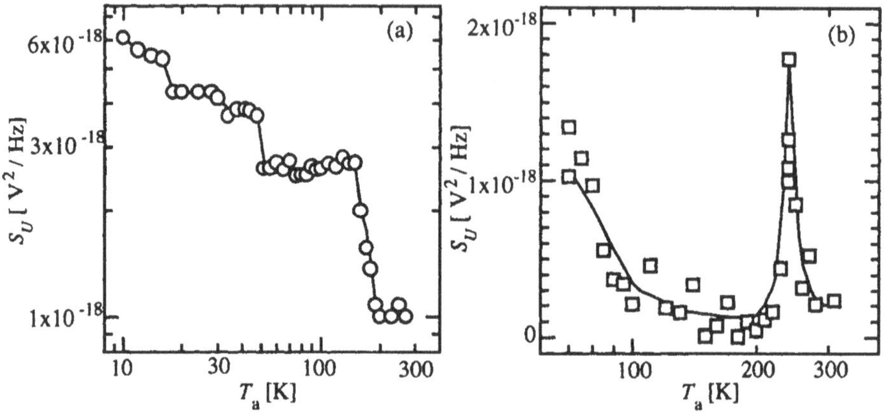

Figure 4: Recovery of noise of irradiated specimens after isochronal anneals for 6×10^2 s at progressively higher annealing temperatures T_a. (a) Recovery of the $1/f$ noise at 1 Hz of an Al film. (b) Intensity of the $1/f^2$ component at 1 Hz of an Au film.

induced increase in noise anneals in discrete steps (Fig. 4a) at the temperatures of the well-known recovery steps of the residual resistivity[14] and is completely recovered after annealing at temperatures above 350 K, as is the irradiation-induced residual resistivity. The steps in noise can therefore be attributed to the annealing of intrinsic atomic defects such as vacancies and self-interstitials.

In the case of irradiated Au an additional component proportional to $1/f^2$ was superimposed on the $1/f$ noise. Measurement of the isochronal annealing was performed at $T_m = 70$ K in order to increase the intensity of the $1/f$ component and thus allow its detection with sufficient accuracy. The irradiation-induced increase of the $1/f$ noise recovered completely in a discrete step at the same temperature as Stage II of the residual resistivity, whereas the intensities of the $1/f^2$ component exhibited pronounced maxima in the annealing stages II and III at approximately 100 K and 250 K (Fig. 4b).

The $1/f^2$ component was correlated with large fluctuations of the voltage across the sample on the time scale of a few seconds. A microscopic explanation might be the formation of metastable defects during the recovery. These defects might enhance the relaxation of mechanical stress or they might fluctuate at 70 K on a time scale of a few seconds. Both mechanisms would lead to a Lorentzian spectrum, whose high-frequency part shows up as $1/f^2$ noise.

4 Conclusions

In the entire temperature range investigated, the $1/f$ noise of Al and Au films showed excellent agreement with the model of Dutta, Dimon, and Horn [7,8]. This indicates that the $1/f$ noise is caused by thermally activated resistance fluctuations. The distribution of activation energies and the attempt frequencies calculated according to the model supports the view that the microscopic origin of the resistance fluctuations is the motion of lattice defects. The results on specimens electron-irradiated at low temperatures shows that noise is increased drastically by the introduction of atomic defects. The annealing of the various defect types, e.g. vacancies and interstitials, resulted in the stepwise recovery of the irradiation-induced noise. The observation of a $1/f^2$ component in irradiated Au specimens with maxima at certain annealing temperatures might be caused by the formation of metastable defects during the annealing.

Acknowledgements

We should like to thank Dr. A. H. Verbruggen, DIMES, TU Delft, The Netherlands, for sample preparation and continuous interest in our work.

References

1. J. H. Scofield, *Rev. Sci. Instrum.* **58**, 985 (1987).
2. A. H. Verbruggen, H. Stoll, K. Heeck, R. H. Koch, *Appl. Phys. A* **48**, 233 (1989).
3. J. Pelz, J. Clarke, *Phys. Rev. Lett.* **55**, 738 (1985).
4. J. Briggmann, K. Dagge, W. Frank, A. Seeger, H. Stoll, A. H. Verbruggen, *phys. stat. sol (a)* **146**, 325 (1994).
5. K. S. Ralls, R. A. Buhrman, *Phys. Rev. Lett.* **60**, 2434 (1988).
6. J. Caro, in: *Noise in Physical Systems and* $1/f$ *Fluctuations*, ed. by V. Bareikis and R. Katilius, World Scientific Publishing, Singapore, pp. 41–46 (1995).
7. P. Dutta, P. Dimon, P. M. Horn, *Phys. Rev. Lett.* **46**, 646 (1979).
8. P. Dutta, P. M. Horn, *Rev. Mod. Phys.* **43**, 646 (1981).
9. N. O. Birge, B. Golding, W. H. Haemmerle, *Phys. Rev. Lett.* **62**, 195 (1989).
10. R. H. Koch, J. R. Lloyd, J. Cronin, *Phys. Rev. Lett.* **55**, 2487 (1985).
11. D. Gupta, K. W. Asai, *Thin Solid Films* **22**, 121 (1974).
12. D. M. Fleetwood, N. Giordano, *Phys. Rev. B* **28**, 3625 (1983).
13. M. Hohenstein, A. Seeger, W. Sigle, *J. Nucl. Mater.* **169**, 33 (1989).
14. J. W. Corbett, in: *"Electron Radiation Damage in Semiconductors and Metals"*, Academic Press, New York and London (1966).

IS 1/F NOISE CAUSED BY MOVING DEFECTS?

A.V.YAKIMOV, S.YU.MEDVEDEV, I.YU.ZVORYKIN
N.Novgorod State University, N.Novgorod 603600, Russia

We try to test the model linked the 1/f noise with the movement of point defects. As the tool the method of spectral-statistical analysis is modified and used. The necessity to re-examine the term "stationarity" of the noise with the account of the limited measuring time was found. We have also found very high sensitivity of the method to the noise non-Gaussianity.

1 Introduction

The nature of the 1/f noise in conducting samples is discussed here. We try to examine the model of Two Level Systems (TLS), see e.g. [1]. Every TLS is associated with a single point defect having two meta-stable states. The noise is stationary if the ensemble of TLS's is fixed. The noise may be considered as Gaussian if the ensemble is large enough. We suppose that defects may diffuse through the sample, changing the current ensemble of TLS's. Thus, the noise loses its stationarity.

Our aim is the test of the noise stationarity and Gaussianity. For this purpose we use the method of the spectral - statistical analysis. Similar investigations were made earlier [2], but results reached are not sufficient for the strict interpretation. This lack is caused by the absence of the confidence interval for measured values.

2 Problems

The first problem we have got was the necessity to re-examine the term "stationarity" of the noise. Process $x(t)$ is stationary (in wide sense) if its statistical moments do not depend on time t, and correlation function $\Phi(t;t+\tau)$ depends only on delay τ. This definition is based on full statistical ensemble of the process. If $x(t)$ is ergodic then the average over ensemble may be replaced by the average over $t\in(-\infty,\infty)$.

In the reality the averaging is made over finite time $t \leq T$. As a result, estimates of all statistical characteristics are random functions of t. If process $x(t)$ is stationary then the variance of these estimates tends to zero when $T \to \infty$. We determine the confidence interval (say, on the significance level 95%) for measuring values. The process is stationary if only negligible part (less than 5%) of estimates is out of this interval. In the opposite case $x(t)$ is considered as non-stationary.

The usage of the method of the spectral - statistical analysis for the reveal of the movement of defects has yielded some unexpected results [3,4]. We performed two experiments: the ***accuracy test***, and the ***correlation test***.

In the first one the accuracy ε_{exp} of the measurement of the filtered noise intensity is estimated. This estimate is compared with the theoretical value $\varepsilon_{th}=\sqrt{(t \cdot \Delta f)}$ found for Gaussian stationary noise. It was assumed, that the noise non-stationarity, caused by the diffusion of defects, is displayed through a reliable deviation of the estimate ε_{exp} from theoretically expected value.

We have found that while the measurement time t is increased then ε_{exp} is decreased. But the increase of the filter band-width Δf may give the rise of the discrepancy between theoretically predicted and experimentally estimated accuracy values.

In the second test the correlation between the noise intensities on outputs of two non-overlapping band-pass filters is estimated. This value for Gaussian stationary noise equals zero, of course, within boundaries of corresponding confidence interval. In experiments both negative and positive correlations were found.

The positive correlation may be explained by instabilities in the measuring set-up (ambient temperature, and voltage supply drift). The negative correlation, as we assumed earlier [3,4], may be explained only by the movement of defects within the sample under test.

The described results have shown the necessity to make the special test of the noise non-Gaussianity influence on the measurement results.

3 Non-Gaussian noise modelling and treatment

The signal was formed by a generator of random numbers in a computer. That gives the uniformly distributed pseudo - random sequence of integers. To reach various degree of affinity to Gaussian process the sum of N_e integers was used:

$$x(t) = (1 / N_e) \, \Sigma x_i (t).$$

In our numerical experiments the values $N_e = 1, 16, 128$ were chosen. When 16 (or more) items are summarised at once, it is usually assumed, that the process is reasonably close to Gaussian one.

For $N_e \geq 16$ the noise ***histogram*** (as the estimate of the *pdf*) is a good "Gaussian" one. At $N_e = 128$ the histogram does practically not differ from Gauss law. The ***spectrum*** was found by *2048* - point FFT, with the usage of the first *800* spectral components of the signal. The deviation from the spectrum of "ideal" white noise corresponds to theoretically expected value, that is about *0.08dB*.

The ***band-pass filtering*** was realised as follows. The signal was Fourier transformed. The intensities of n_f spectral lines were summarised, whose frequencies fall into the chosen band. As a result we got a single sample of the filtered noise intensity. The array of N_T samples was accumulated and statistically treated. The results of the treatment were presented in the following way.

Each result was evaluated and displayed for the ensemble of $n \cdot n_a$ samples; $n = 1 \ldots N$, where $N = 400$ -the total number of the discrete samples on a time; $n_a = 8$ -number of the samples in the intermediate averaging; thus, $N_T = N \cdot n_a$.

The theoretical value of the accuracy was determined in Gaussian approximation: $\varepsilon_{th} = \sqrt{(t\,\Delta f)}$. In the discrete variant the product $(t\,\Delta f) = n \cdot (n_a \cdot n_f)$ is the current number of available non-correlated intensity samples of the spectral lines.

The accuracy ε_{exp} of the measurement was estimated during the data processing. This value was rationed on ε_{th}, that gives the relative estimate of the accuracy of the filtered noise intensity measurement: $\varepsilon_{rel} = \varepsilon_{exp}/\varepsilon_{th}$. If Gaussian stationary noise is analysed, then this relative accuracy gets values $\varepsilon_{rel} \approx 1$, not leaving the frames of the confidence interval. With the account of this circumstance the results of statistical treatment were additionally shifted down and normalised on the half-width of the confidence (with significance level of 95%) interval ΔE:

$$e = (\varepsilon_{rel} - 1)/\,\Delta E = [(\varepsilon_{exp}/\varepsilon_{th}) - 1]/\,\Delta E\,.$$

If more than 95% of data satisfy the condition $|e| < 1$, then the noise is Gaussian and stationary. The stationarity was assumed to be fulfilled in our numerical experiments, the main problem was the test of the noise Gaussianity.

The confidence interval half-width, evaluated in the approximation of delta - correlated Gaussian noise, is as follows [4]

$$\Delta E \approx 4.3 \cdot \varepsilon_{th}\,.$$

This relation was evaluated under the condition $\varepsilon_{th} \ll 1$.

It was assumed, that the noise non-Gaussianity is displayed first of all through the link between spectral lines, whose frequencies are multiple one to another (i.e. between harmonics). Therefore two types of the band-pass filter were chosen.

The intensities of Fourier - components with numbers *2... 101* were summarised in LF filter. That is the filter, covering up to *50* harmonics. In the second case the HF filter, covering lines with numbers *652... 751* was chosen. Harmonically linked spectral lines are completely absent in this band.

The results of numerical experiment for the LF filtered uniformly distributed noise are shown in fig. 1a. Here the numbers of accumulated samples n are pointed on the horizontal axis; the vertical axis contains values of the normalised accuracy e. Results for the HF filter are completely similar, namely, all data tend to be under the bottom of the confidence interval e = -1 while the time is increasing.

Thus, the noise non-Gaussianity is manifested through the link not only between harmonic components, but in more complex way.

In Fig. 1a the vertical line $n=100$ is drawn. Data for $n \leq 100$ correspond to the accuracy available when the filter band-width is decreased four times, that is $n_f=25$. Results for $n_f=25$ are shown in Fig. 1b, where the same frame is used.

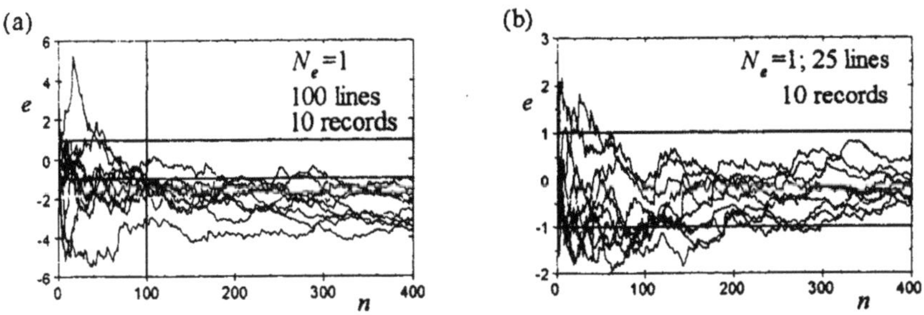

Fig. 1

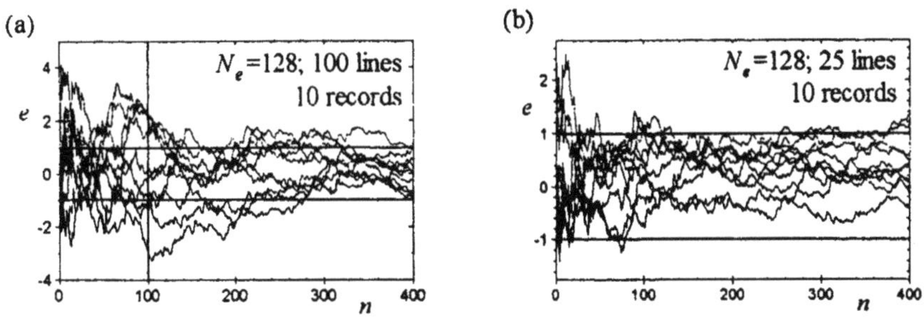

Fig. 2

One can see from Fig. 1b, that 4 times reduction of the filter band-width has resulted the essential reduction of the filtered noise non-Gaussianity. Now the data are practically within the confidence interval restricted by lines $e=\pm 1$.

While summing *16* uniformly distributed items the difference appeared in the results of $n_f = 100$ spectral component summation. Strong data spread was found relatively to the confidence interval limits, with the weak tendency to the reduction while the time of the analysis (the number n) is increased. The similar picture, but with some smaller spread of data for $n_f = 100$, was observed for the sum of *128* items; the corresponding illustration is given in Fig. 2(a, b).

4 Conclusion

Accuracy of the filtered noise intensity measurement is very sensitive to the noise non-Gaussianity. An increase of the time of the analysis explicitly results in the increase of the accuracy. But the increase of the filter band-width may yield the increase of the error in the comparison with value corresponded to "ideal" Gaussian noise. This result agrees with [5], where it is shown, that the accuracy of the intensity measurement is determined by a tri-spectrum of the noise.

Thus, the problems to be solved are seen as follows.

(i) If the 1/f noise is non-Gaussian then what are the type and nature of this non-Gaussianity.

(ii) If the noise is non-stationary then:

a) what are the type and nature of this effect ?

b) how to reveal this non-stationarity on the real time scale?

Acknowledgements

The authors are thankful to Professors F.N.Hooge and L.B.Kiss for the fruitful discussions on considered problems. This investigation was partly supported by The Netherlands Organisation for Scientific Research, Project No. GOS 713-115.

References

[1] Sh.M.Kogan. *Sov. Phys. - Usp.* **28**(2), 170 (1985).

[2] P.J.Restle, M.B.Weissman, and R.D.Black. *J. Appl. Phys.* **54**, 5844 (1983).

[3] A.V.Yakimov *Proc. 7th Vilnius Conf. Fluctuation Phenomena in Physical Systems*, ed. V.Palenskis (Vilnius University Press, 1994. P.81.)

[4] A.V.Yakimov, and F.N.Hooge. *Proc. 13th Int. Conf. Noise in Physical Systems and 1/f Fluctuations*, ed. V.Bareikis and R.Katilius (World Scientific Publishing Co. 1995. P.291.)

[5] G.N.Bochkov, K.V.Gorokhov, and I.R.Konnov. *Pis'ma v JTF* **20**(8), 35 (1994) (in Russian).

ENHANCED 1/f NOISE INDUCED BY ATOMIC RAYLEIGH WAVES IN DISCONTINUOUS PLATINUM FILMS

MIHAI MIHAILA
ICCE - Bucharest, str. E.I. Nicolae 32B , 72996 Bucharest, Romania

ANDREI-PETRU MIHAILA
Electronics Dept., Politechnical Univ. of Bucharest, Bd.. Iuliu Maniu 1-3, 77202 Bucharest, Romania

The temperature dependence of 1/f noise in clean discontinuous platinum films is modeled by phonon density of states superposition method in the range(100-200)K. Our data show that both longitudinal and transversal(vertical) motion of surface atoms generate 1/f noise but enhanced noise level is observed when carriers couple to the vertical motion of surface atoms(Rayleigh waves). A possible connection between the Hooge parameter and the Eliashberg function is revealed.

1 Introduction

For more than seventy years, 1/f noise remains one of the most puzzling phenomenon of solid state physics. Due to its extremely ubiquity, the microscopic origin of 1/f noise is still unsolved. Over the last two decades, evidence has been accumulated proving that defects motion[1]-[12] or some kind of atomic motion[1],[9] could be the microscopic source of 1/f noise in metals. According to some authors[6],[12], experiments ensuing the defect motion models are inconsistent with the model which relies on phonon scattering as source of 1/f noise[13]. Experiments performed near the melting transition[12],[14] showed that the noise intensity decreases upon melting, explicitely recognizing that the lattice is playing a role in generating 1/f noise. Nevertheless, it is concluded that phonons have no influence in the noise mechanism[12]. However, phonon scattering seems to be involved[13] and both bulk[15] and surface phonons[16] were observed in the 1/f noise. The 1/f noise observed in the laser light scattering on quartz[17] and in the current of a tunneling microscope[18] was attributed to phonon number fluctuation. Also, using Phonon Density of States Superposition(PDOS) method, a kind of atomic motion, namely longitudinal motion of surface atoms, has been identified as source of 1/f noise[19] in discontinuous platinum films(DPF). Now we report that in some DPF's the transversal(vertical) motion of the surface atoms generates enhanced 1/f noise. But the vertically polarized surface atomic motion is known as Rayleigh waves[20], thus the atomic Rayleigh wave[21],[22] is identified as source of 1/f noise. Finally, a new physical interpretation of the Hooge parameter is given.

2. Experiments, results and discussion

The data reported here are for clean DPF . Clean means that no desorption peak due to residual gases has been observed during the noise measurements. A number of five samples, evaporated at the same time on a sapphire substrate, were investigated. 1/f noise spectra were observed in all samples. The noise intensity(S_V) followed a quadratic dependence on the applied voltage(V), so as the temperature(T) dependence of the normalized noise intensity(S_V/V^2) has been determined in the (100-200)K temperature range. Figure 1 shows comparatively the dependence of S_V/V^2 vs. T for the sample R_{AB}, with the lowest noise intensity, and for the sample R_{DE} whose noise intensity is enhanced.

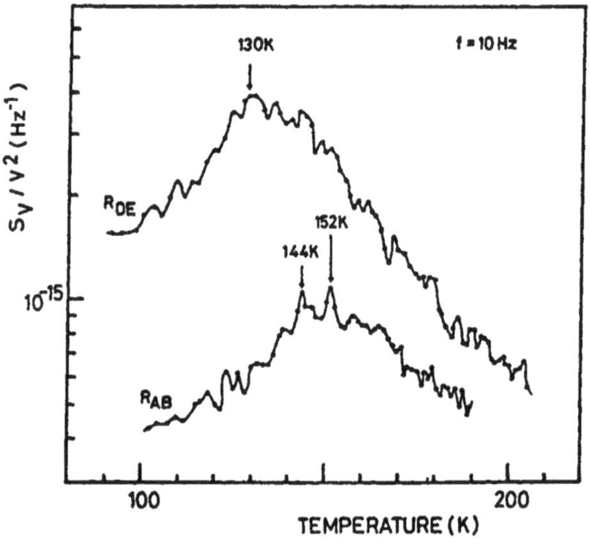

Figure 1: Dependence of the 1/f noise intensity on temperature for the samples R_{AB} and R_{DE} at f=10 Hz; points are experimental data

The most salient feature of the 1/f noise intensity in both samples is the existence of some maxima. While in the sample R_{AB} the maximum is structured with two peaks located at 144K and 152K, respectively, in the case of R_{DE} the maximum is enlarged, down shifted, with a *maximum maximorum* at about 130K. A similar behaviour as in R_{DE} shows the 1/f noise of the other three samples, but in their case 50 times more noise has been found. We were especially intrigued by the observation that all samples having the noise maximum located around 130K showed enhanced 1/f noise. Looking for a physical explanation of this observation, we have resorted to the PDOS method[16],[19] to model the temperature dependence of 1/f noise data. The noise data for the R_{AB}

sample and the function F(ω) obtained by PDOS method are compared in figure 2. The function F(ω) was obtained by the superposition of some surface PDOS and the bulk PDOS for platinum[23]. For surface, only PDOS corresponding to the longitudinal motion of the platinum atoms in the first atomic layer, both in the ΓK and ΓM symmetry directions, were taken from Kern *et al.*[24]. The resulting PDOS was noted F(ω), though the phonon frequency(ω) was converted into an equivalent temperature($T=\hbar\omega/k_B$, where the terms have their usual meaning). The reasonable fit between F(ω) and the noise data points to the longitudinal motion of surface atoms as a source of 1/f noise.

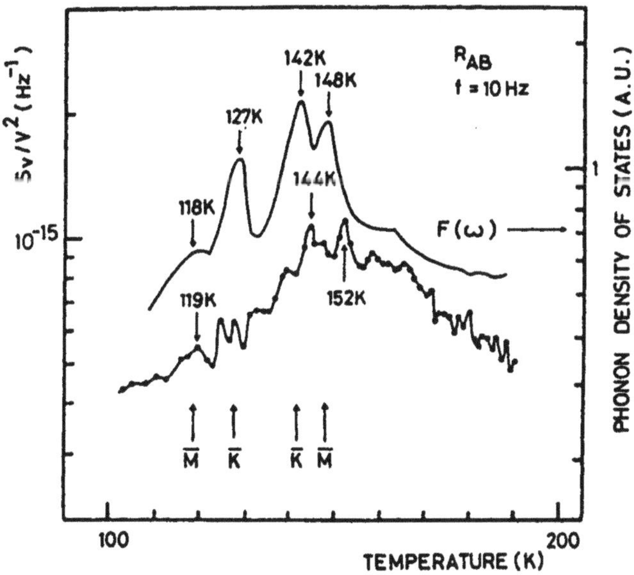

Figure 2: Comparison between the normalized 1/f noise spectral density of the R_{AB} film and the function F(ω); the arrows with the associated letters indicate the symmetry directions in the platinum surface Brillouin zone.

Figure 3 shows the dependence of S_V/V^2 vs. T for R_{DE} and, for comparison, the function F'(ω). In this case, the use of only F(ω) to fit the noise data was in no way satisfactory, let alone the explanation of the maximum at 130K. The way out was to add to F(ω) the PDOS function corresponding to the transversal (vertical) motion of the surface atoms in the first atomic layer, in the ΓK symmetry direction [24]. The result was a new function F'(ω) with a *maximum maximorum* at 129K which can explain the 130K maximum in the noise. These results show that the transversal motion of surface atoms is another source of 1/f noise. The transversal vibrational mode of the surface atoms are known as Rayleigh wave "because Lord Rayleigh in 1885 had shown its existence in a continuous medium; we may liken it to a propagating ripple on the solid surface - it is a motion involved in earthquakes", to quote Myers[25]. Therefore, there is enough reason to consider

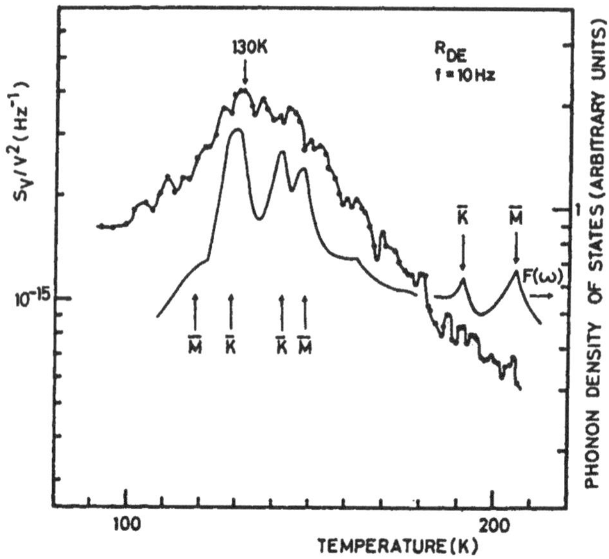

that enhanced 1/f noise observed in R_{DE} film is due to electron scattering by Rayleigh phonons. Corroborated with the Myers' above statement[25], this observation has direct implications on the origin of 1/f noise in solid state physical systems and not only.

Figure 3: Comparison between the normalized 1/f noise spectral density of the R_{DE} film and the function F'(ω); the arrows with the associated letters indicate the symmetry directions in the platinum surface Brillouin zone.

3. Search for the physical significance of the Hooge's parameter

Although widely used to characterize the noisiness of many physical systems, the physical significance of the Hooge parameter[26], α , is still obscure. It is now established that α depends on temperature[5],[11],[14],[27] and scattering mechanisms[13],[27]. Suggestions have also been made that α would have a spectroscopic character[15] From the data on discontinuous platinum films, we have seen that $S_V/V^2 \sim F(\omega)=\Sigma F_i(\omega)$, where $F_i(\omega)$ is the PDOS of the i-th phonon branch. On the Fermi surface, the electrons couple very anisotropically to phonon branches[28]-[31]. Therefore, the participation of $F_i(\omega)$ could be weighted by the associated electron-phonon matrix element(β_i^2), hence $S_V/V^2 \sim \Sigma\beta_i^2 F_i(\omega)$. For some temperature intervals, the anisotropy of β_i^2 could account for the dominance of , say, $F_i(\omega)$ and thus, in the Hooge's form :

$$S_V/V^2 \sim \alpha/f \sim \beta_i^2 F_i(\omega)\ . \tag{1}$$

But the product $\beta_i^2 F_i(\omega)$ of the squared matrix element β_i^2 for the electron-phonon interaction and the phonon density of states $F_i(\omega)$ is known as Eliashberg function. Consequently, the formula (1) suggests a possible connection between the Hooge parameter and the Eliashberg function : $\alpha \sim \beta_i^2 F_i(\omega)$. Such a connection, if any, can explain at least qualitatively many unclear and apparently dissimilar aspects of the 1/f noise in solid state physical systems. Also, it can explain our noise data, if one considers that the electrons couple stronger to the transversal phonon modes, as it happens in noble metals at the necks of the Fermi surface[28]-[32].

4. Conclusion

In conclusion, the transversal motion of surface atoms, also known as Rayleigh wave, has been identified as source of 1/f noise in DPF. It generates more 1/f noise than its longitudinal counterpart. This observation was attributed to a stronger coupling of electrons to transversal surface atomic motion. A possible new connection between the Hooge parameter and the Eliashberg function has been found.

Acknowledgements

One of the authors(M.M.) is indebded to Prof. A. Stepanescu and Prof. A. Masoero for the help in obtaining the experimental data. Thanks are also due to C. Trisca for the help in preparing the manuscript.

References

1. M.B. Weissman, *Rev. Mod. Phys.* **60** , 537(1988).
2. N. Giordano, *Rev. Solid St. Sci.* **3**, 27(1989).
3. Sh. M. Kogan and K. E. Nagaev, *Fiz. Tverd. Tela* **24**, 3381(1982)
4. R. D. Black, P. J. Restle, and M. B. Weissman, *Phys. Rev. Lett.* **51**, 1476(1983).
5. D. M. Fleetwood and N. Giordano, *Phys. Rev.* **B31**, 1157(1985).
6. John H. Scofield, Joseph Mantese and Watt W. Webb, *Phys. Rev.* **B32**, 736(1985).
7. Shechao Feng, Patrick A. Lee and A. Douglas Stone, *Phys. Rev. Lett.* **56**, 1960(1986); *Phys. Rev. Lett.* **59**, 1062(1987).
8. Jonathan Peltz and John Clarke, *Phys. Rev.* **B36**, 4479(1987-I); *Phys Rev. Lett.* **59**,1061(1987).

9. A. H. Verbruggen, R. H. Koch, and C. P. Umbach, *Phys. Rev.* **B35**, 5864 (1987-I)
10. Neil M. Zimmerman and Watt W. Webb, *Phys. Rev. Lett.* **61**, 889(1988); **65**, 1040(1990).
11. C.D. Keener , M.B. Weissman, *Phys. Rev.* **B44,** 9178(1991-I).
12. John H. Scofield, R.W. Epworth, D.M. Tennant , *J. Appl. Phys.* **66**, 2032(1989).
13. F.N. Hooge , L.K.J. Vandamme, *Phys. Lett.* **66A**, 315(1978).
14. G. Gutierrez *et al*, in *Noise in Phys. Syst. and 1/f Noise*, T. Musha, S. Sato, M. Yamamoto(editors.), Ohmsha Ltd.,145(1991)
15. M. Mihaila , *Phys. Lett.* **104A**, 157(1984) ; *Phys. Lett.* **107A**, 465(1985).
16. M. Mihaila, A. Stepanescu, A. Masoero, in *Noise in Phys. Syst. and 1/f Noise*, T. Musha, S. Sato, M. Yamamoto(editors), Ohmsha Ltd., 17(1991) .
17. T. Musha, G. Borbely, M. Shoji , *Phys. Rev. Lett.* **64** , 2394(1990).
18. B. Koslowski, C. Baur, R. Moller, K. Dransfeld , *Surface Science* **280**, 106(1993).
19. M. Mihaila, A. Stepanescu, A. Masoero , in *Noise in Phys. Syst. and 1/f Noise*, V. Bareikis and R. Katilius(editors), World Scientific , 307(1995) .
20. J. W. S. Rayleigh, *Proc. London Math. Soc.* 17, **4**(1887).
21. R. F. Wallis, *Surf. Sci.*, **2**, 146(1964).
22. U. Harten, J. P. Toennies and Ch. Woll, *Faraday Discuss. Chem. Soc.*, **80**, 1(1985).
23. D.H. Dutton, B.N. Brockhouse, A.P. Miiller, *Can. J. of Phys.* **50** , 2915(1972).
24. Klaus Kern *et al*,*Phys. Rev.* **B33**, 4334(1986).
25. H. P. Myers, *Introductory Solid State Physics*, Taylor & Francis, 490(1990).
26. F. N. Hooge, *Phys. Lett.* **29A**, 139(1969).
27. L. K. J. Vandamme, in *Noise in Physical Systems*, M. Savelli *et al.*(editors), Elsevier, 183(1983).
28. J. F. Koch and R. E Doezema, *Phys. Rev. Lett.* **24**, 507(1970).
29. David Nowak and Martin J. C. Lee, *Phys. Rev. Lett.* **28**, 1201(1972).
30. Shashikala G. Das, *Phys. Rev.* **B7**, 2238(1973).
31. L. Dallaire and J. Destry, *Phys. Rev.* **28B** , 2947(1983).
32. A. G. M. Jansen, F. M. Mueller, and P. Wyder, *Phys. Rev.* **B16**, 1325(1977).

NOISE BIASED-PERCOLATION AND ABRUPT-FAILURE OF ELECTRONIC DEVICES

C. PENNETTA
Dipartimento di Fisica, Università di Lecce,
Via Arnesano, 73100 Lecce, Italy

Z. GINGL, L. B. KISS
JATE University, Department of Experimental Physics,
Dom ter 9, Szeged, H-6720 Hungary

L. REGGIANI
Istituto Nazionale di Fisica della Materia
Dipartimento di Scienza dei Materiali, Università di Lecce,
Via Arnesano, 73100 Lecce, Italy

We survey the biased-percolation model based on a two-dimensional resistor network as an approach to understand abrupt failure of electronic devices. Monte Carlo simulations enable us to investigate the evolution of the system including: damage pattern, current distribution, resistance degradation, resistance relative-fluctuations and its power spectrum associated with $1/f$ noise. Unsolved problems related to the development and improvement of the model are discussed.

1 Introduction

Very recently, a new percolation model[1] which promises to be a good candidate for understanding the abrupt failure of electronic devices[2–6] has been introduced. This new model has been called biased percolation, because local Joule heating is assumed responsible for the generation of defects causing percolative breakdown of the device. In this paper we briefly summarize the state of the art of this model and focus onto unsolved problems in the perspective of developing and improving existing findings.

2 State of the art

We consider a thin conductor film as a two-dimensional square lattice network of identical resistors stuck on an insulating substrate at temperature T_0. The lattice is contacted at the left and right hand sides to an external applied voltage U, which is kept constant. The degradation is starting because of the spontaneous creation of insulating defects. In our model, a defect corresponds to an infinite value of the resistance of an element (open-circuit model). The total degradation of the network implies the existence of at least one continuous path of defects between the upper and lower side of the lattice and therefore the measured resistance becomes infinite in this case. We take the probability W_α of creation of a local defect indexed by α as:

$$W_\alpha = exp(-\frac{E_o}{K_B T_\alpha}) \tag{1}$$

where E_0 is an activation energy characteristic of the defect, K_B the Boltzmann constant, and T_α the local temperature at the resistor α :

$$T_\alpha = T_o + A r_\alpha i_\alpha^2 \tag{2}$$

A is the characteristic parameter responsible for the coupling between current and device degradation, its value, measured in (K/W), depends on the efficiency in the thermalization by the substrate of each resistor r_α, i_α being the current flowing in it. It is interesting to note that this model provides a spontaneous symmetry breaking: at the beginning, when there is no defect yet, the probability of defect creation is homogeneous in space. By the accidental creation of the first defect at a random location, the original symmetry is spontaneously broken and the further defects will grow with a higher probability around the first one. When A is zero, the corresponding probability W_α^0 is the same for all resistors, which corresponds to a degradation governed by the standard percolation model.[7,8]

Monte-Carlo simulations are carried out using square networks with sizes $N \times N$ up to $N = 100$. Starting from the perfect lattice, the defects are generated according to the probability W_α. All local currents i_α are then recalculated, the probability W_α is updated, and applied to generate new defects. As realistic parameters we take $T_0 = 300\ K$, $E_0/K_B = 3000\ K$, $r_\alpha = 1\ \Omega$, $U = 1\ V$ and $A = 100 \times N^2\ K/W$ (for purposes of convenience we choose A scaling as N^2). As indicators of degradation we have studied the evolution of the damage-pattern, resistance degradation and its fluctuations. The main results are summarized below.

(i) The damage exhibits a filamentary pattern perpendicular to the direction of the current flow, as illustrated in Fig. 1. In particular, only a few clusters of defects are found in the device when it breaks down (infinite resistance), most of the material still remaining defect free.

(ii) The evolution of the network resistance R and of its normalized variance $< \delta R^2 > /R^2$ exhibit a sharp transition to failure as illustrated in Figs. 2 and 3 respectively. This latter quantity is evaluated assuming that each resistor is fluctuating in time independently from each other, with a variance

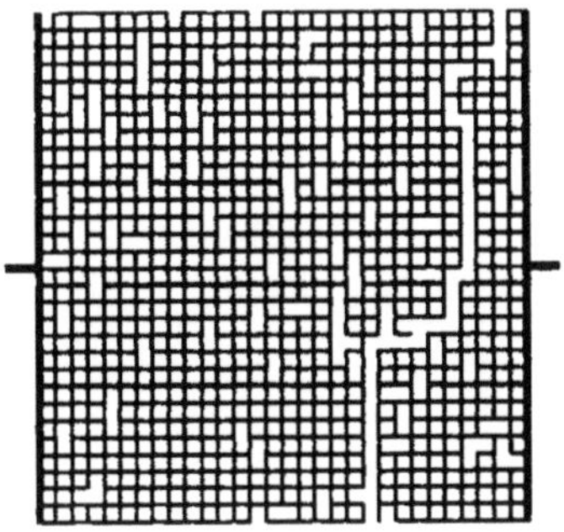

Figure 1: Lattice with size 30 × 30 close to complete failure.

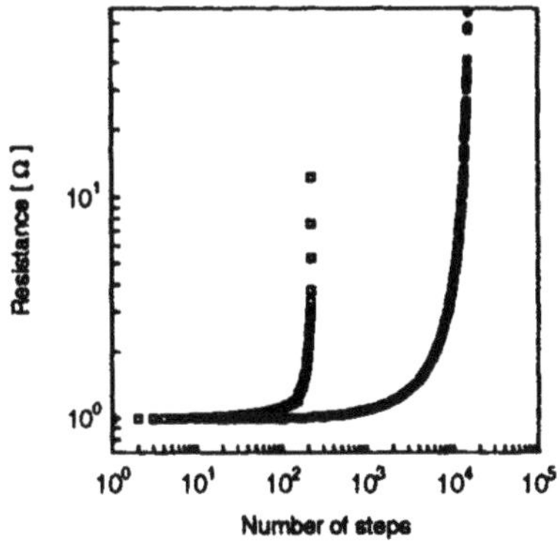

Figure 2: Evolution of the lattice total resistance for the standard (circle) and biased (square) percolation models, respectively. Curves refer to a sample with sizes 100 × 100.

$< \delta r_\alpha^2 >= \rho^2$, the same for all resistors, so that[8]:

$$\frac{< \delta R^2 >}{R^2} = \frac{\rho^2}{r^2} \frac{\sum_\alpha i_\alpha^4}{(\sum_\alpha i_\alpha^2)^2} \tag{3}$$

where we account for the fact that all $r_\alpha = r$ and $\rho^2/r^2 \ll 1$. We remark that, contrary to standard percolation, the increase of the resistance and of $< \delta R^2 > /R^2$ is not significant up to the last few percent of the device lifetime, where the resistance is diverging very steeply.

(iii) Pure $1/f$ noise spectra exhibit a colored transition near the abrupt failure of the device as illustrated in Fig. 4. Within our model the noise spectrum of resistance fluctuations $S_{\delta R}(f)$ at frequency f is:

$$S_{\delta R}(f) = \frac{R^2}{r^2(\sum_\alpha i_\alpha^2)^2} \sum_\alpha s_{r\alpha}(f) i_\alpha^4 \tag{4}$$

where $s_{r\alpha}(f)$ is the spectral density of the α-th resistor assumed of Lorentzian form:

$$s_{r\alpha}(f) = \frac{4\rho^2 \tau_\alpha}{1 + (2\pi\tau_\alpha f)^2} \tag{5}$$

with the correlation time of the α-th resistor, τ_α, hyperbolically distributed between 10^{-6} and 1 s. In this way, the spectral density of total resistance at the initial step exhibits a $1/f$ spectrum over several decades of frequencies.

Overall, the above features show valuable agreement with available experiments[4–6] and offer interesting possibilities to test the reliability of electronic devices.[2–3]

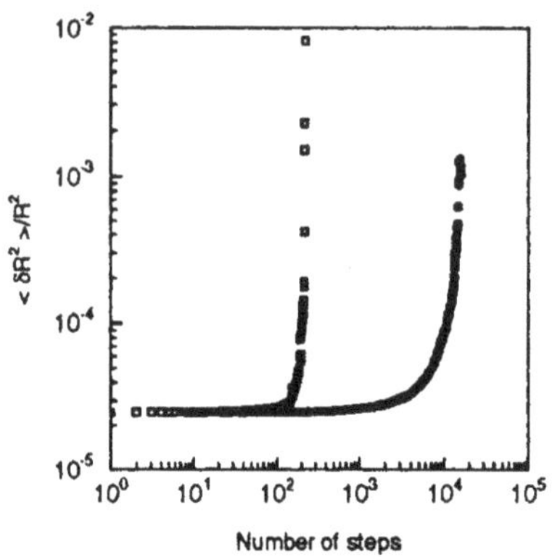

Figure 3: Evolution of normalized variance for standard (circle) and biased (square) percolation models, respectively. Curves refer to a sample with size 100 x 100.

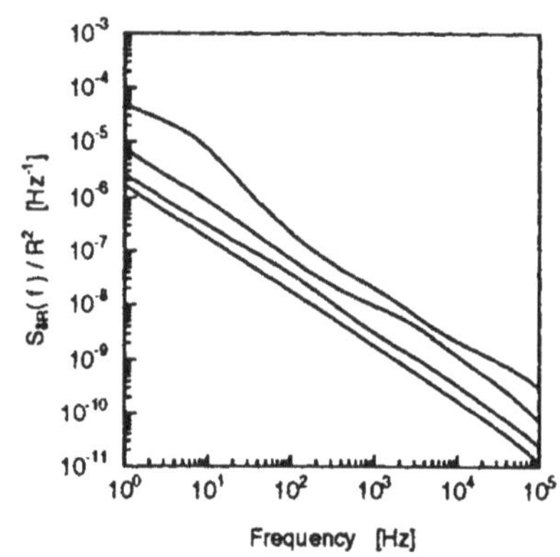

Figure 4: Normalized spectral density of total resistance fluctuations calculated from the biased percolation model for a lattice with size 100 x 100. The lowest curve is for $R = 1.0\ \Omega$. The others follow in increasing order of R, respectively $R = 1.5\ \Omega$, $R = 3.4\ \Omega$, $R = 5.0\ \Omega$.

3 Unsolved problems

Unsolved problems of this new model are summarized as follows:

(i) The normalized variance of the resistance R scales with R : $\Delta R^2/R^2 \propto R^\gamma$ with γ real. In the case of standard percolation this scaling exponent γ is universal. Is this universality also present in the case of biased percolation ?

(ii) What is the role of different biasing conditions ? What happens for different values of the biasing quantity ?

(iii) What is the effect of contact geometry and boundaries ? For example we may have a point-contact, a grating contact, and so on.

(iv) What could be the effect of dishomogeneities in the efficiency of cooling by the substrate (due, for example, to the quality of the adesion or to other dishomogeneous features of the substrate) ?

(v) What is the effect if the sample is non-ohmic ?

(vi) Is it possible to define an effective noise temperature for the network as a good indicator to monitor abrupt failure ?

(vii) In which way the lifetime of the device can be correlated to the evolution provided by Monte Carlo simulation. In other words which relation can exist between the iteration number and the time step ?

(viii) In which way the network sizes can be correlated to the geometrical and structural characteristics of the real device ? In particular, how the granularity of the film influences its noise characteristics ?

(ix) The model can be upgraded to provide a different sensitivity for noise and resistance degradation ?

(x) Can the present model be implemented by introducing others probabilities of generating defects, or different kinds of defects ? In particular, can this model be usefull for studing conductor - superconductor [9] transitions ?

Acknowledgments

This work has been performed within the European Laboratory for Electronic Noise (ELEN) supported by CEC contract ERBCHRXCT920047, PECO EAST ELEN ERBCHRXCT920047. Partial support from Hungarian grant OTKA F14309, TO14290, TO16342

References

1. Z. Gingl *et al.*, *Proc. Workshop on Noise and Reliability of Semiconductor Devices*, ed. by J. Sikula and P. Schauer (Univ. Brno Press, Brno, p. 151, 1995).
2. T. M. Chen and A. M. Yassine, *IEEE Trans. on Electron Devices* **41**, 2165 (1994).
3. L. K. J. Vandamme, *IEEE Trans. on Electron Devices* **41**, 2176 (1994).
4. B. K. Jones *et al.*, *Microelectron. Reliab.* **35**, 13 (1995).
5. K. Dagge *et al.*, *Appl. Phys. Lett.* **68**, 1198 (1996).
6. *Reliability and Degradation - Semiconductor Devices and Circuits*, ed. by M. J. Howes and D. V. Morgan, (John Wiley & Sons, New York, 1986).
7. D. Stauffer and A. Aharony, *Introduction to Percolation Theory* (Taylor & Francis, London, 1994).
8. R. Rammal *et al.*, *Phys. Rev.* A **31**, 2662 (1985).
9. L. B. Kiss and P. Svedlindh, *Phys. Rev. Lett.* **71**, 2817 (1993).

NOVEL DYNAMIC APPROACH TO THE 1/f NOISE PROBLEM IN SOLIDS VIA SHORT-LIVED LARGE ATOMIC ENERGY FLUCTUATIONS IN NANOMETER REGIONS

YU.L. KHAIT, I.B.SNAPIRO
Solid State Institute, Technion-Israel Institute of Technology,
Haifa 32 000,Israel

A solution of one of the main unsolved problems of $1/f$ noise related to the noise origin is suggested. We propose random short-lived large energy fluctuations(SLEF's)of small numbers $N_o \geq 1$ of atoms and SLEF-generated transient point dynamical defects(TPDD's) of lifetime $\Delta\tau = 10^{-13} - 10^{-12}s$ to be a source of $1/f$ noise in solids. The SLEF's and TPDD's having exponentially broad relaxation times cause random (in space and time) trapping(localization) and strong scattering of carriers.This produces stochastic dynamical fluctuations in the carrier density and mobility which possess $1/f$ spectral component.The Hooge-like relation with the parameter $a_H = 10^{-2} - 10^{-4}$ dependent on material characteristics has been obtained for metals and semiconductors through the SLEF-based approach.

1 Introduction

An extremely broad range of solids exibits so-called $1/f$ noise or *flicker* noise characterized by the spectral density (SD) $S \propto 1/f^{\gamma}$ that makes $1/f$ noise almost a universal phenomenon [1-22]. Hence one can expect the origin of flicker noise to have the same universal nature common for all solids. However, at present, the problem of the origin of $1/f$ noise is *unsolved* in spite of over sixty years of investigations, a great amount of experimental data accumulated, numerous models proposed and impressive achievements in the analysis of experimental data and properties of possible sources of $1/f$ noise. The concept of defect-fluctuators with an exponentially broad range of relaxation times $\tau = \tau_o exp(E/kT)$ has been put forward for the interpretation of experimental data[1-5,17,20,21]. Nevertheless, the problem of the the nature of the fluctuators *has not yet been solved.*

In this paper we propose a novel stochastic dynamic source of $1/f$ noise, based on the kinetic many-body electron-related theory of short-lived large energy fluctuations(SLEF's) of small numbers $N_o \geq 1$ of atoms (of lifetime $\Delta\tau = 10^{-13} - 10^{-12}$s) and SLEF-induced rate processes in solids[23-41]. We suggest that the random dynamical SLEF's take the role of the fluctuators generating $1/f$ noise. The SLEF's have exponentially broad relaxation times. The SLEF theory applied successfully to numerous processes and materials shows that SLEF's generate transient point dynamic defects(TPDD's) of nanometric size and lifetime $\Delta\tau_R \approx \Delta\tau$. Each of the TPDD's breaks up the local material order, symmetry and stability and changes dramatically atomic and electronic properties in the nanometer region during $\Delta\tau$. The SLEF's and TPDD's existing simultaneously and permanently in the solid generate a strong electron-lattice interaction which can cause trapping (localization) of mobile carriers , strong carrier scattering and upward and downward electron transitions in the SLEF-affected nanometric regions[23a,25-30,33-36,39]. In other words, each of the SLEF's induces transient local conductor-insulator transitions (TLCIT's) in the SLEF-affected material region. The SLEF's and TLCIT's appear spontaneously in random places at random instants of time. The random persistent sequences (generations) of simultaneously occurring SLEF's,TPDD's and TLCIT's which form random dynamic arrays exist permanently in solids[23a,31-36]. They cause stochastic dynamical fluctuations in the carrier number and mobility and in the material resistance. These fluctuations contain the $1/f$ noise component,and thus they generate $1/f$ noise.

2 The Origin of $1/f$ Noise - Statement of the Problem

The fact that $1/f$ noise has been observed in virtually all kinds of materials [1-22] motivates one to expect the origin of $1/f$ noise to be common for all materials, although specific properties of different materials can cause some observed differences in particular noise features found in different kinds of materials. There exists a widespread opinion that $1/f$ noise is generated by ensembles of defect-fluctuators possessing a broad distribution of relaxation times τ [1-5,17-21]. However the *nature* of the fluctuators and thus that of the *origin* of $1/f$ noise is one of the key *unsolved problems* in the field. Observations of $1/f$ noise inversely proportional to the sample volume V are usually attributed to the bulk nature of the noise. In many experimental situations the famous Hooge empirical relation between SD S_I of current fluctuations, the total number N_c of carriers and frequency ω takes place [9] $S_I/I^2 = \alpha/N_c\omega$. Here the parameter α is of the order of 10^{-3} in many cases, but it is not a "universal constant" and its value in some cases can deviate strongly from the "normal" value of $\alpha_o \approx 10^{-3}$. Two $1/f$ noise components exist simultaneously in *metals*[1]: a) a strongly temperature-dependent noise, which is independent of the substrate. The amplitude of this noise is proportional to $\exp(-E_M/kT)$ at temperature $T < 350K$,where $E_M \approx 0.1$ to 0.15 eV and b) a weak temperature-dependent component which, however, depends strongly on the substrate nature. In *semiconductors* strong experimental evidence of the relation of $1/f$ noise to lattice scattering has been presented. Hooge and coauthors [2] have concluded that "it is as if the density of phonons fluctuates with a $1/f$ spectrum". The concept of fluctuators having the exponentially broad distribution of relaxation times attributes this distribution to some thermally activated processes caused by fluctuators which generate $1/f$ noise. We shall see below that the proposed SLEF-based model of $1/f$ noise which includes SLEF-induced thermally activated processes as a source of $1/f$ noises, introduces SLEF-generated fluctuators having relaxation times $\tau \propto exp(\Delta G/kT)(\Delta G >> kT)$ but does not produce thermally activated particles jumps over energy barriers. Let us first consider some limitations imposed by the conventional theory of solids which are associated with the following well known assumption [42-44]: 1) *Average atomic displacements* (or oscillation amplitudes) σ_A are *small* compared to the average interatomic spacing d, i.e. $\sigma_A << d$. 2) *The electron motion can be separated from the atomic motion* through the Bohr-Oppenheimer adiabatic principle due to the above assumption $\sigma_A << d$. 3) The collective atomic motion can be described in terms of *harmonic* (quasiharmonic) approximation and of phonons when $\sigma_A << d$. 4) Relatively small electronic perturbations are taken into account. 5) The electron-lattice interactions are usually described in terms of the electron-phonon interactions. 6) Only conventional long-lived defects (vacancies, etc.) and their interactions with carriers are taken into account. 7) The solid is usually assumed to be characterized by its equilibrium atomic, electronic and electromagnetic parameters (if some external factors do not force it to deviate from equilibrium). The above assumptions, although proven extremely effective in various fields, impose strong limitations on the consideration of a broad range of processes in solids which are associated with large transient atomic displacements (LTAD's) $\Delta q_o = |q_o - \bar{q_o}| \approx d >> \sigma_A$ from the mean positions $\bar{q_o}$ (such as atomic and defect diffusion, etc.) This was first stressed by Frenkel [45] about 50 years ago. The theory of SLEF's and SLEF-induced processes [23-41] frees one from the above limitations through the consideration of large local non-equilibrium fluctuations hypotized by Boltzmann 100 years ago[46], SLEF-induced LTAD's and strongly interrelated atomic and electronic phenomena in nanometric regions which break the aforementioned assumptions. These SLEF-induced phenomena take place in all kinds of materials. The SLEF-based electron- related theory has been applied successfully to the consideration of various rate processes, phase and structural transformations in metals [23a,24], metal binary alloys [35], crystalline and amorphous semiconductors [23a-25,27-29,31], semiconductor lasers and light emitting diodes [30]), metal-semiconductor interfaces[36], semiconductor surfaces [26], high- T_c and low-T_c superconductors [33] and superconducting ceramics at $T > T_c$ [39], semiconductor superlattices of nanometer periods [37] and semiconductors affected by light [38]. In this work we extend the SLEF theory to the problem of the origin of $1/f$ noise.

3 A Dynamic Stochastic Model of the SLEF-generated Picosecond Nanometric Fluctuator

The theory of SLEF's and SLEF-related phenomena is well documented in literature[23-40]. Here we summarize,mainly, some key points of the theory related to SLEF-induced LTAD's and TPDD's able to localize (trap) carriers for a short while and scatter them strongly.This causes random fluctuations $\Delta N_c(t) = N_c(t) - \overline{N_c}$, $\Delta n(t) = n(t) - \overline{n}(t)$, $\Delta\mu = \mu(t) - \overline{\mu}$ in the carrier number $N_c(t) = Vn(t)$ in the sample volume V, concentration $n(t)$ and mobility $\mu(t)$ around their mean values $\overline{N_c} = V\overline{n}$,$\overline{n}$ and $\overline{\mu}$. We use the SLEF theory that considers the dynamics of SLEF-induced correlated many-body picosecond atomic and electronic phenomena which occur during $\Delta\tau \approx 10^{-13} - 10^{-12}s$ in nanometer regions of solids(or their surfaces). Each SLEF of lifetime $\Delta\tau = \tau_1 + \tau_2$ emerging in random places at random instants of time generates $N_o \geq 1$ hyperthermal fluctuating atoms(HFA's). During the SLEF formation time $\tau_1 \approx 0.5\Delta\tau$ the HFA's receive the energy $\varepsilon_{op} >> kT$ only from the causal nanometer HFA vicinity of radius $R_1 \approx c_s\tau_1 \approx (2-3)d \approx 10^{-7}cm$ and volume $\Omega_1 \approx 4R_1^3$(containing $\Delta N_1 \approx 30 - 100$ atoms) due to the causality condition. The HFA gives the "borrowed" energy back to the HFA surroundings of radius $R_2 \approx c_s\tau_2 \approx R_1$ during the SLEF relaxation time $\tau_2 \approx \tau_1$ (here c_s is of the order of the sound velocity). During $\Delta\tau$ the SLEF perturbs the HFA vicinity of radius $R_A \approx c_s\Delta\tau \approx 2R_1 \approx (4 - 10)d$ and volume $V_A \approx 4R_A^3 \approx 30R_1^3$ containing $N_A \approx (3-10)10^2$ atoms.Mobile electrons moving with the velocity $v_e = 10^7 - 10^8 cm/s$ can be affected by a single SLEF at distances not larger than $r_e \approx \Delta\tau v_e \approx (10^2 - 10^3)d$ [23,25-30,33,35-40].SLEF's can correlate with one another at distances of the order of $R_A = 10d$ on the atomic level and at distances $r_e \approx (10^2 - 10^3)d$ on the electronic level.The HFA's experience LTAD's $\Delta q_o >> \sigma_A$,which can be seen directly in molecular dynamic simulations[31,32].They are associated with large picosecond correlated atomic and electronic distortions in the HFA nanometer neighbourhood. The SLEF-induced LTAD's and a local transient disordering form TPDD's of lifetime $\Delta\tau_d \approx \Delta\tau = 10^{-13} - 10^{-12}s$ which differ qualitatively from the traditional long-lived point defects(vacancies,interstitials,etc.)[23-28,30-37]. The SLEF's generate a local non-equilibrium state and fluctuations in the phonon density in the HFA's vicinity. The picosecond correlated motion of the HFA's and many surrounding particles in the HFA nanometer causal vicinity during the SLEF lifetime is described by the coupled kinetic integro-differential equations[23,41].The SLEF probability and rate coefficients are calculated from a solution of the SLEF kinetic equations[23,41]. The SLEF-induced TPDD's create strong transient local atomic and electronic distortions and instabilities that destroy, for a short while, the material symmetry,order and stability and generate local transient non-equilibrium states in the nanometric volumes.The LTAD's and TPDD's breaking the Born-Oppenheimer adiabatic approximations induce large local transient electronic rearrangements and the motion and crossing of electronic levels(surfaces) and also non-adiabatic electron transitions in the HFA vicinity;new local moving electron levels are also created by SLEF's.This causes strong carrier-lattice interaction effects[23a,25-30,35-40].

The simultaneously existing SLEF's and TPDD's of various $\varepsilon_{op} >> kT$ form a dynamic random array (the SLEF and TPDD generation) of the lifetime of the order of $\Delta\tau = 10^{-13} - 10^{-12}s$. The persistent random sequences of the picosecond SLEF (and TPDD) generations exist permanently in the sample. At any fixed instant of time t' the TPDD random array distorts randomly the spatial distribution $U_p(r,t')$ of the energy potential for carriers which form(at fixed t) a 3-dimensional random field. This potential contains random transient traps for carriers of picosecond lifetime.When t varies, the random array of $N_f(t) = V\eta(t)$ coexisting SLEF's and TPDD's changes randomly the spatial potential distribution $U_p(r,t)$which forms (3+1)-dimensional spatial-temporal random function $U(r,t)$. Here $\eta(t)$ is the random cubic density of simultaneously oc curring SLEF's and TPDD's. Thus every picosecond generation of SLEF-induced TPDD's produces the random potential which,in turn,creates random localization(trapping)of mobile carriers during $\Delta\tau$. This generates random fluctuations $\Delta n(t) = n(t) - \overline{n}$ and $\Delta N_c(t) = V\Delta n(t) = N_c(t) - \overline{N}_c$ in the carrier density $n(t)$ and number $N_c(t) = Vn(t)$ around their mean values $\overline{n}$ and $\overline{N}_c = \overline{n}V$ in the sample volume.In *semiconductors* the average number(per SLEF and time interval $\Delta\tau$) of localized carriers is $\overline{g}_{ts} = \overline{n}\overline{v}_e\Sigma_t\Delta\tau$ [23a,25-30,33,35,36].Usually this number is $\overline{g}_{ts} << 1$;eg $\overline{g}_{ts} = 10^{-4} - 10^{-2}$ for $\overline{n} = 10^{15} - 10^{17}cm^{-3}$,cross section $\Sigma_t \approx 10^{-13}cm^2$ and $v_e = 10^7cm/s$. In *metals* where the average carrier concentration is $\overline{n} \approx N_A\gamma_v \approx 3 * 10^{22}cm^{-3}$ every SLEF produces picosecond localization of $\overline{g}_{tm} \approx (N_o+\chi_1)\gamma_v to(N_o+\chi_1+\chi_2)\gamma_v$ valence electrons. These electrons belong to $(N_o + \chi_1)to(N_o + \chi_1 + \chi_2)$ atoms which include N_o HFA's, the χ_1 or/and χ_2 nearest HFA's and next to the nearest HFA neighbours;γ_v is the number of valence electrons per atom. The $(N_o + \chi_1)to(N_o + \chi_1 + \chi_2)$ atoms which occupy volume $\delta V_1 = (N_o + \chi_1)d^3$ to $\delta V_2 = (N_o + \chi_1 + \chi_2)d^3$ constitute the most distorted "core" of every TPDD. Therefore one can expect SLEF-induced picosecond localization(per TPDD) of $g_{tm} \approx (N_o + \chi)\gamma_v$ electrons; at $N_o = 1,\chi = 12$ and $\gamma_v = 1 - 3$ one finds $g_{tm} = 13 - 40$. During $\Delta\tau$ the TPDD core ceases to be a conductor since carriers are unable to pass through the strongly distorted nanometer material region. This means that every SLEF induces the transient local conductor-insulator transition(TLCIT) in the nanometer region of volume $\delta V_1$to δV_2. Random TLCIT's produced by sequences of random arrays of simultaneously occurring SLEF's and TPDD's generate a random dynamic current redistribution similar, in a sense, to that discussed in [21].Besides,every SLEF (and TPDD) causes scattering on the average of $\overline{g}_s = \overline{n} v_e \Sigma_s \Delta\tau$ external carriers located in the TPDD vicinity of radius $r_e \approx v_e \Delta\tau$. Consider now the probability of SLEF's and TPDD's to occur, taking into account that the TPDD's can cause Δn^{up} upward or/and Δn^d downward electron transitions (between levels with average energy separation $< \delta e >$) occurring in the HFA nanometer vicinity during the SLEF lifetime $\Delta\tau$.The SLEF's of$\varepsilon_{op} \geq \Delta E >> kT$ and TPDD's occur with the probability(per second) [23,25-30,33,35-40]

$$W = \Delta\tau^{-1} exp(-\Delta G/kT) \tag{1}$$

calculated from the kinetic consideration. Here $\Delta G = \Delta E + (| < \delta e > | - kT)\delta n$ is the effective SLEF free energy,$\delta n = \Delta n^{up} - \Delta n^d$, $\delta G = \delta E - T\delta S$, $\delta E = \delta n | < \delta e > |$ and $\delta S = k\delta n$.From eq.(1) one finds the exponentially broad distribution of relaxation times of SLEF-induced phenomena

$$\tau = W^{-1} = \Delta\tau exp(\Delta G/kT), \tag{2}$$

From eq.(2) one finds the distribution $P(\tau) = W^{-1} = 1/\tau$. The following two kinds of SLEF's are considered.The first kind of "irreversible"SLEF's forms thermally activated rate processes(eg. atomic diffusion, etc.) each of which has its SLEF threshold energies $\Delta E = E + < \delta e > \delta n$ taking the role of measured activation energies [23a,25-30,34-40].The second kind of SLEF's termed as "reversible" generates HFA's,LTAD's and TPDD's, but these SLEF's do *not* produce atomic transitions over energy barriers,as can be seen in computer simulations[31,32]; nevertheless these SLEF's are described by Arrhenius-like eqs.(1) and(2)[23a,31,32-35] and can cause experimentally observed effects[35]. The average cubic density of simultaneously occurring independent SLEF's of $\varepsilon_{op} \geq \Delta E >> kT$ and the mean distance between them are[23a,33-35]

$$\overline{\eta}(\Delta E) = d^{-3} exp(-\Delta G/kT), \overline{r}_f(\Delta E) \approx \overline{\eta}^{-1/3} \tag{3}$$

Consider now the component of $1/f$ noise associated only with SLEF-induced localization(trapping) of carriers.This component results from SLEF-induced fluctuations $\Delta n(t) = \overline{g}_t \Delta\eta(t)$ and $\delta N_c(t) = V\overline{g}_t \Delta\eta(t)$ in the carrier density and number. Here $\Delta\eta(t) = \eta(t) - \overline{\eta}$ is the fluctuation of the SLEF random density $\eta(t)$ around $\overline{\eta}$. For a given ΔE the autocorrelation functions for $\Delta N_f(t,\Delta E) = V\Delta\eta(t,\Delta E)$ and $\Delta N_c(t,\Delta E)$ are

$$K(\Delta t,\Delta E) = \overline{[\Delta N_f(\Delta E)]^2} exp[-\Delta t/\tau(\Delta E)]$$
$$\varphi_c(\Delta t,\Delta E) = \overline{g}_t^2 K(\Delta t,\Delta E)$$
$$\overline{[\Delta N_c(\Delta E)]^2} = \overline{g}_t^2 \overline{[\Delta N_f(\Delta E)]^2} \tag{4}$$

Hence one finds the corresponding spectral densities $S_f[\omega,\tau(\Delta E)] = \overline{[\Delta N(\Delta E)]^2}\tau[1+(\omega\tau)^2]^{-1}$ and $\varphi_c(\omega,\tau(\Delta E)) = \overline{[\Delta N_c(\Delta E)]^2}\tau[1+(\omega\tau)^2]^{-1}$. Integrating φ_c over τ with the weight $P(\tau) = 1/\tau$ and using the theorem about mean value one finds for the total spectral functions $\widetilde{S_c}(\omega)$ and $\widetilde{S_I}(\omega)$ the relations

$$\widetilde{S_c}(\omega)/\overline{N}_c^2 = \widetilde{S_I}(\omega)/\overline{I}^2 = \overline{(\Delta N_c^2)}/(\omega N_c^2) \tag{5}$$

Taking $\overline{(\Delta N_c^2)} = \beta_c \overline{N}_c$ and $\overline{(\Delta N_f^2)} = \beta_f \overline{N}_f$ we find the Hooge-like relations

$$\widetilde{S_c}(\omega)/\overline{N}_c^2 = \widetilde{S_I}(\omega)/\overline{I}^2 = a_H/(\omega N_c) \tag{6}$$

where

$$a_H \approx \beta_f \overline{g}_t^2 \overline{\eta}/\overline{n} \tag{7}$$

4 Discussion and Conclusions

A dynamical SLEF-based approach to one of the *unsolved* problems of $1/f$ noise origin and the fluctuators nature discussed in the previous sections leads to experimentally verifiable conclusions. Eq.(7) enables one to estimate the parameter a_H for different materials.Consider first $1/f$ noise generated by reversible SLEF's which produce LTAD's $\delta \approx 0.2d$ of the order of the Lindemann value not inducing HFA jumps over energy barriers.Local disordering effects of such LTAD's are confirmed by the Mossbauer spectroscopy in binary metal alloys[35].The mean cubic density of such SLEF's $\overline{\eta} = d^{-3}\exp[-\Delta G(\delta)/kT]$ entering eq.(7) contains the activation free energy $\Delta G(\delta = 0.2d) = \Delta E(\delta) - T\delta S(\delta)$. The energy $\Delta E(\delta)$ is calculated from the relation[28,33,35]

$$\Delta E(\delta) = (\delta/d)^2 B\Omega_o \tag{8}$$

and $\delta S = k\ln(D_o/D_{onorm})$[23a,25,27,28].Here B is the bulk modulus,Ω_o is the mean volume per atom,D_o is the self-diffusion Arrhenius pre-exponential factor and $D_{onorm} \approx d^2/6\Delta\tau \approx 5*10^{-4}cm^2/s$.Calculate now the Hooge parameter a_H for metals at $T = 300K$,eg. for Cu,using $D_o(Cu) \approx 0.7cm^2/s$, eqs(7) and (8) and taking $\chi_1 = 12$,$N_o = 1$, $\gamma_v = 1$ and standard values for B and

Ω_o. Then one finds $a_H(Cu) \approx 4 * 10^{-3}$ that is in reasonable agreement with experimental observations.

The strong temperature dependence of $a_H(Cu) \propto exp(-\Delta G/kT)$ can be expected below the temperature T_p at which the SLEF's of $\Delta G(\delta \approx 0.2d) \approx 0.25d$ eV experience a percolation-like process. This percolation,similar to those considered for other SLEF-related processes[23a,33,34,35], satisfies the condition $r_{fA} \approx A_A d \approx 7d$. Hence one finds $T_p = \Delta G(klnA_A)^{-1} \approx 500K$, that is in agreement with observations[3].At higher $T > T_p \approx 500K$ the temperature dependence is expected to be weaker or even can be replaced by some reduction in a_H since the SLEF's of $\Delta G(\delta)$ start to correlate negatively to one another that can reduce the SLEF probability. These conclusions are also in reasonable agreement with observations. Calculate now the Hooge parameter a_{HS} for typical semiconductors at $T = 300K$.Consider n-Ge with $\overline{n} = 10^{15}cm^{-3}$, $g_{ts} \approx 3 * 10^{-4}$,$D_o \approx 1cm^2/s$ and $\delta \approx 0.2d$. In this case one finds from eqs. (3), (7) and (8) $\overline{\eta} \approx 7 * 10^{-5}d^{-3}$, $r_f \approx 24d$.Hence one obtains $a_H S \approx 5 * 10^{-4}$ that is in agreement with observations.Here one can expect a relatively weak temperature dependence(compared to that in metals) since $\overline{r}_f < \lambda_B = h/(m^* * v_e) \approx 80d$ and $4\lambda_B^3\overline{\eta} \approx 35$. Thus every carrier wave function "feels" simultaneously many SLEF-induced random TPDD's associated with LTAD's $\delta \approx 0.2d$. A further increase in T does not change this situation substantially. However a substantial reduction in T below $T_1 \approx 200K$ which satisfies the condition $r_f(T_1) \approx \lambda_B$ can cause a stronger temperature dependence in a_{HS}.

The proposed SLEF-based approach to the *unsolved* problem of the $1/f$ noise origin suggests the following two pairs of ω_{jmin} and $\tau_{jmax}(j = 1, 2)$ of the lowest frequencies and longest relaxation time.The first pair associated with SLEF-induced electronic perturbations in the region of radius $r_e \approx v_e\Delta\tau = (10^2 - 10^3)d$ is $\omega_{1min} = 2\pi/\tau_{1max} = 2\pi(A_e^3\Delta\tau)^{-1} = 2(10^7 - 10^4)s^{-1}$ and $\tau_{1max} = A_e^3\Delta\tau = (10^6 - 10^9)\Delta\tau$ for $A_e = r_e/d$,$v_e = 10^7 - 10^8 cm/s$ and $\Delta\tau = 3 * 10^{-13}s$.The second pair is $\omega_{2min} = 2\pi/\tau_{2max} = 2\pi(A_L^3\Delta\tau)^{-1}$ and $\tau_{2max} = A_L^3\Delta\tau$. These parameters are associated with the SLEF-induced electromagnetic perturbations of frequency$\approx 2\pi/\Delta\tau \approx 2(10^{13} - 10^{12})s^{-1}$ which propagate with the velocity c_L at distance $R_L = c_L\Delta\tau$ during $\Delta\tau$;$A_L = R_L/d$.For semiconductors where $c_L = 3 * 10^{10}/n_R \approx 10^{10}cm/s$ one finds $A_L \approx 10^5$,$\tau_{2max} \approx 3 * 10^2 s$ and $\omega_{2min} \approx 2 * 10^{-2}s^{-1}(n_R = 2 - 4)$. In conclusion, a stochastic dynamical SLEF-based approach to the *unsolved* problem of the origin of $1/f$ noise and its applications to metals and semiconductors is proposed. A resonable agreement with observations has been found.

We gratefully acknowledge useful discussions with Dr.B.Ashkinadze of the Israel Institute of Technology.

References

1.P.Dutta and P.M.Horn,Rev.Mod.Phys.53,497(1981)
2.F.N.Hooge,T.G.M.Kleinpenning,and L.K.J.Vandamme,Rep.Prog.Phys.44, 479(1981)
3.Sh.M.Kogan, Usp.Fiz.Nauk 145,285(1985)[Sov.Phys.Usp. 28,170(1985)]
4.M.B.Weissman,Rev.Mod.Phys.60,537(1988)
5.N.V.D'yakonova,M.E.Levinshtein,and S.L.Rumyantsev,Fiz.Tekh.Poluprovodn. 25,2065-2104(1991)
6.M.Savelli,et al ,Noise in Physical System and 1/f Noise(North-Holland,Amsterdam,1988)
7.S.Feng,in:Mesoscopic Phenomena in Solids,eds.B.L.Al'tshuler,et al (Elsevier,Amsterdam,1991)
8.M.Giordano,in:Mesoscopic Phenomena in Solids,eds.B.L.Al'tshuler,et al (Elsevier,Amsterdam,1991)
9.F.N.Hooge,Phys.Lett.A29,139(1969)
10.R.L.Stratonovich,Non-Linear Non-Equilibrium Thermodynamics (Springer,Berlin,1992)
11.AIP Conf. Proceed.285.Noise in Physical Systems and 1/f Fluctuations,
Ed's P.H.Handel and A.L.Chung(Amer.Inst.Phys.,New York,1993)
12.N.B.Lukyanchikova,Fluctuational Phenomena in Semiconductors and
Semiconductor Devices,in Russian (Moskow Press:House Radio and Communication,1990)
13.A.van der Ziel,Noise in Measurements(Wiley New York 1975)
14.N.Y.Chen,R.Jonker,V.C.Matijasevic,H.M.Jaeger,and J.E.Mooij, Appl.Phys.Lett.67,133(1995)
15.S.Scouten,Yisi Xu,B.H.Moeckly,and R.A.Buhrman,Phys.Rev.B50,16121(1994)
16.C.T.Rogers and R.A.Buhrman,Phys.Rev.Lett.55,859(1985)
17.A.V.Bobyl at al.,Physica C 247,7(1995)
18.J.Bernamont,Ann.Phys.(Leipzig)7,71(1937)
19.Sh.M.Kogan and K.E.Nagaev,Solid State Commun. 49,387(1984)
20.L.B.Kiss and P.Svendlindh,Phys.Rev.Lett.71,2817(1993)
21.C.T.Seidler, S.A.Solin and A.C.Marley, Phys.Rev.Lett.76,3049(1996)
22.C.Parman and G.Kakalios,Phys.Rev.Lett.67,2529(1991)
23.Yu.L.Khait:(a)Phys.Reports 99,237(1983);(b)In Recent Progress in Many Body Theories,
eds.A.Y.Kallio et.al.(Plenum,N.Y.,1988)
24.Yu.L.Khait,PhysicaA,103,1(1980)
25.Yu.L.Khait,R.Beserman,Phys.Rev.B38,8107(1988),Yu.L.Khait, et al Phys.Rev.B38,6107(1988)
26.Yu.L.Khait,R.Weil,R.Beserman,W.Beyer and W.Wagner,Phys.Rev.B42, 9000(1990)
27.Yu.L.Khait,Semiconductor Sci.Techn.6,C84(1991)
28.Yu.L.Khait,R.Beserman,D.Show and K.Dettmer,Phys.Rev.B50,14893(1994)
29.Yu.L.Khait,and V.Richter,Journ.Phys.(U.K.)D:Appl.Phys.Rap.Comm.26, 8306(1993).
30.Yu.L.Khait,J.Salzman and R.Beserman,Appl.Phys.Lett.53,2135(1988); ibid 55,1170(1989)
31.Yu.L.Khait,A.Silberman,R.Weil and J.Adler,Phys.Rev.B44,8308(1991)
32.Yu.L.Khait,Yu.Kurskii and J.Adler,Bulletin of Israel Physical Society,39,119(1993)
33.Yu.L.Khait,Zs.Phys.B71,8(1988)
34.(a)Yu.L.Khait,PhysicaB 139,237(1986);(b)Yu.L.Khait,Phys.Stat.Solidi B131,K19(1985)
35.Yu.L.Khait,I.Snapiro and H.Shechter,Phys.Rev.B52,9392(1995)
36.Yu.L.Khait and R.Weil,J.Appl.Phys.78,6504(1995)
37.K.Dettner,W.Freiman,M.Levy,Yu.L.Khait and R.Beserman,Appl.Phys.Lett 66,2376(1995)
38.I.Abdulhalim,R.Beserman,Yu.L.Khait and R.Weil,Appl.Phys.Lett.51, 1898(1987)
39.Yu.L.Khait,Nucl.Instr.Meth.in Phys.Res.B 1996(in Press)
40.Yu.L.Khait ,in Plasma Properties Deposition and Etching ,Material Sci. Found.,
ed's J.J.Pouch and S.A.Alterovitz (Trans.Tech.Publ, Switzerland,1993)
41.The transient correlated motion of $N_o \geq 1$ HFA's and of many surrounding particles
in the nanometer FA vicinity is governed by coupled kinetic integro- differential equations obtained
from the Liouville equation(in Ref.23a,b).
42.C.Kittel, Quantum Theory of Solids (Wiley New York,1975)
43.The Physics of Metals,ed. J.N.Ziman (Cambridge Univ.Press., Cambridge,1969)
44.N.W.Ashcroft and N.D.Mermin, Solid State Physics (Holt, Rinehart and Winston, New York,1975)
45.Ja.Frenkel, Kinetic Theory of Liquids (Claredon,Oxford,1946)
46.L.Boltzmann, Vorlesungen uber Gasteorie (Barth,Leipzig,1896)

ON THE NATURE OF 1/F NOISE IN SEMICONDUCTORS AT HIGH ELECTRIC FIELDS

E.B. KISLITSYN, S.A. KORNILOV
St.-Petersburg State University of Telecommunications, Moika 61, 191186, St.-Petersburg, Russia

The experimental study of illumination influence upon 1/f current noise in GaAs avalanche diodes is described. The main results are: - in 1/f noise formation participate traps; - there are experimental indications to an additive 1/f noise source existence.

1 Introduction

In [1] a physical model of the bulk 1/f noise has been developed. This model, like the "surface" model of McWhorter, attributes 1/f noise to occupancy fluctuations of traps. According to [1], 1/f-like spectrum is formed if capture cross-section of electrons (for n-type materials) exponentially decreases with the energy distance from the bottom of the conduction band. This assumption seems to be realistic.

A review of experiments, supporting the mentioned model, in [2] has been published. The most important seem to be experiments, demonstrating illumination influence upon 1/f noise spectrum. It was found that the spectrum modification is produced by the photons with energy close to the energy gap magnitude. These experiments were carried out on donor doped GaAs and Si specimens; at that the illumination practically did not affect free electrons concentration. So the illumination could influence 1/f noise only by means of the generation of holes: the holes, captured by traps, change their occupancy and thus influence 1/f noise of the electron current. In essence light experiments directly demonstrate a participation of traps in 1/f noise formation.

Let us note two details: a) experiments were carried out at comparatively weak electric fields; b) at high temperature the influence of illumination upon 1/f noise disappears [2].

In [3] the model, proposed in [1], has been used to calculate the magnitude of the current 1/f noise in avalanche diode. However, operation conditions of avalanche diodes are so much specific (very high electric field, up to hundreds kV/cm, velocity saturation, base depletion, impact ionisation of traps), that arises a doubt about a possibility of the adaptation of any 1/f noise model, checked at well less fields.

Aside from the considered publications, the problem of 1/f noise in semiconductors at high electric fields in [4-14] has been discussed. The models, developed or used in [4-8,10-12], are based on a mobility fluctuations concept. In [4,8,10] the primary physical source of 1/f noise was not mentioned, in [5-7] it was attributed to the scattering of electrons on acoustic phonons. In [12] 1/f noise of hot electrons was considered as a result of mobility fluctuations arising due to the scattering of electrons on neutral metastable centres. But the influence of traps in [13] was not taken into account. In [13] results of the analysis of 1/f noise in IMPATT diodes and oscillators are presented. This work is based on a model, including two independent 1/f noise sources: charge fluctuations of traps and fluctuations of drift velocity due to the scattering of electrons on neutral metastable centres. It was found that theoretical results agree with experimental data. But the methodology, developed in [13], is rather complex and results, got in this work, demand an independent confirmation.

So, the problem of 1/f noise at high electric fields is not yet clear and demands new experimental facts. In this paper we communicate some experimental results concerning 1/f noise sources in GaAs at very high fields.

2 Experimental results and discussion

The principal object of experiments was to bring out a role of traps in 1/f noise formation at high electric fields. The most adequate and direct way to do that is to investigate a response of 1/f noise to illumination of a specimen under study [2].

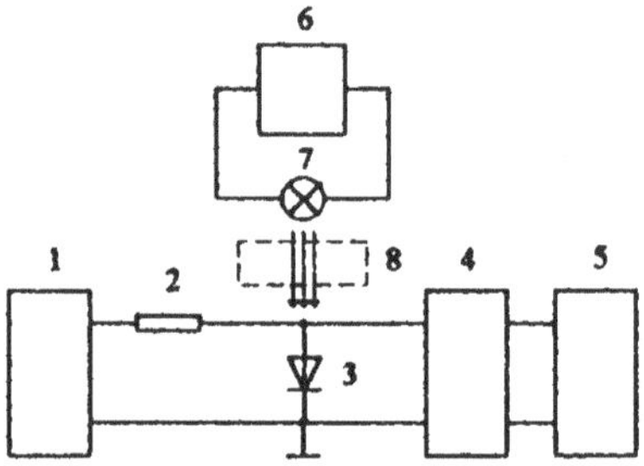

Figure 1: The experimental set-up. 1 - power supply of the avalanche diode 3; 2 - ballast resistance 600 Ohm; 4 - low-noise transistor amplifier; 5 - spectrum analyser; 6 - power supply of the incandescence lamp 7, 8 - accessory optical filter.

Experiments were carried out on GaAs diodes by means of the methodology, described in [3,4]. Me-n-n$^+$ uniformly-doped structures without punch-through were investigated. The diode parameters are: donor concentration $N_D = 10^{16}$ cm^{-3}, depletion layer and multiplication layer lengths $l = 3$ μm and $l_a = 0.9$ μm, structure diameter $D = 150$ μm. The diodes were operating in developed breakdown mode, the maximal strength of electric field (in multiplication layer) was about 400 kV/cm. To illuminate the structure a quartz window in diode package was provided. As a light source an incandescence lamp and optical filters were used. The experimental set-up is shown in Fig. 1.

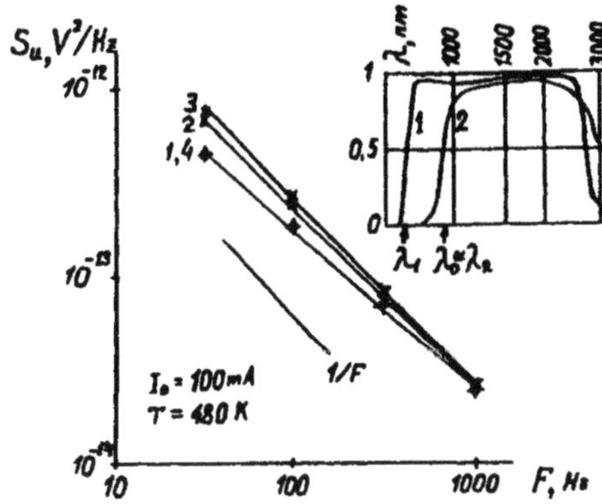

Figure 2: Noise voltage spectra and characteristics of optical filters (insert).

Fig. 2 demonstrates the illumination influence upon 1/F-noise spectra. The insert shows spectral characteristics of optical bandpass filters used in experiments. Here λ_1 is the wavelength corresponding to the high-frequency boundary of the filter No. 1, $\lambda_0 = hc/\Delta E_g$ is the wavelength of photons having energy equal to the energy gap ΔE_g. The wavelength λ_2, corresponding to the high-frequency boundary of the filter No. 2, is very close to λ_0.

The spectrum 3 in Fig. 2 was measured without illumination. The spectrum 4 was measured by broadband illumination (without filters). One can see that the light makes the spectrum more flat, suppressing 1/f noise at low frequencies. To find out what part of light spectrum is responsible for this effect, the measurements with optical filters were performed. In the case of the filter No. 2 the suppression effect is almost absent (spectrum 2). On the contrary installation of the filter No. 1 brings to the same effect as by the illumination without filters (spectra 1 and 4 practically coincide). These facts mean that the 1/f noise suppression is produced by photons having energy E falling in the range $hc/\lambda_1 > E > hc/\lambda_0$, i.e. by photons having energy more than ΔE_g, but close enough to it. The measurement of the diode resistance has shown that its magnitude changes by illumination no more than by 0.2 % . It makes impossible to relate the 1/f noise suppression by illumination with free electron concentration change.

These results indisputably prove a participation of traps in 1/f noise formation process. In the main they agree with whose which were got at lower fields [2]. So the trap mechanism of 1/f noise, developed in [1] and outlined in Introduction, can be extended, at less qualitatively, to the case of very high electric fields.

Let us discuss another interesting question. Experiments, carried out at low electric fields, have shown that the effect of illumination on 1/f noise depends on temperature. Being well pronounced at low and room temperature, it weakens with temperature rise, disappearing for GaAs at 430 K ($N_D = 10^{15}$ cm^{-3}) [2] and 540 K ($N_D = 10^{17}$ cm^{-3}) [14]. Our experiments were carried out at 480 K ($N_D = 10^{16}$ cm^{-3}). One can see from Fig. 2 that in our case illumination suppresses noise level at low frequencies, but this effect is rather weak (near 3 dB).

A simple explanation of these facts can reside in existence of an additional, insensitive to illumination, 1/f noise source [14]. In line with two-noise source model described in [13] (see Introduction), this unknown 1/f noise source can be associated with velocity fluctuations arising due to the scattering of electrons on neutral metastable centres. It is, certainly, only a suggestion. However there is an interesting way to clarify the question, at least in part.

To distinguish, is 1/f current noise related with concentration fluctuations or with mobility (velocity) fluctuations, the magnetoresistance methodology was used [2]: experiments at low temperature have shown that 1/f noise in GaAs manifests itself as concentration fluctuations. This result supports the trap model of 1/f noise. But does it held at high temperature, when 1/f noise does not respond to illumination? It is clear, that extension of mentioned methodology to the case of high temperature could give an important information.

3 Conclusions

Experiments, carried out on n-GaAs avalanche diodes and the discussion of results have brought to conclusions:

1. At high electric fields, just as at low, the participation of traps in 1/f noise formation is indisputable.

2. Some experimental results point to the existence of an additional 1/f noise source which is insensitive to illumination and prevails at high temperature. This noise source can be associated with drift velocity fluctuations, arising due to the scattering of electrons on neutral metastable centres.

3. To prove the additional 1/f noise sources existence and to establish its physical nature we need more experimental data. We must find answers to the questions:

(i) Does manifest itself 1/f noise at high temperature in the form of velocity fluctuations? The magnetoresistance methodology seems to be suitable to get the answer, at least at low field conditions [2], but as against [2] the experiment should be carried out at high temperature.

(ii) If so, then: Is the scattering on neutral metastable centres the principal primary source of 1/f velocity fluctuations? This question seems to be a difficult one. Its discussion hardly make sense till the answer to the first question will be found.

Acknowledgements

We are grateful to S.L. Rumyantsev and M.E. Levinshtein for valuable discussion and contribution. We thank also V.M. Vald-Perlov for providing experimental avalanche diodes.

References

1. N.V. Dyakonova, M.E. Levinshtein, *Sov. Phis. Semicond.* **23**, 175 (1989).
2. N.V. Dyakonova, M.E. Levinshtein, S.L. Rumyantsev, *Sov. Phis. Semicond.* **25**, 1241 (1991).
3. N.V. Dyakonova, M.E. Levinshtein, *Sov. Phis. Semicond.* **23**, 743 (1989).
4. T.G.M. Kleinpenning, *Physica* B **103**, 340 (1980).
5. T.G.M. Kleinpenning, *Physica* B **113**, 189 (1981).
6. Th. van de Roer, *Sol. St. El.* **23**, 695 (1980).
7. Th. van de Roer, *Physica* B **168**, 53 (1991).
8. T.G.M. Kleinpenning, *Physica* B **142**, 229 (1986).
9. M.E. Levinshtein, S.L. Rumyantsev, *Sov.Phis.Semicond.* **19**, 1015 (1991).
10. S.A. Kornilov, K.D. Ovchinnikov, V.M. Pavlov, *Izu. vuz. Radiophysica* **28**, 607 (1985) (In Russian).
11. S.A. Kornilov, K.D. Ovchinnikov, *Noise in oscillators, amplifiers and frequency multipliers on IMPATT-diodes* (GUT, St-Petersburg, 1993) (In Russian).
12. V.B. Orlov, A.V. Yakimov, *Physica* B **154**, 175 (1989).
13. S.A. Kornilov, K.D. Ovchinnikov, I. Corbella, *Proc. ICNF'95, 319* (World Scientific, Singapore, 1995).
14. M.E. Levinshtein, S.L. Rumyantsev, *Techn.Phis.Lett.* **19**(4), 247 (1991).

THE $1/f^{1/5}$ PROBLEM

N. V. DYAKONOVA, M. E. LEVINSHTEIN, S. L. RUMYANTSEV
Ioffe Institute, St. Petersburg, 194021, Russia.

J. W. PALMOUR
Cree Research, Inc., 4022 Stirrup Creek Dr. Suite 322, Durham NC 27713, USA

1 Introduction

The low-frequency noise spectrum in semiconductors and semiconductor devices is usually represented well as a superposition of $1/f$ noise and of one or several Lorentzians. However, for GaAs MESFETs it is often observed, that the low-frequency noise is a superposition of Lorentzians and of the noise with the spectrum of the type $1/f^{\gamma}$, where γ lies in the range 1.3–1.5 [1–4]. It is customary to associate the $1/f^{1.5}$ noise with the diffusion noise [5], or with the surface thermal noise [3].

However two different cases should be distinguished in the $1/f^{1.5}$ problem.

On the one hand the $S \sim 1/f^{1.5}$ dependence is observed in condition of the weak temperature dependence of the noise. This case may be observed either for volume or surface noise sources. The models discussed in Refs. [3,5] can be applied for this cases.

On the other hand noise with the spectrum $1/f^{\gamma}$, where $\gamma = 1.3 - 1.5$, is often observed in the temperature range, where generation-recombination (GR) noise due to a local level is a dominant component of the low-frequency noise. In this paper we will demonstrate that this phenomenon can be caused by the impurity level broadening which arises in any real crystal because of local stresses, lattice defects, doping inhomogeneities etc. However in order to explain the experimental dependences one has to suppose very unusual form of the level broadening, exponential temperature dependence of the capture cross section ($\sigma = \sigma_0 \exp(-E_1/kT)$ and linear dependence of E_1 on level position. By this means the validity of the model, proposed in this paper, must be supported or argued by other experiments.

2 Results and discussion

Fig. 1 shows the temperature dependences of the spectral density of channel resistance fluctuations, $S = S_R/R^2$ in a GaAs MESFET. The equilibrium electron concentration in the channel n_0 is $10^{17}\,\mathrm{cm}^{-3}$, and the channel volume V is

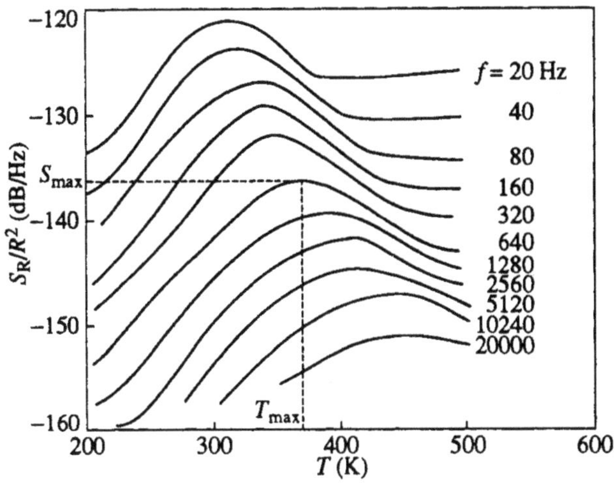

Figure 1: Temperature dependences of the spectral density of resistance fluctuations for a GaAs sample.

$2 \times 10^{-9}\,\mathrm{cm}^3$. For details of sample parameters and experimental conditions, see [6]. It easy to check that in temperature range $200 < T < 300\ K$ the frequency dependences of noise closely follow the $1/f^{1.5}$law. Fig. 2 shows temperature dependences of the spectral density of channel resistance fluctuations for SiC FET. The equilibrium carrier concentration in the channel is $10^{17}\,\mathrm{cm}^{-3}$, the channel volume at the gate voltage $U_g = 0$ is equal to $5 \times 10^{-10}\,\mathrm{cm}^3$. The drain-source voltage $U_{ds} = 1$ V corresponds to the ohmic regime. Details of the sample parameters are given in [7]. It easy to check again that at temperatures corresponding to the left (low-temperature) edges of the peaks in Fig. 2 ($440 < T < 500$ K) the frequency dependences of the noise follow the $1/f^{1.5}$ law. Similar behavior is observed for 4H-SiC FETs at high temperatures $T \geq 640$ K [8].

In Ref. [9], $S(T)$ curves have been presented for Si JFET. The value of γ calculated for the low-temperature peak edges is 1.4.

Strikingly similar results obtained for such different materials as GaAs, SiC, and Si led us to think that the $1/f^{1.5}$ noise has its origin in some general properties of local levels in semiconductors.

Consider first data for SiC since in this case the GR noise due to the local level can be clearly separated from other noise sources. The full curves in Fig. 3 show $S(f)$ deduced from Fig. 2 for four frequencies. These curves are plotted after substraction of the background noise (broken curves in Fig. 2).

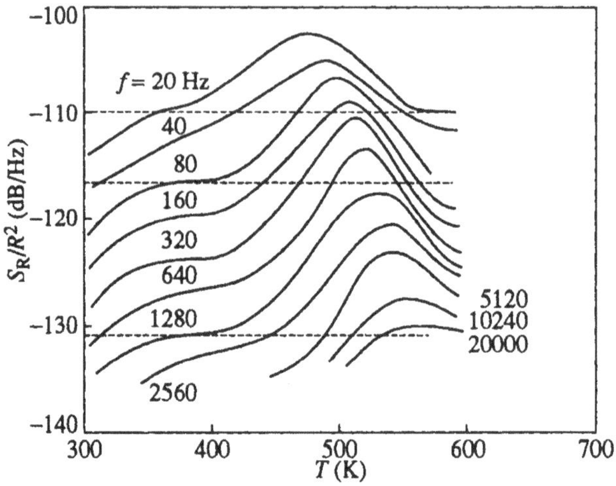

Figure 2: Temperature dependences of the spectral density of resistance fluctuations for a 6H-SiC sample [7].

Fig. 3 clearly demonstrates that the peaks in Fig. 2 are due to GR noise caused by a single local level.

With the "best fit" method the following values were determined from the experimental curves $S(T)$ shown in Fig. 4: $E_0 = 0.3$ eV, $E_1 = 1.1$ eV, $\sigma_0 = 5 \times 10^{-11}\,\mathrm{cm}^2$, $N_t = 5 \times 10^{15}\,\mathrm{cm}^{-3}$. The broken curves in Fig. 3 show the results of computer simulations of the GR noise, using to standard expression [10]. The exponential temperature dependence of the capture cross-section σ is taken into account. It is seen that the calculated and the measured dependencies $S_{\max}(f)$ and $T_{\max}(f)$ are in a good agreement. However for low-temperature edges of the peaks the slope of the experimental curves is less than that of the calculated curves at any frequency. It is this difference that leads to the frequency dependence $S(f) \sim 1/f^{1.5}$instead of $S(f) \sim 1/f^2$ for steeper calculated curves.

The observed broadening of the $S(T)$ peaks can be accounted for by the broadening of the local level.

The broadening of impurity levels is often invoked to explain experimental DLTS data [11,12]. However, the effect of level broadening on the shape of noise spectra $S(T)$ has not been previously discussed.

When the broadening effects have to be taken into account, the impurity band with distribution function $\rho(E)$ must be considered. The density of states $\rho(E)$ providing the best agreement between the simulation and the

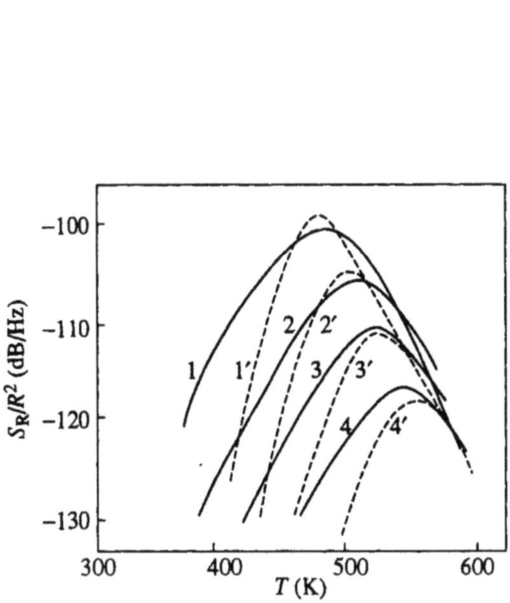

Figure 3: Temperature dependences of the noise spectral density for the same sample of SiC as in figure 2. Full curves reperesent the experimental data. Broken curves show the results of computer simulations. Frequency (Hz): $1, 1'-20$, $2, 2'-80$, $3, 3'-320$, $4, 4'-1280$.

experimental results is shown in Fig. 4 (inset).

The dependences $S(f, T)$ were calculated by integrating standard expression weighted with the density of states $\rho(E)$ over energy.

The results show that if the activation energy E_1 is independent on energy E, there is no accordance with the experiment for any $\rho(E)$.

We used the simplest model assumption of linear relation between E_1 and E with a single fitting parameter β [14]:

$$E_1 = E_{10} + \beta(E - E_0) \tag{1}$$

where E_{10} is the activation energy for $E = E_0$.

Assuming the above dependence $E_1(E)$ we succeed in fitting well the results of the calculation to the experiment. Fig. 4 shows the experimental $S(T)$ curves for SiC (the same as in Fig. 3) and the best fitting results of calculation.

Similar results were obtained for GaAs [10]. However the very special form of the line and unusual dependence of the activation energy E_1 on level position in the forbidden gap have to be assumed to explain the experimental results. Hence the independent experiments with low-temperature optical impurity spectroscopy of high resolution have to be provided to confirm or argue against this phenomenological model.

1. J. Graffeuil and J. Caminade, Electr. Lett. **10**, 266(1974)
2. J. R. Hellum and L. M. Rucker, Solid-State Electron. **28**, 549(1985)

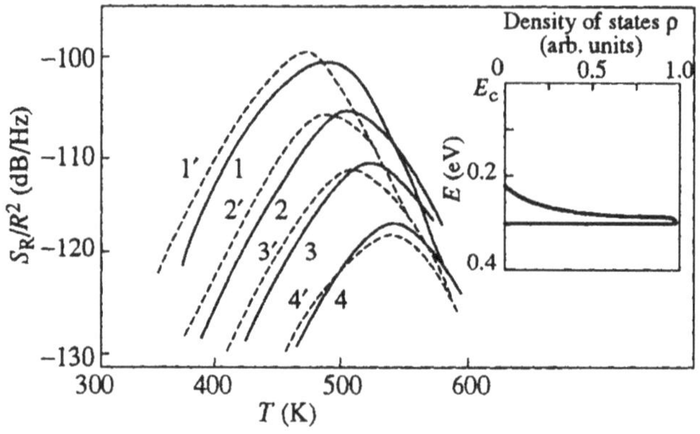

Figure 4: Comparison of experimental (full curves) and calculated (broken curves) $S(T)$ dependences for SiC. Frequency (Hz): $1, 1'-20$, $2, 2'-80$, $3, 3'-320$, $4, 4'-1280$. The inset shows $\rho(E)$ used in the calculations.

3. M. Pouysegur, J. Graffeuil and J. L. Cazoux, IEEE Trans. Electron Devices **34**, 2178(1987)
4. M. E. Levinshtein and S. L. Rumyantsev, Tech. Phys. Lett. **19**, 247(1993)
5. K. M. Van Vliet and G. R. Fasset, Fluctuation Phenomena in Solids (ed. by Burgess, Academic Press,New York), p 267(1965)
6. M. E. Levinshtein and S. L. Rumyantsev, Tech. Phys. Lett. **19** 247(1993)
7. M. E. Levinshtein, J. W. Palmour and S. L. Rumyantsev, Semicond. Sci.Techn. **9** 2080(1994)
8. M. E. Levinshtein, J. W. Palmour, S. L. Rumyantsev and D. B. Slater Jr., J.Appl.Phys., submitted for publication.
9. R. A. Spaulding, Proc. IEEE **56** 886(1968)
10. N. V. Dyakonova, M. E. Levinshtein and S. L. Rumyantsev, Semicond. Sci.Techn. **11** 177(1996)
11. J. R. Kirtley, T. N. Theis, P. M. Mooney, and S. L. Wright, Appl.Phys. **63** ,1541 (1988)
12. P. Omling, L. Samuelson and H. G. Grimmeiss, J. Appl. Phys. **54** 5117(1983)
13. V. N. Abakumov, V. I. Perel and I. N. Yassievich, Nonradiative Recombination in Semiconductors (Modern Problems in Condensed Matter Science vol 33) (Amsterdam: North-Holland) 1991

TEMPERATURE DEPENDENCE OF 1/f NOISE

X. Y. CHEN
Dept. of Electrical Engineering, Eindhoven University of Technology
5600 MB Eindhoven, the Netherlands

We discuss here the temperature dependence of 1/f noise in semiconductors. The experimental results from InP and GaAs epitaxial layers are presented. Previous measurements on semiconductors like InSb, Si and Ge are reviewed. No single model can explain all the findings.

1 Introduction

In homogeneous samples of semiconductors the 1/f noise can be described by equation (1).[1] The 1/f noise is characterized by a parameter α

$$\frac{S_R}{R^2} = \frac{S_V}{V^2} = \frac{\alpha}{fN}. \tag{1}$$

Various trends in the temperature dependence of α have been observed. The problem is: the existing models that work well at a constant temperature do not explain the temperature dependence of α.

In this contribution, we present recent results of the temperature dependence of α measured on InP and GaAs. For a review of the problem, all results from the last 15 years are also cited here. We may have some possible explanation for each specific situation. However, we do not understand the general results.

2 Experimental results of α (T).

2.1 The results for InP

We measured the noise in unintentional doped InP grown by chemical beam epitaxy (CBE). A series of samples are used. Gradually, we have succeeded in improving the quality of the samples. Figure 1 shows the temperature dependence of α.

In fig. 1 we also plotted the results reported by Tacano et al.[2] who measured the noise in heavily doped so-called FIB InP samples. The noise level is more than two orders lower than in our samples.

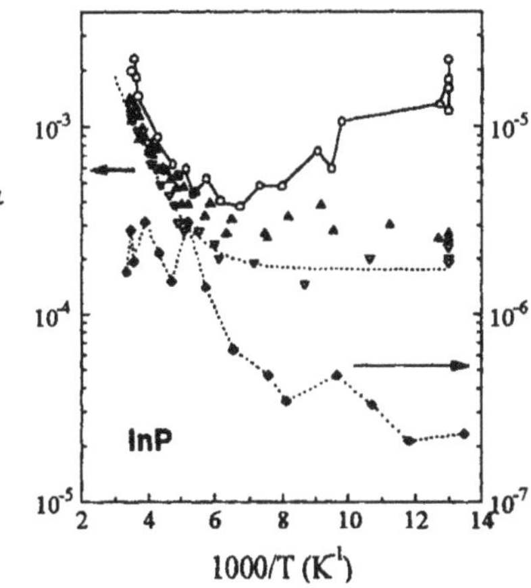

Fig. 1. α (T) in InP.
◆ : Tacano's InP.[2] The others are CBE InP, ○ : the purest CBE InP.

2.2 The results for GaAs

When studying the annealing effect on the proton damaged GaAs, we measured the temperature dependence in MBE (molecular beam epitixy) GaAs. The results are given in figure 2. The damage by proton irradiation introduces the excess 1/f noise that can be reduced by annealing. Annealing can repair the lattice to a certain extent.[3] In figure 2 we plotted the results for GaAs reported by Ren et al.,[4] and for InGaAs hetero - structure by Tacano et al.[5]

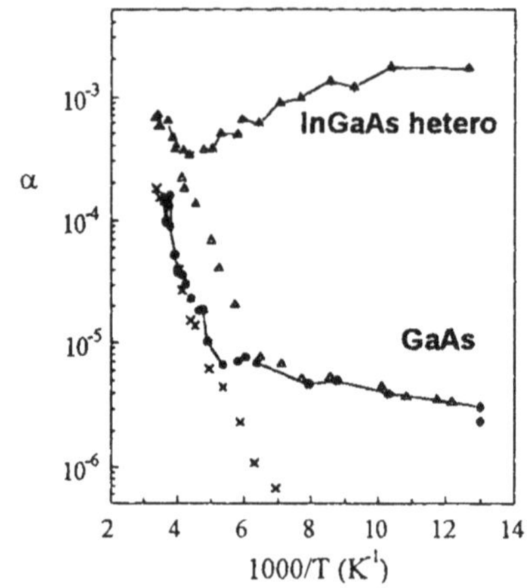

Fig. 2. α (T) in GaAs and InGaAs[5]
●: MBE GaAs; △ ×: Ren;[4] ▲: Tacano.[5]

2.3 The results for InSb

Alekperov et al.[6, 7] reported the results for single crystal InSb. The results based on n-type InSb are shown in figure 3. When the distribution of impurities in the sample was inhomogeneous, the noise becomes high and weakly dependent on temperature like date series of 1. For p-type InSb, the original paper showed the relative voltage fluctuation S_V / V^2 versus temperature instead of α. The noise level increases with decreasing temperature.[7]

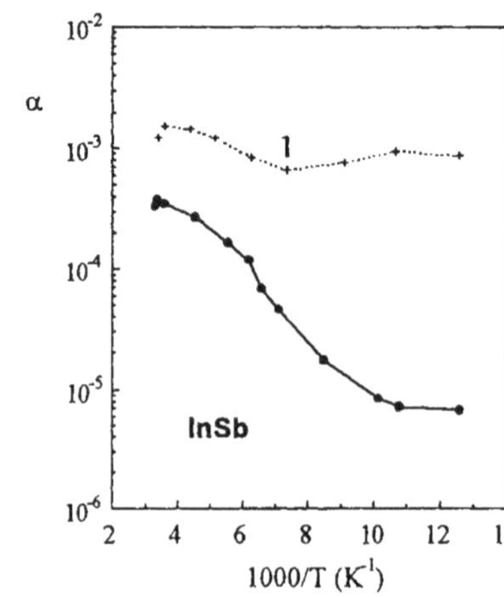

Fig. 3. α (T) in InSb (Alekperov[6])
+: With inhomogeneous distribution of impurities; ●: purer InSb.

2.4 The results for Si and Ge.

The temperature dependence of the 1/f noise in Si has been intensively studied. Clevers[8] reported many data from single crystal Si. It was shown that in all cases the noise was bulk noise. Various trends in the temperature dependence of α have been observed. There do not seem to be any reproducible results for different samples with different doping levels and different structures in geometry. He concluded that the different ways in which the samples were prepared might create different temperature dependences. Figures 4, 5 and 6 show the results reported by Palanskis et al.,[9] Luo et al.,[10] and Bisschop et al..[11]

One can conclude:

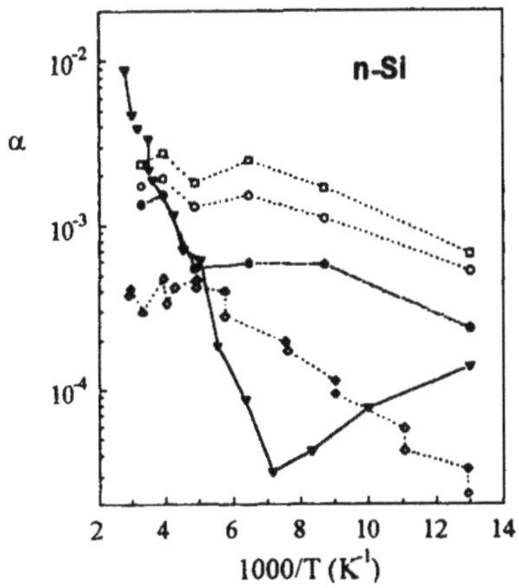

Fig. 4. α (T) in crystal n-Si.
◆: Palenskis;[9] □○●: Luo;[10] ▼: Bisschop.[11]

a) Various trends of α(T) exist, even in the same material.

b) The value of α scatters on a large scale in the same material with different doping levels.

3 Discussion.

3.1 Temperature dependence

Usually the 1/f noise increases with increasing temperature. The dependence of the 1/f noise on temperature could be anything if there is non-bulk 1/f noise. Therefore we would rather concentrate on samples where the bulk 1/f noise dominates. For these samples a promising common trend of temperature dependence of 1/f noise can be found. Generally in a good quality material like MBE, CBE growing semiconductors and pure crystal semiconductors, the temperature dependence of α composes of two branches, (i) a branch independent of temperature at lower temperatures and (ii) another branch with strong dependence on temperature at higher temperatures. The latter can be fitted either by an activated process or by a power law as expressed in equation (2) and (3). The activation energy ΔE or the power constant γ is different in different cases.

$$\alpha = \alpha_0 + b \times \exp\left(\frac{-\Delta E}{KT}\right), \tag{2}$$

$$\alpha = \alpha_0 + \beta T^{\gamma}, \tag{3}$$

where α_0, b and β are temperature independent.

In a narrow temperature range it is very hard to distinguish between the curves of equation (2) and (3), because for each value of ΔE one can always choose a proper value of γ that makes the curve of equation (3) very close to the curve of equation (2). ΔE was found between 0.1 and 0.2 eV. However, what that activated process is, is a mystery. γ is then found between 3 and 6 around room temperature.

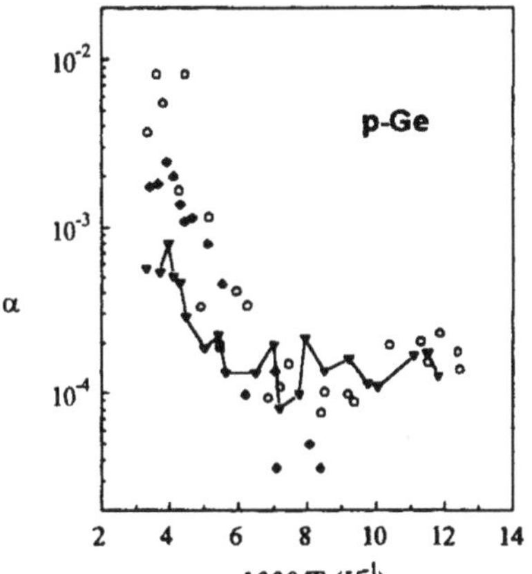

Fig. 6. α (T) for Ge from Bisschop.[11]

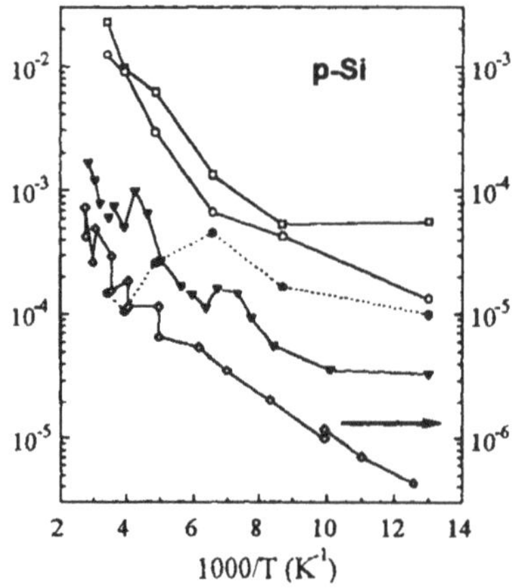

Fig. 5. α (T) in crystal p-Si.
◆: Ref. 9 □○●: Ref. 10 ▼: Ref. 11.

Luo et al.[10] also offered an alternative explanation by assuming that the 1/f noise originates in a thin surface layer via carrier trapping at the interface to the oxide layer or in the layer itself. Tacano et al.[2, 5] observed the temperature dependence of α in III-V compounds. They interpreted it by the Handel's quantum theory.

3.2 Wide range of α values

The wide range of α values was not only found in the experiments discussed above, it is quite common. To solve this problem, Hooge and Vandamme[12] proposed a model in which only lattice scattering shows 1/f noise when several scattering mechanisms are present. Then α can be expressed by

$$\alpha = \left(\frac{\mu_{meas}}{\mu_{latt}}\right)^2 \times \alpha_{latt}, \qquad (4)$$

where μ_{latt} is the lattice scattering mobility, μ_{meas} the measured mobility and α_{latt} a constant depending on the material. This model can explain the wide spread of α in good quality samples of the same material. Each material can be characterized by a value of α_{Latt}, but α_{Latt} has different values in different materials. At room temperature, $\alpha_{Latt} \approx 5\text{-}7 \times 10^{-4}$ for GaAs,[4, 13, 14] $\alpha_{Latt} \approx 3\times10^{-3}$. for InP,[15] $\alpha_{Latt} \approx 2\times10^{-3}$ for Si.[11] However, this model does not work so well at 77 K where we do not always find α proportional to μ_{meas}^2.[2, 8, 11] Except for the recently published results for InP at 77 K,[15] there is no evidence for the correctness of equation (4) at 77 K.

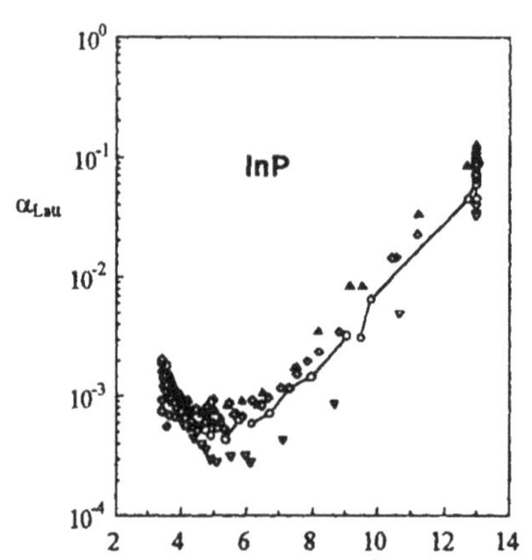

Fig. 7. α_{Latt} (T) for CBE InP.
○ : the purest sample.

The temperature dependence of α_{Latt} can be determined from equation (4) for a material. Palenskis et al.[9] reported that the temperature dependence of α in a Si sample follows the temperature dependence of $(\mu_{meas}/\mu_{latt})^2$, and hence that α_{latt} is a constant. Later Bisschop et al.[11] measured the noise from a series of samples with different impurity scattering, the model $\alpha \propto \mu_{meas}^2$ holds at temperatures down to 150 K, not at temperatures below 150 K. He proposed that when a sample is pure enough the measured α is α_{Latt}. He concluded a thermally activated process as in equation (5) for α_{Latt} in Si and Ge where $\Delta E \approx 0.1eV$. However, we are left with the problem why the model of the 1/f noise in lattice scattering can not explain the results for highly doped Si at low temperatures. More recently, Ren et al.[4] used a series of n-GaAs samples with different contributions from impurity scattering. Their results again showed that the model holds above 150 K. He ended the discussion by simply taking α_{Latt} equal to a constant 7×10^{-5} (=A) at lower temperatures. α_{Latt} is also described by equation (5) with $\Delta E = 0.13eV$.

$$\alpha_{Latt} = A + B\exp\left(-\frac{\Delta E}{kT}\right). \tag{5}$$

Our low temperature results for the purest InP sample are not consistent with the results given above. While cooling down the purest InP samples, the α value passes through a minimum and then increases again with decreasing temperature. In the other InP samples, α is temperature dependent and more or less follows equation (2). All results, including Tacano's, fit equation (4) at 300 K and 77 K.[15] By using equation (4) we obtain a temperature dependence of α_{latt} for all InP samples. The temperature dependence of α_{Latt} plotted in figure 7 is similar to that of α from the purest sample. We do not understand why α_{latt} depends on temperature in this way. Such temperature dependence of α_{Latt} was also found with other pure semiconductors. Tacano et al. measured the noise in heterostructure devices where the 1/f noise comes from two-dimensional electronic gas in the undoped channel. The noise increases with decreasing temperature below 150 K (see Fig. 2). Bisschop obtained such a trend in his purest Si sample too (see Fig. 6).

4 Conclusions and questions

Bulk 1/f noise of semiconductors has various temperature dependences. In a good quality sample the 1/f noise increases with increasing temperature. At high temperatures, around room temperature, the 1/f noise can either be described by a thermally activated process or by a power law. It is not understood what causes this dependence. The model with 1/f noise in the lattice scattering can successfully explain most results at higher temperatures, but not all at lower temperatures. α_{Latt} in pure samples increases with decreasing temperature. We do not understand that at all.

REFERENCES

1. F.N. Hooge, Phys. Lett. A 29, 139 (1969)
2. M. Tacano, H. Tanoue and Y. Sugiyama. Jpn. J. Appl. Phys. 31, L316(1991)
3. X. Y. Chen, V. Aninkevicius. X. Y. Chen, V. Aninkevicius. in *Proceedings of the 7th Vilnius Conf. Fluctuation Phenomena in Physical Systems, Palanga, 1994,* edited by V. Paleaskis (Vilnius University, Vilnius, 1994), p. 260.
4. L. Ren and F.N. Hooge, Physica B 176, 209 (1992).
5. M. Tacano, in *Proc. of the 13th Inter. Conf. "Noise in Phys. Systems and 1/f Fluctuations",* edited by V. Bareikis, R. Katilius. (World Scientific, New Jersey, 1995), p. 279
6. S. A. Alekperov, N. Ya. Guseinov, Ch. O. Kadzhar, and E. Yu. Salaev, Sov. Phys. Semicond. 20, 973 (1986)
7. S. A. Alekperov and F. L. Aliev, Sov. Phys. Semicond. 24, 578 (1990)
8. R. H. M. Clevers, Physica B 154, 214 (1989)
9. V. Palenskis and Z. Shoblitskas, Solid State Commun. 43, 761 (1982)
10. J.Luo, W.F. Love, and S.C. Miller, J. Appl Phys. 60, 3196 (1986).
11. Bisschop J., and Cuijpers J.L., Physica B+C 123, 6 (1983)
12. F.N. Hooge and L.K.J. Vandamme, Phys. Lett. 66A, 315 (1978)
13. L. Ren and M. R. Leys, Physica B 172, 319(1991)
14. X. Y. Chen, M. R. Leys and F. W. Ragay. Electron. Lett. 30(7), 600(1994)
15. X. Y. Chen, M. R. Leys, Solid-State Electronics 39, 1149 (1996)

1/f NOISE IN SEMICONDUCTOR MATERIAL INTERPRETED AS MODULATED GENERATION-RECOMBINATION NOISE

F. GRÜNEIS
Institut für Angewandte Stochastik, Fr.-Herschelstr.4, 81679 München, Germany
email: Ferdinand.Grueneis@t-online.de

M.I. TÖRÖK
JATE, Dom Ter 9, 6720 Szeged, Hungary
email: mitorok@physx.u-szeged.hu

It is assumed that fluctuations of some GR-levels in a semiconductor material are modulated by an underlying 1/f noise process with unknown physical origin. In accordance with empirical findings the Hooge coefficient α_H can be related to the lifetime τ of charge carriers.

1 Introduction

In context with a homogeneous semiconductor resistor, Hooge [1] experimentally found the relation

$$S_{1/f}(f) = (I_0^2 / N_0)(\alpha_H / f) \tag{1}$$

Herein $S_{1/f}(f)$ is the power spectral density of a fluctuating current I(t); I_0 is the mean of I(t). N_0 is the mean number of free charge carriers and α_H is the Hooge coefficient.

Recently, in semiconductors Lukyanchikova *et al.* [2] found empirically that the 1/f spectrum is closely related to the spectrum of GR-noise (GR=generation-recombination). The power spectral density of GR-noise is given by

$$S_{GR}(f) = 4(I_0^2 / N_0)(\langle \Delta N^2 \rangle / N_0)\, \tau / \{1+(2\pi f \tau)^2\} \tag{2}$$

Herein, $\langle \Delta N^2 \rangle$ is the variance of fluctuations originating from some GR-levels; τ is the mean lifetime of charge carriers. For finding whether 1/f noise is proportional to $\langle \Delta N^2 \rangle$ and/or τ, the temperature dependence of α_H was investigated. As a result, α_H was found to be only temperature dependent on τ and

$$\alpha_H = \beta \tau \tag{3}$$

Lukyanchikova [3] investigated this relation for different semiconductor materials and found that (3) is valid for τ ranging from 10^{-4} to 10^{-10} s. Since β is ranging only from $0.5*10^2$ to 10^3 s^{-1}, ß was suggested as a better measure for 1/f noise.

For interpreting empirical results, it is assumed that 1/f noise essentially is GR-noise which is modulated by a slowly varying time function. In this context, a mathematical model where 1/f noise is described as a modulated noise process is appropriate. In the next chapter, such a model, the clustering Poisson process is presented as a possible candidate and is generalized for applying also to a GR-process. Here the unsolved problems are: how can the cluster model be modified to apply to semiconductor physics? What are the predictions of this model? Is the Hooge formular still valid? How is about the temperature dependence?

2 1/f noise interpreted as modulated random noise

It is wellknown that shot noise is due to random occurrence of elementary current impulses. The random occurrence is described by a homogenous Poisson process for which the mean rate of impulses is constant in time. The spectrum is white up to a frequency inverse to the lifetime of impulses and is denoted $S_{shot}(f)$.

The case, the mean rate of impulses is no longer constant but is modulated in time by an underlying fractal noise process, has been investigated by Grüneis *et al* [4,5]. Such a process is constructed by a random succession of clusters of electrons and was denoted the clustering Poisson process (=CPP). Electrons in the cluster are supposed to occur at random; let λ be the time between electrons in the clusters. Denote m the number of electrons in a cluster and p_m the probability of finding m electrons in a cluster with m = 1,2,..N_{max}; N_{max} is a maximum cluster size. For $p_m \sim m^{-2}$, a pure 1/f shape is obtained. In this case, a mean cluster size $<m> = \Sigma\, mp_m \cong 0.6 \ln N_{max}$ and a mean cluster duration $\tau_c = <m><\lambda>$. For the CPP the spectrum is within the scaling region expressed by

$$S(f) = S_{shot}(f)\ \{\ 1 + (\ f_c / f\ )^b\ \} \tag{4}$$

In addition to shot noise one obtains a 1/f noise component which is due to cluster formation. b is the fractal exponent. f_c is the frequency where 1/f noise is equal to shot noise. It was shown that

$$f_c = 1 / 3\tau_c \tag{5}$$

and thus is equivalent to the mean correlation time of 1/f noise. Since τ_c is only dependent on the cluster parameters, f_c is a quantity independent of shot noise. The second term in (4) can formally be compared to Hooge's relation (1) resulting in $\alpha_H = 2f_c\tau$. For this reason, the CPP is a possible candidate for an interpretation of empirical results of (3) which is presented in the following.

3 How can the cluster model be applied to semiconductor physics?

As was shown by Van der Ziel [6], GR-noise can be understood as being caused by fluctuations in the rate of generation and recombination and the theory of shot noise applies to both. Regard some GR-level between the c-band and valence band; here any levels as concerns traps, donors and recombination centers may be taken into account. Assume this GR-process is modulated as described by the CPP. By analogy to (4), the spectrum is expressed by

$$S(f) = S_{GR}(f)\ \{\ 1 + (\ f_c / f\ )^b\ \} = S_{GR}(f) + S_{1/f}(f) \tag{6}$$

Denote by $<\Delta N^2{}_{1/f}>$ the variance of fluctuations of some GR-levels which are supposed to be responsible for 1/f noise. When this GR-level is identified $<\Delta N^2{}_{1/f}>$ can be calculated as described by Burgess [7]. Then, according to (2), the first term of (6) is the GR-spectrum which writes as

$$S_{GR}(f) = 4\ (I_0{}^2 / N_0)\ (<\Delta N^2{}_{1/f}> / N_0)\ \tau\ /\ \{\ 1+(2\pi f\tau)^2\} \tag{7}$$

The second term $S_{1/f}(f)$ is due to cluster formation which is modulating the GR-process. In combination with (7), the second term of (6) is expressed as

$$S_{1/f}(f) = S_{GR}(f)\ (\ f_c / f\ )^b = 4\ (I_0{}^2/N_0)\ (<\Delta N^2{}_{1/f}>/N_0)\ \tau\ \ (\ f_c / f\ )^b\ \ /\ \{\ 1+ (2\pi f\tau)^2\} \tag{8}$$

For $b = 1$ and $f < 1/2\pi\tau$, this is compared with (1) giving rise to

$$\alpha_H\ =\ \ 4\ f_c\ (<\Delta N^2{}_{1/f}>/N_0)\ \ \tau \tag{9}$$

Inserting (5), α_H can be given an even more informative form

$$\alpha_H\ \cong\ (<\Delta N^2{}_{1/f}>/N_0)\ \ (\ \tau\ /\ \tau_c) \tag{10}$$

In this form, Hooge's constant is related to the ratio between the mean lifetime τ of charge carriers and the mean duration τ_c of clusters and to the contribution of fluctuations $<\Delta N^2_{1/f}>/N_0$ of the GR-level which is supposed to be responsible for 1/f noise.

4 Discussions and Unsolved Problems

The results of this paper are based on the supposition that electrons in some GR-levels are generated and/or recombined in clusters. As a consequence, there are two contributions in the spectrum: the usual GR-spectrum which is due to the overall occurrence of electrons and a 1/f noise term which is due to cluster formation. For a clustered GR-process, in principle any level between valence and c-band may be taken into account. The results of the model are:

- the proposed model is a number fluctuation model,
- the Hooge formular is still valid,
- in accordance with empirical findings of (3), α_H is found to be proportional to τ,
- the Hooge coefficient α_H of (10) is essentially given by the quotient of two time constants: lifetime of charge carriers and mean cluster duration; since α_H is a dimensionless constant this is a reasonable result,
- α_H is also proportional to $<\Delta N^2_{1/f}>$ describing the contribution of the unknown GR-level responsible for 1/f noise. This is not in contradiction to the empirical results of Lukyanchikova *et al.* [2], however the levels contributing to their $<\Delta N^2>$ have to be excluded as possible levels giving rise to 1/f noise,
- at sufficiently high frequencies, 1/f noise is followed by a plateau as given by (7); this plateau however, may be buried in thermal noise or GR-noise of other origin,
- often α_H is measured as a function of temperature T; relation (3) suggests to separate the dependence on τ and to measure $\beta(T) = \alpha_H(T) / \tau(T)$.

An unsolved problem is the formation of clusters of electrons in the GR-process which may be due to a physical process of unknown origin. Among many other possibilities one may ask:

- is there some tiny correlation between electrons?
- is there the possibility of recombination via many levels of dislocations which may lead excitation of electrons in shallow traps?

Acknowledgements

The authors are very grateful for inspiring discussions with N.Lukyanchikova, Kiev and L.B.Kiss, Szeged.

References

1. F.N. Hooge, Phys Lett A29, 139 (1969).
2. N.B. Lukyanchikova, M.V. Petrichuk, N.P. Garbar, A.P. Sasciuk and D.I. Kropman, Physica B 167, 201-207 (1990).
3. N.B. Lukyanchikova, Physics Letters 180A, 285-288 (1993).
4. F. Grüneis, Physica 123A, 149-160 (1984).
5. F. Grüneis and T. Musha, Jap J Appl Phys 25,No.10, 1504-1509 (1986).
6. A. Van der Ziel, Fluctuation Phenomena in Semi-Conductors, Butterworths Scientific Publications, London 1959, p.28.
7. R.E. Burgess, Proc Phys Soc Lond, B68, 1020-1027 (1956).

ANOMALOUS ADDITIONAL LOW-FREQUENCY NOISE OF SURFACE ORIGIN GENERATED IN THIN GaAs AND InP LAYERS

P. GOTTWALD, ZS. KINCSES

Faculty of Electronic Engineering of the Technical University of Budapest, H-1521, Hungary

B. SZENTPÁLI

Research Institute for Technical Physics of the Hungarian Academy of Sciences, Budapest P.O.B. 76, H-1325 Hungary

Pure $1/f$ noise enhancement in the Low-Frequency Noise (LFN) spectrum of planar unpassivated InP and GaAs resistor samples has been measured under the influence of surface contamination. Furthermore, noise enhancement reminiscent of a resonant curve is superimposed on the LFN spectrum of surface passivated GaAs planar resistor samples. The physical background of these phenomena is presently unknown.

1 Introduction

Previous investigations on the low-frequency noise (LFN) behaviour of GaAs and InP have shown that the surface related additional noise components are by no means negligible. Furthermore, it is also clarified that the additional noise is generally due to process induced defects. Thus, the LFN measurement technique is an effective inspection tool in semiconductor technology.

The most common form of the additional LFN is the generation-recombination (G-R) noise having a Lorenzian spectral distribution and characteristic temperature dependence. Furthermore, additional $1/f$ noise is mostly a characteristic of semiconductor crystals of poorer perfection. Clearly, $1/f$ noise is mainly a bulk effect. Nevertheless, under the influence of controlled contamination on the free semiconductor surface, additional noise of surprisingly pure $1/f$ spectral distribution was observed. It was experimentally proved that the additional $1/f$ noise is of surface origin, and was dominantly observed for InP.

In other cases, if certain passivation technologies were used, an unusual selective enhancement of the LFN spectrum has been observed, reminding one of a resonant curve. The frequency of the enhancement was strongly temperature dependent. This effect is certainly caused by an unknown flaw in the passivation technology and has been observed - very rarely - for GaAs.

2 Experiment

In this work LFN measurements have been carried out on surface sensitive planar resistors. The length and the line width of the resistors were 800 and 40 μm respectively. The resistor chips were fabricated by wet chemical MESA etching on a thin (~300 nm) n-doped ($n = 2x10^{17}$ cm^{-3}) MOCVD InP or VPE GaAs layer, grown in both cases on a semiinsulating substrate. The layout of the patterns was the same as published in [1]. High quality Ohmic contacts were fabricated on the samples [2]. The samples were either exposed in a N_2 atmosphere to a saturated vapour of water, methanol, acetone or chloroform acting as contaminants, or were passivated by different techniques as will be detailed later.

To convert the resistance fluctuation to LFN voltage an inspection DC current was applied. Depending on the sample resistance this current was set to a value which caused a DC voltage drop of 8 V across the resistor. Thus, a longitudinal electric field strength of 100 V/cm was present in the sample. Both of the DC resistance and the noise spectra (frequency range: 2 Hz..20 kHz) were carefully monitored during the experiments.

In the case of the unpassivated samples, the LFN and the DC resistance were monitored during the exposure time. After each contamination cycle the surface was cleaned by a special technique [1], using UV light.

In the other case, the selective enhancement in the LFN spectrum was measured after the following passivation technologies has been applied to the GaAs samples:

a.) Passivation by RF-sputtered Si_3N_4 (Target-temperature: 30 °C, N_2 pressure: $5.2x10^{-3}$ mbar, acceleration voltage: 1.4 kV, process-time 10 min.)
b.) Hg sensitised photo-CVD of SiO_2 layer [3]. (Target-temperature: 150 °C, thickness: 100 nm)

The passivated samples were investigated by measuring the temperature dependence of the LFN spectra and the DC sample resistance in the temperature range of 0 - 80 °C.

3 The unsolved problems

a.) 1/f-type additional noise. - While exposing the samples to the contaminating atmosphere, the noise and resistance started to increase for both of the GaAs and InP. After a maximum value has been reached, the noise started to fall moderately [1], [4]. An excellent *1/f*-type additional noise of high intensity was generated in InP, and the maximum noise enhancements were about 40, 25, 6 and 2 dB for water, methanol, acetone and chloroform, respectively. Fig. 1. shows the noise enhancement for InP under the influence of water. For GaAs on the other hand, smaller and only nearly 1/*f*-like additional noise has appeared in case of water and acetone [1]. The maximum enhancement caused by water was only about 19 dB.

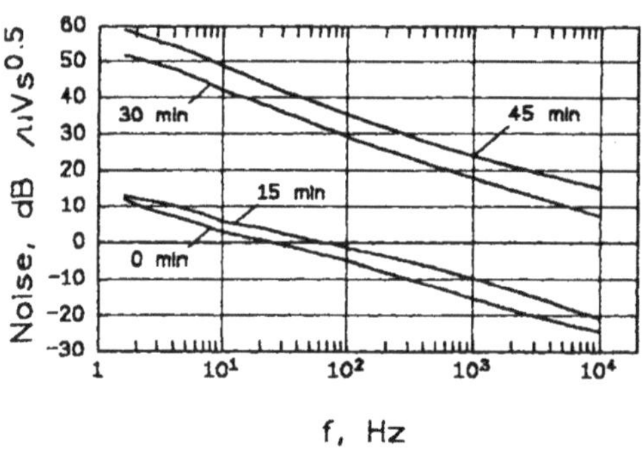

Fig.1. LFN spectra measured on unpassivated InP resistor samples, while exposing it to saturated vapour of water in a N_2 atmosphere. The lowest noise was measured at the beginning, then the noise was monotonically growing. The spectra have been measured after 15, 30 and 45 min contaminating exposure.

The shapes of the resistance transient curves [4] were monotonically growing for the organic solvents, but a local minimum appeared for water in the InP samples. The total change of the resistance was in all cases less than 10%. (The resistance change of the InP samples was 7% for methanol and water and only about 3% for acetone and chloroform.)

The mechanism of the generation of such a perfect 1/*f* noise due to surface contamination is not yet understood completely. Namely, it cannot be explained by the decrease of the mean carrier number caused by the widening of the surface depletion region [1], since this change is too small. Clearly, a 10 % increase of the sample resistance indicates a decrease of the mean carrier number in the same order, which would only cause about 0.41 dB increase in the 1/*f* noise, due to Hooge's empirical equation. Unfortunately, other models based on fluctuating defect states [5] are also insufficient to explain such a high intensity 1/*f* additional noise.

Furthermore, the extra noise can be cancelled by UV illumination ($\lambda \sim 400$ nm) and by simultaneous moderate heating of the sample in a pure N_2 atmosphere [1]. It should be noted that the heating alone does not cancel the noise. This fact gives evidence that the additional 1/*f* noise is caused by the adsorption of the contaminating materials. Probably they adhere to the semiconductor surface in the form of radicals. Regarding the photon energy of the recovering UV light, the attributed formation energy may be at about 2 eV. The noise enhancement and the accompanying resistance increment monitor this process. However, it is not clear while the additional noise is pure 1/*f* - type in many cases?

b.) Additional noise with resonant enhancement. - Certain passivation technologies for GaAs, as detailed previously, result in an unusual LFN enhancement, which looks like a resonant curve. Fig. 2. demonstrates that the frequency of the peak enhancement f_r is temperature dependent, for both of the passivations by sputtered Si_3N_4 and by the photo-CVD of SiO_2..

Furthermore, there is a decrease of the frequency of the peak enhancement, if the DC voltage on the sample increases. This decrease is only slight for the samples passivated by Si_3N_4, but considerable greater for the passivation by photo-CVD of SiO_2. Below a DC field strength of about 60 V/cm in the sample, the anomalous enhancement has been hidden in both cases.

Additionally, if the samples were illuminated by visible light, the resonant enhancement has vanished by an accompanying moderate increase of the $1/f$ noise.

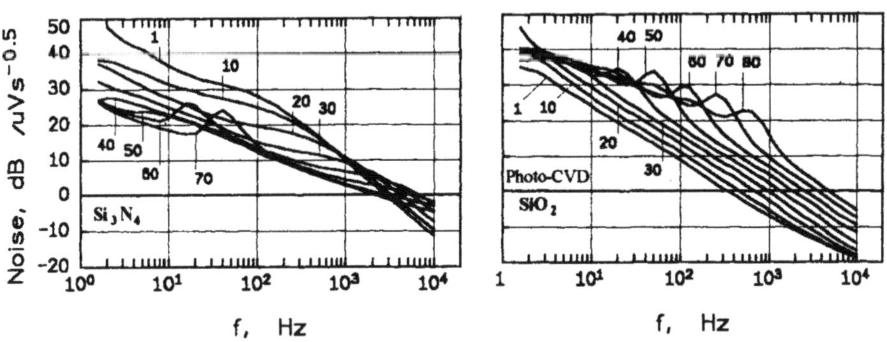

Fig. 2. Anomalous peak enhancement in the LFN spectra of surface passivated GaAs resistor samples in the temperature range of 10 - 80 °C, measured just after processing. Results for passivation by sputtered Si_3N_4 and for passivation by photo-CVD of SiO_2 layer are given.

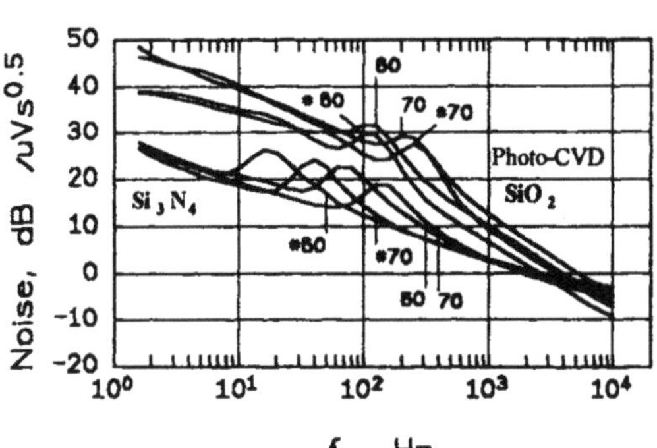

Fig. 3. Long-term drift of the noise spectra. Shift of the resonant enhancement at 60 and 70 °C in a time interval of 3 years.
- Curves #60 and #70 were measured in 1992 while curves 60 and 70 were measured in 1995
- Upper curves: passivation by photo-CVD of SiO_2 layer
- Lower curves: passivation by sputtered Si_3N_4.

Moreover, a long-term drift of the shape of the spectra has been observed in a 3 years interval, and the frequency of the peak enhancement f_r has shifted toward higher frequencies with the time. This is illustrated in Fig. 3.

Due to the temperature dependence of f_r an Arrhenius plot of (τT^2) has been constructed for the different cases (Fig 4.). Note that $\tau = (2\pi f_r)^{-1}$. Calculation of the activation energy resulted in a well-defined level at about 600 meV for the passivation by photo-CVD of SiO_2 just after processing. For the samples passivated by Si_3N_4 on the other hand, a level at about 880 meV has been found after processing, but 3 years later a lower value of 600 meV has been determined.

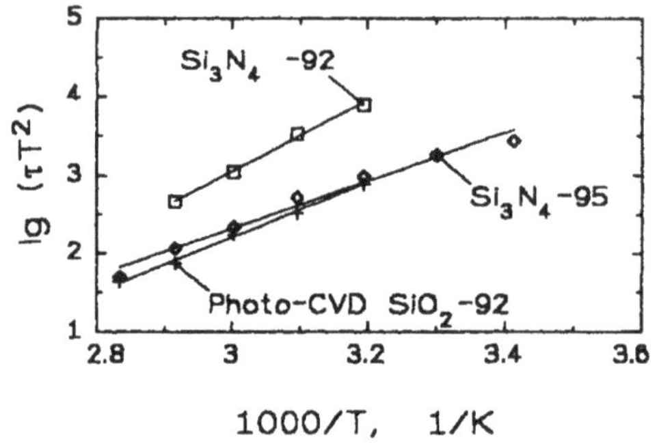

Fig. 4. Arrhenius plot for passivation by photo-CVD of SiO_2 just after the processing (-92). For the samples passivated by Si_3N_4 the plot is constructed after processing (-92) and 3 years later (-95).

These observations suggest that slow surface states (deep levels) are formed at the interface of the semiconductor and the deposited dielectric layer. However, the spectrum of the additional noise is very different to the Lorenzian one. Clearly, the derivative of a spectrum consisting of a $1/f$ spectrum and one or more Lorenzian components can never be positive. Due to the field-strength dependence and the long-term drift phenomenon the physical picture is further complicated and is not yet clear. Moreover it is not understandable why the two basically different passivation technologies can lead to the same unusual type of the additional noise.

4 Conclusion

Summarising the observations it should be stated that additional noise of unknown origins has been generated in special compound semiconductor structures. Without passivation these have often the form of $1/f$ -type LFN, thus complicating the separation of noise components of bulk and surface origin. Otherwise, some passivation technologies - however under unknown conditions - are able to generate very disturbing additional noise reminding us of a resonant enhancement in the spectrum. Additionally, this latter effect shows a long-term drift ranging even after several years. The exact physical background is presently unknown.

Acknowledgements

Thanks are due to Dr. E. Kuphal (German TELEKOM) for supplying the InP wafer material. The research was supported by the National Scientific Research Foundation (OTKA)/Hungary (contract No.: 773 and TO 15612), and also by the European Scientific Co-operation (contract No.: CP94/01180 /*Copernicus*)

References

1. A. Ambrózy et al.: Proc. *Noise in Physical Systems and 1/f Fluctuation, ICNF'91*, 23, Kyoto, (1991).

2. C. Heedt at al. : Proc. *Fourth Int. Conference on Indium Phosphide and Related Materials*, IEEE, 238, Newport (1992).

3. H. Kräutle: Passivation of InP for optoelectronics, *Seventh International Symposium on Passivity*, Clausthal, Germany, (1994).

4. P. Gottwald et al.: Proc. *Noise in Physical Systems and 1/f Fluctuations ICNF'95*, 299, Palanga, Lithuania (1995).

5. S. Borrello et al.: Solid-State Electronics **36**, 407 (1993).

THERMAL NOISE, FLUCTUATION-DISSIPATION

QUANTUM VACUUM FLUCTUATIONS, ZERO POINT ENERGY AND THE QUESTION OF OBSERVABLE NOISE

D. ABBOTT, B.R. DAVIS
Centre for High Performance Technologies and Systems (CHiPTec), EEE Department, University of Adelaide, SA 5005, Australia
email: dabbott@eleceng.adelaide.edu.au

N.J. PHILLIPS
Department of Physics, De Montefort University, Leicester LE7 9SU, England

K. ESHRAGHIAN
Faculty of Engineering, Edith Cowan University, Joondalup, WA 6027, Australia

In this paper we review the unsolved problem surrounding the exact relationship between noise, zero point energy and vacuum fluctuations. We survey the unresolved debate highlighting marked differences of opinion in the literature. Much of the uncertainty is shown to be due to unresolved fundamental issues in quantum mechanics.

1 Introduction

If we consider the thermal noise across a resistor R, loaded by a capacitor C, we can classically calculate noise over the total bandwidth. This has been carried out [1] for the various limiting cases of R and C and the results are displayed in Table 1.

Table 1: Thermal noise over infinite bandwidth for different cases of limiting R and C.

	Classical			Quantum		
	$\langle v_n^2\rangle$	$\langle i_n^2\rangle$	$\langle q_n^2\rangle$	$\langle v_n^2\rangle$	$\langle i_n^2\rangle$	$\langle q_n^2\rangle$
$R\to 0$ Shorted Cap.	0	0	0	0	0	0
$R\to\infty$ Open Cap.	$\frac{kT}{C}$ (dc)	0	kTC (dc)	$\frac{kT}{C}$ (dc)	0	kTC (dc)
$C\to\infty$ Shorted Res.	0	∞	∞	0	$\frac{2}{3hR}(\pi kT)^2$	∞
$C\to 0$ Open Res.	∞	0	0	$\frac{2R}{3h}(\pi kT)^2$	0	0

If we examine the classical solutions in Table 1, the most obvious problem with thermal noise formula $\langle v_n^2\rangle = 4kTR\Delta f$ is that it classically predicts infinite noise voltage for $C \to 0$ and infinite noise current for $C \to \infty$. This is an analogous situation to the black-body radiation problem where the Rayleigh-Jean's law suffers from the so-called ultraviolet catastrophe – the divergent black-body curve having infinite area over all frequencies. Anticipating this, Nyquist [2] in 1928 suggested replacing kT with the one-dimensional form of Planck's law

$$\frac{hf}{e^{hf/kT}-1} \tag{1}$$

which reduces to kT as $f \to 0$ and rolls off for $hf > kT$. This roll-off conveniently imposes a physical limit on the bandwidth and we see for this quantum case, in Table 1, that the infinities in question disappear. The remaining infinity for noise charge, in the quantum case, is not a breakdown of quantum theory but is due to $C \to \infty$ becoming an infinite store of charge [1]. $C \to \infty$ can be thought of as being modeled by an ideal voltage source and note, furthermore, that the noise process becomes non-stationary.

So far so good, Nyquist's quantum term successfully removes the unwanted infinities, however introduces a new set of problems. Firstly, this quantum term, alone, is obviously inadequate as it predicts that we can communicate with noiseless channels if $hf > kT$ (ie. in the Tera Hertz band). This is no longer an academic question as gallium arsenide resonant tunneling quantum electronic devices now operate in the THz domain. Gallium arsenide detectors[3] and sources [4] of THz radiation have been reported.

A second problem is that the quantum term, in Eqn. 1, predicts zero energy at $T = 0$ which is a violation of the Uncertainty Principle. As we shall see the solution to this creates a further conundrum.

2 The Quantum Energy Catastrophe

During 1911-12, Planck's 'second theory' produced the following modification to the quantum term [5]

$$\frac{hf}{e^{hf/kT}-1} + \frac{hf}{2} = hf \coth\left(\frac{hf}{2kT}\right). \tag{2}$$

The extra $hf/2$ term is called the *zero-point energy* (ZPE) and in this case, at $T = 0$, the Uncertainty Principle is not violated. This creates a further conundrum in that $hf/2$ is infinite when integrated over all frequencies, which is an apparent return to the type of 'catastrophe' problem we saw in the classical case. One can only assume that Nyquist accordingly did not suggest this form and probably would have been aware of Planck's own misgivings concerning the experimental objectivity of $hf/2$. The inclusion of $hf/2$ in standard noise texts only became popular after 1951 following the classic work of Callen & Welton [6] that produced the $hf/2$ ZPE term as a natural consequence of their generalized treatment of noise in irreversible systems using perturbation theory.

The solution to the catastrophe problem is that $hf/2$, in fact, turns out to be the *ground state* of a quantum mechanical oscillator. If n is the quantum number, which is a positive integer, then the allowed energy states for a quantum oscillator are $(n + \frac{1}{2})hf$ and thus the ground state is given when $n = 0$. As there is no lower energy state than the ground state, there is no energy level transition available to release the ZPE. Therefore it can be argued that $hf/2$ should be dropped before integration of the quantum expression. This procedure is an example of *renormalization*, which basically redefines the zero of energy. Renormalization is a significant area of quantum field theory and is usually presented in a more formal manner. The problem of renormalization is an open question in the theory of gravitation where there is the apparent catastrophe of *total* energy becoming infinite. For most laboratory measurements there is no catastrophe as we are only interested in energy *differences*. It is rather vexing that many basic texts herald quantum theory as removing the classical catastrophe, without admitting to the new set of catastrophe type problems it introduces such as in gravitation – a modern fully covariant theory of renormalization [7] resolves some problems, but the case is not yet fully closed.

The fact that the ground state energy, which we call ZPE, cannot be released means that texts that quote the Callen & Welton $hf/2$ term as an observable noise component are not strictly correct. However, by coincidence it turns out that the mean square of the zero point fluctuation (ZPF) also has the $hf/2$ form [8]. The mean square does not vanish with renormalization, of course, and this ensures the Uncertainty Principle survives renormalization. The mean square fluctuation is a detectable quantity and represents the magnitude of the ZPF. This noise starts becoming significant, just when the thermal noise begins to roll-off, in the THz band, thus preventing the possibility of noiseless communication.

Each mode contributes $hf/2$ towards the mean square fluctuation and, for an infinite number of frequencies, the magnitude is infinite. It is considered that this infinity is not fundamental, since the measurement conditions have not been specified. It can be shown [8] that for any finite observation bandwidth and volume of space the magnitude of the fluctuations of a quantum field is finite – if either the bandwidth is infinite or the measurement is evaluated at a *point* in space then the fluctuations become infinite.

3 The Steak Grilling Debate

In 1982, Grau & Kleen expressed the view that $hf/2$ is both unextractable and unobservable, adding their memorable rejoinder in the *Solid-State Electronics* journal that $hf/2$ is not "available for grilling steaks" [9]. Uncannily, about the same time Koch, Van Harlingen & Clarke (KVC) published noise measurements in superconductors reporting to have observed ZPF [10]. Over the next 3-4 years a number of independent superconductor papers followed, all nonchalantly quoting the KVC interpretation of ZPF as standard. In reply, Kleen (1987) essentially restated his case pointing out an unanswered question in the superconductor measurements [11]. As far as we are aware there has been no published KVC reply. This debate epitomizes the tension in schools of thought between $hf/2$ merely producing a measurement artifact (school of Kleen) and $hf/2$ being a real noise power (school of KVC).[a]

The orthodox position, is that the effects of ZPF are observable such as in the Casimir effect [12]. ZPF also has an orthodox status in explaining the observations of Mullikan [13], Lamb [14] and the nature of liquid helium [15]. On the other hand, consensus is not total as the school of Kleen has some support [16,17], the commonly supposed link between spontaneous emission and ZPF has been criticized [18] and the overall understanding of ZPF is also questioned as expressed, for example, in the following quote [19]:

> "The obvious question, then, is whether the zero-point energy and the vacuum fluctuations are one and the same thing. If they are, why is it that the former can be eliminated from the theory? The answer is not yet clear, and a deeper significance has yet to be discovered. Therefore, we will adopt the view that the zero-point energies are to be formally removed from the theory...., and all physical effects of the type.... discussed are to be ascribed to quantum fluctuations of the vacuum.... It must be admitted that the vacuum is not completely understood, neither physically nor philosophically. Whether or not the vacuum fluctuations are intimately related to the (unobservable) zero-point energy remains an open question."

where the expression "vacuum fluctuations" is an alternative term for ZPF. The view that ZPF cannot give rise to a detectable noise power itself, but can indirectly modulate or induce a detectable noise power has been expounded by Senitzky [20]. As for grilling steaks, the debate still sizzles but has shifted away from electrical noise theory. Controversial attempts to harness ZPE are underway using the concept of system self-organization [21] and presupposing the idea that the ground state is not the actual source of energy but is a 'pipeline'

[a] It is curious to note that KVC consistently always refer to the term 'ZPF' in their papers, whereas Kleen always uses the term 'ZPE' – hence there is the added confusion of semantics entangled with valid points of disagreement.

into some universal background source [22]. In an enterprising decade where there have been controversial attempts to consider superluminal velocity [23] and quantum information theory (promising two bits of information from one physical bit [24] and a form of teleportation [25]), there is no doubt that we have not heard the last of ZPE research. It remains to be seen what concrete results are produced and, if any, what the implications are to noise theory. Until further evidence, the quantum zero-field should be regarded as a conservative field as far as the extraction of energy is concerned. We can illustrate this using the thought experiment of a pair of parallel plates being pulled together by the Casimir effect – we can imagine one of the moving plates attached to a cord over a pulley with a miniscule mass on the end. As the mass is raised, the plate therefore does work and hence a small amount of energy is extracted from ZPF. However, external energy must be put into the system, to separate the plates to restart the process. Hence we have a conservative field. It could be argued that the ZPF is merely releasing externally introduced energy, stored by the system, and this may be a mechanical analogy of Senitzky's view [20].

On the other hand, Jaynes has pointed out [26] that the energy density of the Lamb shift, in a hydrogen atom, caused by ZPF, would give rise to a Poynting vector about three times the power output of the sun. This had led to a view that ZPF has no reality [27]. Hence the level of reality of ZPF, in this example, is in tension with the previous example. This also reflects the tension between KVC and Kleen.

Another consequence of a literal view of ZPE is that via the $E = mc^2$ relation and general relativity, this energy can also act as the source of a gravitational field – call this energy density in space W. Then the Kepler ratio for a planet with mean distance R from the sun and period T is proportional to $m_{\text{sun}} + (\frac{V}{c^2})W$, where V is the volume of the sphere of radius R. To agree with observed ratios for the planets the upper frequency cutoff for W can be no higher than optical frequencies [28]. But any attempt to account for the Lamb shift with ZPF requires a cutoff thousands of times higher, at the Compton wavelength [28]. This gravitational energy would not only disturb the above ratios, but it would radically disrupt the solar system. This *ad hoc* selection of frequencies for the operation of ZPF for the convenience of explanation is problematic.

4 Quantum Cut-Off Experimental Status

Fig. 1 shows a theoretical plot of the quantum term for different temperatures. The $hf/2$ term is plotted to illustrate that at normal working frequencies and temperatures it is vanishingly small, so for these conditions it can be

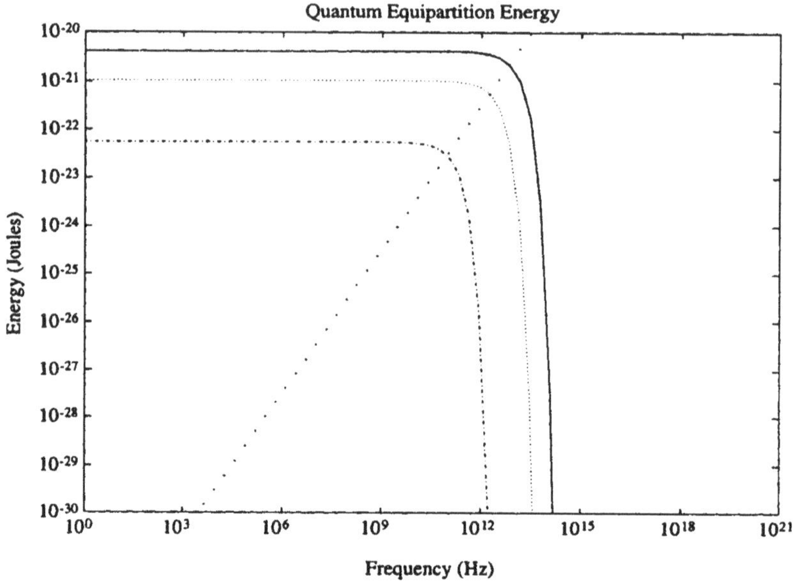

Figure 1: Quantum equipartition energy versus frequency for 300K, 77K and 4K. The line represents the hf/2 term plotted separately.

neglected regardless of the status of debate. It can be seen from the Fig. 1 plot that experimental verification of the quantum cut-off point for electrical noise is rather difficult due to the Tera Hertz frequencies. If the temperature is reduced, to reduce the cut-off frequency, we see that the maximum energy of the curves falls, thus making noise detection more difficult. In 1981, van der Ziel [29] proposed to make measurements in this region, at 100 GHz using Hanbury-Brown Twiss circuits; unfortunately, this research effort was never completed. The only curves we have today, for electrical noise, appear to be those of the type of KVC, which show no cut-off due to ZPF becoming significant. Therefore, as far as we are aware, there are no measurements that directly demonstrate the quantum cut-off for *electrical* thermal noise, to this day. Although the cut-off region, for electrical noise, has so far been obscured by ZPF it may become possible in the future to view at least part of this region, without violation of the Uncertainty Principle, if somehow the concept of ***squeezed states*** can be successfully employed for the electrical case (eg. [30]).

5 Conclusion

We have reviewed the debates surrounding the objectivity of the influence of ZPF on electrical noise. Although in the literature, terminology is not standard, we suggest to prevent confusion that the unextractable and unobservable groundstate is called ZPE, whereas the vacuum fluctuations themselves are called ZPF. We noted that the mean square fluctuation of ZPF has the form hf/2 and ZPE also has the form hf/2. This has caused some consternation in the literature and we highlighted that these quantities are different. ZPE can be removed by renormalization, whereas the effects of ZPF can be seen in a number of physical phenomena. It is clear that noise measurements are affected by an hf/2 law, as seen experimentally, otherwise communication channels would be noiseless above a certain frequency. However unresolved debate surrounds whether this represents a real noise power or is some quantum disturbance of a measurement (with no power to grill steaks). Also, Senitzky proposed a third option that ZPF cannot do work, but can modulate power from an outside source. All these views have problems: (1) insistence on a measurement artifact, with no work done seems to deny the reality of other observed ZPF effects, (2) whether power is produced or modulated, as per Senitzky, still leaves the problem of potential indefinite increase by hf/2. It seems that vacuum fluctuations are still not fully understood. Solutions could come either from developments in quantum physics or alternatively there is an opportunity to further tackle the problem from the point of view of noise.

Acknowledgments

Thanks are due to Dr. Murray W. Hamilton, University of Adelaide; Prof. B. L. Hu, University of Maryland; Prof. Laszlo B. Kiss, JATE, Hungary; Dr. Roger H. Koch, IBM Watson Research Center; Prof. Matt McIrvin, Harvard University; Prof. Tony Siegman, Stanford University; Prof. Paul C. W. Davies, University of Adelaide, and Prof. W. Tom Grandy, Jr., University of Wyoming for helpful communications on the subject of ZPE/ZPF. Funding from the Australian DEET TIL program is gratefully acknowledged.

References

1. D. Abbott *et al, IEEE Trans. Edu.* **39**, 1 (1996).
2. H. Nyquist, *Phys. Rev.* **32** (1928).
3. C.J.G.M. Langrak *et al, Appl. Phys. Lett.*, **6**, 23, (1995)

4. D. Birkedal *et al, Proc. 9th Int. Conf. on Ultrafast Phenomena*, California, (1994)
5. T. S. Kuhn in *Black-Body Theory and the Quantum Discontinuity 1894-1912* (Oxford, 1978).
6. H. B. Callen and T. A. Welton, *Phys. Rev.* **83**, 1 (1951).
7. N. D. Birrell and P.C.W. Davies, *Quantum Fields in Curved Space* (Cambridge Univ. Press, 1982).
8. W. H. Louisell in *Quantum Statistical Properties of Radiation* (Wiley, 1973).
9. G. Grau and W. Kleen, *Solid-State Elec.* **25**, 8 (1982).
10. R. H. Koch, D. J. Van Harlingen and J. Clarke, *Phys. Rev. Lett.* **47**, 17 (1981).
11. W. J. Kleen, *Solid-State Elec.* **30**, 12, (1987).
12. C. Itzykson and J-B. Zuber, *Quantum Field Theory* (McGraw-Hill, 1980).
13. R. S. Mullikan, *Nature* **114**, (1924).
14. T. A. Welton, *Phys. Rev.* **74**, 9 (1948).
15. C. J. Adkins, *Equilibrium Thermodynamics* (McGraw-Hill, 1975).
16. A. H. W. Beck, *Statistical Mechanics, Fluctuations and Noise* (Edward Arnold, 1976).
17. L. B. Kiss, *Solid State Comm.* **67**, 7 (1988).
18. A. E. Siegman, *Microwave Solid-State Masers* (McGraw-Hill, 1966).
19. W. T. Grandy, *Introduction to Electrodynamics* (Academic Press, 1970).
20. I. R. Senitzky, *Phys. Rev. A* **48**, 8, (1993).
21. M. B. King, *26th Proc. IECE* **4**, 1991.
22. H. E. Puthoff, *Phys. Rev. A* **40**, 9 (1989).
23. G. C. Giakos and T. K. Ishii, *IEEE Microwave & Guided Wave Lett.* **1**, 12, 1991.
24. C. Bennett and S. Wiesner, *Phys. Rev. Lett.* **69**, 20 (1992).
25. C. Bennett *et. al.*, *Phys. Rev. Lett.* **70**, 13 (1993).
26. E.T. Jaynes in *Complexity, Entropy and the Physics of Information*, ed. W.H. Zurek (Addison-Wesley, MA, 1990).
27. E.T. Jaynes in *Physics and Probability*, ed. W.T. Grandy and P.W. Milonni (Cambridge University Press, 1993).
28. W.T. Grandy, *Private Communication*, 1996.
29. A. van der Ziel, *Sixth Int. Conf. Noise Phys. Sys.* April 6-10, (1981).
30. B. Baseia and A. L. De Brito, *Physica A* **197**, (1993).
31. H. E. Puthoff, *Spec. Sci. and Tech.* **13**, 4, (1990).

QUANTUM NOISE IN TRANSPORT RESISTIVE SYSTEMS IS IT DETECTABLE ?

L. REGGIANI

Istituto Nazionale di Fisica della Materia, Dipartimento di Scienza dei Materiali, Università di Lecce, Via Arnesano, 73100 Lecce, Italy

C. PENNETTA

Dipartimento di Fisica, Università di Lecce, Via Arnesano, 73100 Lecce, Italy

V. GRUŽINSKIS, E. STARIKOV, P. SHIKTOROV

Semiconductor Physics Institute, Goštauto 11, 2600 Vilnius, Lithuania

L. VARANI

Centre d' Electronique et de Micro-optoélectronique de Montpellier (CNRS UMR 5507), Université Montpellier II, F-34095 Montpellier, France

The problem of the generalization to the quantum case of Nyquist formula is rivisited. We address its applicability to a semiclassical resistor and point out the possibility to provide a crucial test in favour of the presence and/or absence of the zero-point contribution.

1 Introduction

The aim of this communication is to revisit some fundamental questions concerning the quantum formulation of the Nyquist theorem [1] as applied to a semiclassical resistor where carriers behave similarly to a Brownian motion. Nyquist theorem states that at thermal equilibrium the spectral density of current fluctuations at frequency f of a given two-terminal device as measured in the outside short circuit, $S_I(f)$, is given by:

$$S_I^{cl}(f) = 4Re[Y(f)]K_BT \tag{1}$$

where K_B is the Boltzmann constant, T the lattice temperature and $Re[Y(f)]$ the real part of the device admittance. Equation (1) refers to the original *classical* form of Nyquist theorem [1]. We remark the possibility to write Eq. (1) for the spectral density of voltage-fluctuations $S_V(f)$ as measured at the terminals of the open circuit by replacing $Y(f)$ with its reciprocal quantity, the device impedance $Z(f)$. The extension of Eq. (1) to the *quantum* case, i.e. by substituting the Planck spectrum to the classical Rayleigh-Jeans value

K_BT, was already suggested by Nyquist himself as:

$$S_I^{qu}(f) = 4Re[Y(f)]\frac{hf}{exp(\frac{hf}{K_BT}) - 1} \tag{2}$$

h being the Planck constant. A quantum derivation of Nyquist formula provided in a seminal paper by Callen and Welton[2] suggested a further generalization of (2) which includes the zero-point energy of the harmonic oscillator thus giving:

$$S_I^{qu,zp}(f) = 4Re[Y(f)]hf[\frac{1}{exp(\frac{hf}{K_BT}) - 1} + \frac{1}{2}] \tag{3}$$

The classical Eq. (1) has been verified experimentally and numerically through simulative techniques (e.g. with the Monte Carlo method[3]). The quantum Eq. (2) has been indirectly verified by well established measurements of the spectrum of black-body radiation by absorption. However the problem remains open for the case of a semiclassical resistor where still it can be asked: (i) which of the form (2) and (3) is more appropriate to generalize Eq. (1) ? (ii) Is any direct detectable way to provide a crucial test in favour of (2) or (3) ?

2 State of the art

Concerning question (i) Callen and Welton[2] provided a first principle derivation of Eq. (3), succesively Landau-Lifshitz[4] and Kubo et al[5] confirmed this result. However, several perplexities have been arosen about both the theoretical derivation and its applicability to real cases[6].

MacDonald[7] has doubted about the inclusion of zero point term in expressions for Brownian mouvements and Robinson[8] provided as basic arguments on the same subject the fact that the notion of resistance and dissipation in quantum mechanics is a delicate problem to be treated as explicitely statistical processes describing the coupling between one quantum system, the circuit, and another quantum system, the heat sink. Indeed dissipation will violate ultimately quantum mechanics since all mechanical motion will eventually cease contrary to the uncertainty principle.

Bell[9] argued that zero point applies to the average power but cannot take place in any exchanges and is not to be included in the available noise power. In favour of this statement theoretical results of Bogoliubov and Shirkov[10] and Kubo[11] giving an average energy of a harmonic oscillator not including the zero point contribution are claimed.

Gupta[12], by using a Weber[13] argument, argued that the zero-point energy of the harmonic oscillator should not be included in the available noise-power because it cannot be extracted from the oscillator. In so doing, the oscillator would remain with zero energy thus violating the uncertainty principle.

Kiss [14] concluded that the presence of zero-point term would contradict basic results of quantum mechanics.

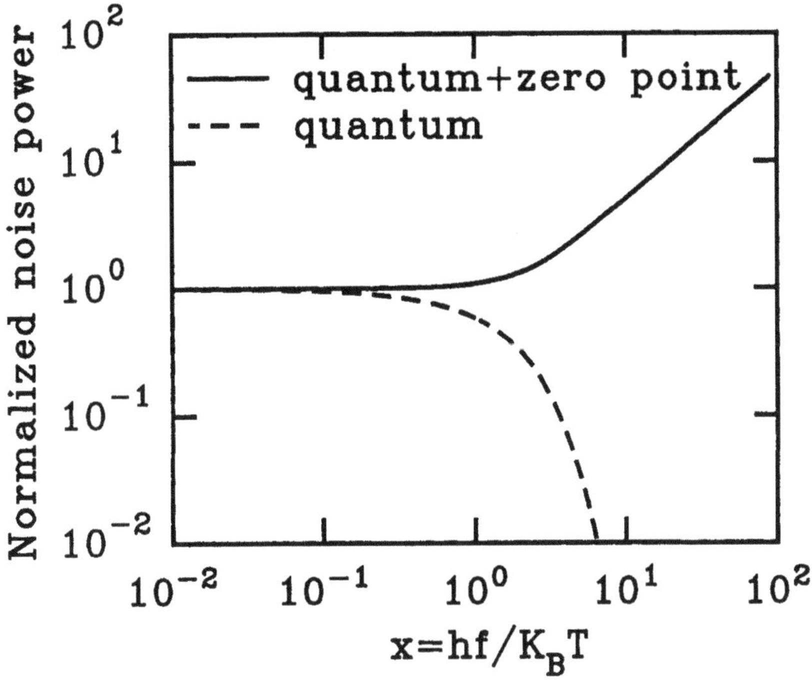

Figure 1: Normalized noise power as a function of normalized frequency. Dashed curve refers to the quantum case, continuous curve to the quantum plus zero-point energy case, respectively.

In any case, the possibility to confirm from a macroscopic measurement Eq. (3) is still under controversy, despite of a measurement made by Koch et al [15] who claimed evidence of it in a pioneer experiment in resistivity shunted Josephson junctions.

3 Unsolved problems

According to the discussion of the previous section, we conclude that the validity of Eq. (2) and/or (3) for the case of a semiclassical resistor represents still a challenge for researchers. From the point of simulative techniques it implies the introduction of the coupling between the carrier dynamics and the electromagnetic field in a self consistent way inside the kinetic equations. This step has never been solved to the authors knowledge and should be an interesting field for future research. From the point of view of experiments we offer below a tentative scheme whose practical realization is left to the skilfulness of experimentalists. By introducing for convenience the dimensionless frequency $x = hf/(K_BT)$, we report in Fig. 1 the cut-off region of the current spectral density of Eq. (2) and (3) normalized to the classical one. This numerical quantity represent the available noise power per unit bandwidth normalized to K_BT. For $x = 4$ we obtain the values 0.07 and 2.1 for quantum and quantum with zero point spectra, respectively. This difference should be experimentally detectable in a low temperature (i.e. 1 K) noise-power measurement at high frequency (i.e. 80 GHz) of a resistor at equilibrium. Here, despite of being at 1 K the noise temperature of the resistor under test should be of 0.07 K or of 2.1 K according to the validity of Eqs. (2) or (3).

Acknowledgments

This work has been performed within the European Laboratory for Electronic Noise (ELEN) supported by CEC contract ERBCHRXCT920047, PECO EAST ELEN ERBCHRXCT920047. Partial support from Ministero dell' Universitá e della Ricerca Scientifica e Tecnologica (MURST) Italiano is acknowldged.

References

1. H. Nyquist, Phys. Rev. **32**, 110 (1928).
2. H.B. Callen and T.A. Welton, Phys. Rev. **83**, 34 (1951).
3. C. Jacoboni and L. Reggiani, Rev. Mod. Phys. **55**, 645 (1983).
4. L.D. Landau and E.M. Lifshitz, Statistical Physics, (Pergamon, London, 1958).
5. R. Kubo, M. Toda and N. Hashitsume "Statistical physics II" Springer Verlag (Berlin, 1985).
6. C.W. Gardiner "Quantum noise" Springer Verlag (Berlin, Heidelberg, 1991).
7. D.K.C. MacDonald, Noise and fluctuations: an introduction, (Wiley, New York, 1962).
8. F.N.H. Robinson "Noise and fluctuations in electronic devices and circuits" Clarendon Press (Oxford, 1974).
9. D.A. Bell, Noise and solid state, (Pentech, London, 1985).

10. N.N. Bogoliubov and D.V. Shirkov, Introduction to the theory of quantized fields, (Interscience, London, 1959).
11. R. Kubo, Rept. Prog. Phys. **29**, 255 (1966).
12. M.S. Gupta, Proc. IEEE **70**, 788 (1982).
13. J. Weber, Phys. Rev. **90**, 977 (1953).
14. L.B. Kiss, Solid State Commun., **67**, 749 (1988).
15. R.H. Koch, D.J.Van Harlingen and J. Clarke, Phys. Rev. **B26**, 74 (1982).

QUENCHING THE THERMAL NOISE DOWN TO THE QUANTUM LIMIT

G. CAGNOLI, L. GAMMAITONI[a]
Dipartimento di Fisica, Universita' di Perugia
I-06100 Perugia, Italy

J. KOVALIK, M. PUNTURO
Istituto Nazionale di Fisica Nucleare, Sezione di Perugia,
I-06100 Perugia, Italy

F. MARCHESONI
Dipartimento di Fisica, Universita' di Camerino
I-62032 Camerino, Italy

Thermal noise in mechanical suspension systems is presently the most severe limitation to the sensitivity of the new generation of interferometric gravitational wave detectors, like VIRGO and LIGO, in the frequency range between 5 and 500 Hz. For this reason, a few experimental groups around the world are challenging the strategic goal of queching thermal noise effects.
Here we address present strategies, present some experimental data and discuss theoretical and experimental implications of such a task by pointing out present limits and still unsolved problems.

1 Introduction

Thermal noise is the name commonly given to fluctuations affecting a physical observable of a macroscopic system in thermal equilibrium with its environment.

The internal energy of a macroscopic apparatus at thermal equilibrium is shared between all its degrees of freedom or, equivalently, between all its normal modes each carrying an average energy kT, where k is the Boltzmann constant and T the equilibrium temperature. This is true also for such modes as the oscillations of springs, pendula, needles, etc. Such an energy manifests itself as a random fluctuation of the relevant observable experimentally perceived as the noise affecting its measured value.

Thermal noise is ubiquitous and is one of the unavoidable limits to the precision of mechanical measurements.

[a]Electronic mail address: Gammaitoni@perugia.infn.it

2 Sensitivity limit to the detection of gravitational waves

It has been estimated that thermal noise poses a severe limit to the sensitivity of the new generation of interferometric gravitational wave detectors, like VIRGO [b] and LIGO, in the frequency range between 5 and 500 Hz, due to the fluctuations in the position of the suspended elements (test masses and optics) of the interferometer and to the internal modes of the mirrors. It has been remarked that the whole suspending structure (super attenuator) can be treated as a multi-stage pendulum. Due to the action of thermal noise the position of each element of the pendulum chain will fluctuate in time. This is particularly true for the position of the optical components (mirror, beam splitter or suspended bench) located at the last stage of the chain. Such fluctuations combine with the gravitational wave induced displacement, thus setting a lower limit to the antenna sensitivity fig. 1.

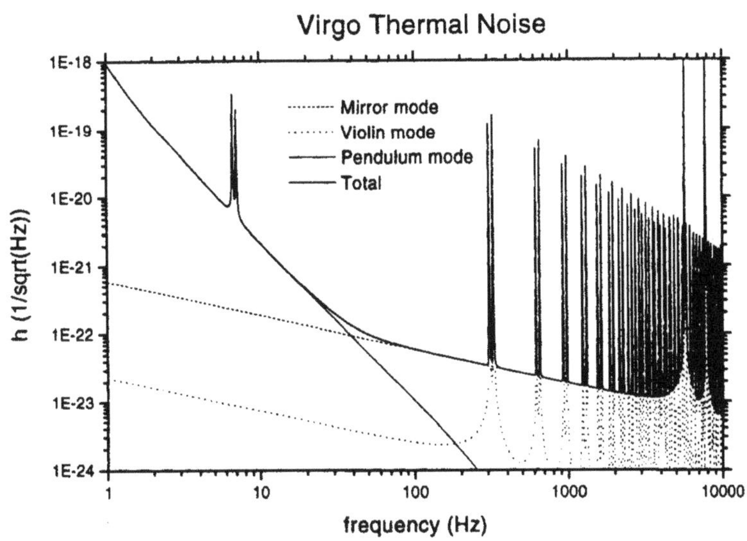

Figure 1: Thermal noise contribution to the VIRGO sensitivity curve (estimate).

It is clear that the determination of the spectral properties of the thermal noise affecting the antenna is an important and crucial task. It has been well

[b] For a description of the VIRGO project see the VIRGO central web site at "http://www.pg.infn.it/virgo".

known, since the fifties, that the spectral character of thermal fluctuations can be connected to the dissipative properties of the monitored observable (*Fluctuation-Dissipation theorem* (FDT)[1]): the equilibrium fluctuations (thermal noise) of a macroscopic system can be probed by applying a weak (linear response) external perturbation, which couples to the observable of interest. The system absorbs energy from the perturbation in a way which is completely determined by the spectral properties of the fluctuations, in the absence of the perturbation. [2]

This suggests an experimental way to study the spectral properties of thermal noise affecting a macroscopic observable: let us call $x(t)$ a physical observable which quantifies the response of a system to an applied external force $f(t)$. The response function $H(\omega)$ (called "generalized susceptivity" in the physical literature) is a complex quantity defined by:

$$X(\omega) = H(\omega)F(\omega); \tag{1}$$

with $H(\omega) = H'(\omega) + iH"(\omega) = |H(\omega)|e^{i\phi(\omega)}$. Here capital letters denote the Fourier transform of the corresponding time dependent quantities. The FDT sets a relation [1] between the power spectral density of $x(t)$, $< x(\omega)^2 >$, and $H"(\omega)$, i.e.

$$< x(\omega)^2 >= -2\ kT\ \frac{H"(\omega)}{\omega} \tag{2}$$

Most notably, the quantity $H"(\omega)$ can be accessed experimentally and the fluctuation power spectral density $< x(\omega)^2 >$ can be obtained accordingly. The r.m.s. value $\sqrt{< x(\omega)^2 >}$ can be obtained through the relation [1]:

$$\sqrt{< x(\omega)^2 >} = \sqrt{\frac{1}{\pi}\int_0^\infty < x(\omega)^2 > d\omega} = kT\ H'(0) \tag{3}$$

The detector sensitivity curve can thus be estimated by measuring the dissipation properties of the suspension structure or, to be more precise, the imaginary part of the response function of the observable of interest $x(t)$ (i.e. the position of the mirror center of mass) to the conjugate force $f(t)$. The response function $H(\omega)$ is obtained from 1 by recording the time series of the force and the relevant displacement and, then, taking their Fourier transforms. Finally, the fluctuation spectral density (thermal noise) is computed from the $H(\omega)$ imaginary part, $H"(\omega)$, by using the FDT in 2.

3 How to reduce thermal noise effects

Once the expected spectral noise properties of the thermal noise affecting the system are known, a crucial task is the quenching of thermal noise effects in the frequency band of interest, down to the quantum limit. This is usually accomplished by acting on the dissipation properties of the mechanical structures. After reducing external losses, one has to address the internal friction effects. An effective quenching strategy is mainly based on a proper choice of the material and on a careful characterization of the intrinsic dissipation mechanism.

Such a characterization is performed phenomenogically through the study of the material loss angle ϕ. This empirical quantity which should account for a number of different dissipation sources resists to a general theoretical treatment [3,4]. Moreover simple "ad hoc models" generally fail to reproduce experimental data.

The detailed knowledge of the dependence of the loss angle on load and frequency is still an open problem, a pre-condition for the realization of highly sensitive mechanical devices. Such a strategy for the reduction of thermal noise effects is commonly implemented by designing mechanical oscillators with high quality factor (see fig. 2).

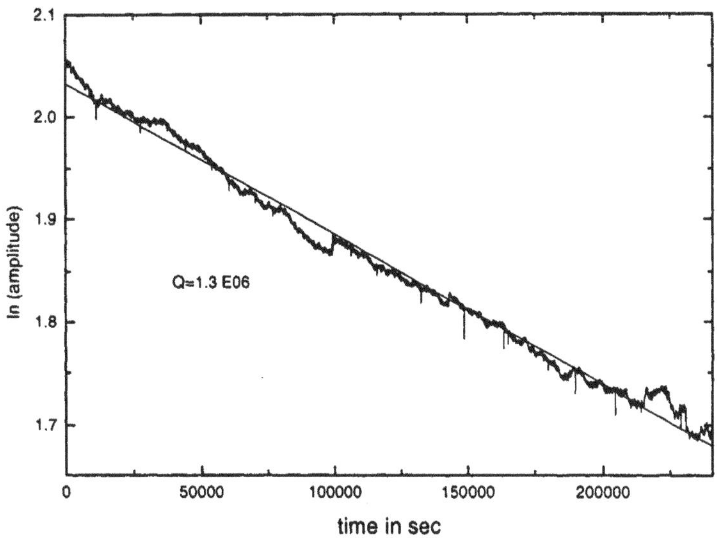

Figure 2: Logarithmic relaxation of a full scale last stage suspension system prototype, as measured in Perugia. $Q = 1.3 \pm 0.1\ 10^6$ at $\nu_0 = 0.6\ Hz$.

The effectiveness of such a strategy relies much on the assumption that the loss angle is independent from the frequency. While this assumption is far from being generally verified by experiments, we still miss a viable approach to frequency dependent losses. In particular, the problem of measuring losses in the low frequency regime, for a wide frequency band, is still unsolved.

Acknowledgments

Work supported by the VIRGO project, an INFN-CNRS collaboration.

References

1. R. Kubo, M. Toda and N. Hashitsume, Statistical Physics II (Springer, Berlin, 1985) and L.D. Landau, E.M. Lifsits, Statistical Physics (Mir, Moscow, 1976).
2. L. Reichl, A Modern Course in Statistical Physics ,(Univ. of Texas press, Austin, 3rd ed. 1988).
3. A.S. Nowick and B.S. Berry, Anelastic Relaxation in Cristalline Solids (Academic Press, New York, 1972).
4. F. Marchesoni, G. Cagnoli and L. Gammaitoni, Phys. Lett. A 187, 359 (1994) and G. Cagnoli, L. Gammaitoni, J. Kovalik, F. Marchesoni, M. Punturo, Phys. Lett. A 213, 245 (1996).

A NEW CIRCUIT THEORY PARADOX IN THE NOISE ANALYSIS OF A 2-STAGE RC LADDER

D. ABBOTT, B.R. DAVIS
Centre for High Performance Technologies and Systems (CHiPTec),
EEE Department, University of Adelaide, SA 5005, Australia
email: dabbott@eleceng.adelaide.edu.au

K. ESHRAGHIAN
Faculty of Engineering, Edith Cowan University,
Joondalup, WA 6027, Australia

A new paradox is described where the correlation between the thermal noise voltages in a 2-stage RC ladder behaves unexpectedly at limiting component values. A future resolution to this unsolved problem may possibly uncover a limitation in the circuit theory formalism for handling noise. There may be be a connection between this problem and Penfield's motor paradox proposed in 1966.

1 Introduction

For the first time, we present an unsolved paradox in a simple two stage RC ladder (Fig. 1). In the Appendix, we solve the relevant complex integral for the circuit showing that correlation between the two capacitor noise voltages is zero, ie. $\langle v_1 v_2 \rangle = 0$. However, we also show that if $R_2 \to 0$ or $R_1 \to \infty$ then $\langle v_1 v_2 \rangle$ suddenly becomes non zero! For low R_2 we obtain some correlation, whereas for large R_1 we get anticorrelation. In practice, zero correlation will not be observed as we impose a limited measurement bandwidth. Zero correlation is only obtained when we consider the total frequency band. Nevertheless, the predicted crossover from correlation to anticorrelation, as resistor values change, is a surprise result.

This dilemma is unsolved and may highlight a limitation in the circuit theory formalism for noise. If the capacitors are replaced by inductors, it may be that the problem has similar roots to Penfield's motor paradox[1,2].

It would be interesting to recalculate the correlation terms if we replace kT with the one dimensional form of Planck's law[3], to impose the quantum limit to bandwidth. Closed form solutions of the resulting integrals appear to be exceedingly difficult and could probably be expressed in terms of the χ function (the derivative of $\ln \Gamma(z)$). However, as $R_1 \to \infty$, the integration anomalies occur near $f = 0$, so the quantum form would not affect the result in this particular case.

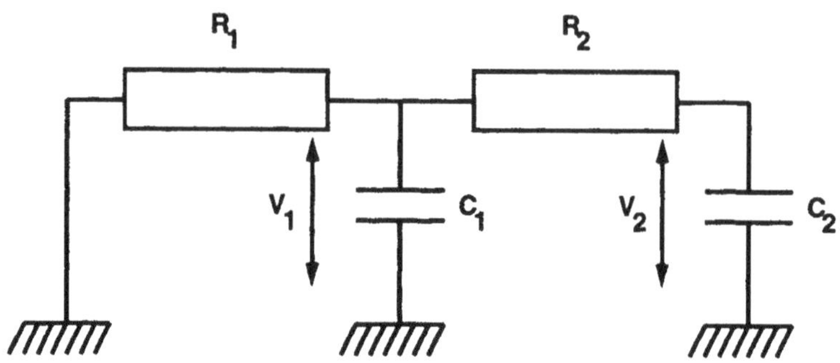

Figure 1: Two stage RC ladder – a source of an unsolved paradox.

2 Discussion

It may be that there are anomalies introduced by letting $R_1 \to \infty$ or $R_2 \to 0$ *before* we do the integration. It would therefore be instructive to take limits after the integration, for comparison. However, we do not get the opportunity to do this because $\langle v_1 v_2 \rangle = 0$ and thus there are no variables in this expression to manipulate. To overcome this problem, we write a new variable $\langle v_{ij} \rangle$ which is the voltage noise across capacitor C_i due to resistor R_j. So we can separate the noise contributions from the two resistors as

$$\langle v_1 \rangle = \langle v_{11} \rangle + \langle v_{12} \rangle \tag{1}$$

and

$$\langle v_2 \rangle = \langle v_{21} \rangle + \langle v_{22} \rangle \tag{2}$$

hence the correlation can be written as

$$\langle v_1 v_2 \rangle = \langle v_{11} v_{21} \rangle + \langle v_{12} v_{22} \rangle. \tag{3}$$

It can then be easily shown that

$$\langle v_{11} v_{21} \rangle = \frac{kT}{C_1 + C_2 + \frac{R_2 C_2}{R_1}} \tag{4}$$

and

$$\langle v_{12} v_{22} \rangle = -\frac{kT}{C_1 + C_2 + \frac{R_2 C_2}{R_1}}. \tag{5}$$

Therefore the sum of these two terms, which is the total correlation, is zero as before. Now if we let $R_1 \to \infty$ or $R_2 \to 0$, the total correlation is still zero. This contradicts our initial non-zero results, when the limits were taken *before* the integration. This basically summarizes the dilemma. The differences *can be* mathematically explained: for instance if $R_1 \to \infty$, before integration, the positive part of the curve becomes a delta function and is no longer captured by the integral. Hence we effectively integrate under the negative portion of the curve and the correlation becomes negative. The unsolved problem is the physical interpretation, in terms of noise, of the ordering of the limits.

3 Conclusion

We have outlined an unsolved problem regarding thermal noise correlation in a 2-stage RC ladder. A solution may improve understanding of the circuit theory formalism. If the capacitors are replaced by inductors, there may be some similarities with Penfield's paradox posed in 1966, though this requires further investigation.

Acknowledgments

Funding from the Australian DEET TIL program is gratefully acknowledged.

Appendix

General Complex Integral for Capacitor Problem

We need to solve the integral of the general form:

$$I = \frac{1}{2\pi j}\int_{-j\infty}^{+j\infty} \frac{a_0 + a_1 s + a_2 s^2}{(b_0 + b_1 s + b_2 s^2)(b_0 - b_1 s + b_2 s^2)} ds \tag{6}$$

Let $b_0 + b_1 s + b_2 s^2 = b_2(s - s_1)(s - s_2)$, so by factorizing the denominator and taking a contour integral we have,

$$I = \frac{1}{2\pi j}\oint_C \frac{a_0 + a_1 s + a_2 s^2}{b_2^2(s - s_1)(s - s_2)(s + s_1)(s + s_2)} ds. \tag{7}$$

Taking the sum of the residues,

$$I = \frac{a_2 s_1 s_2(s_1 - s_2) - a_0(s_1 - s_2)}{2 s_1 s_2(s_1 - s_2)(s_1 + s_2) b_2^2} \tag{8}$$

and using $s_1 s_2 = b_0/b_2$ with $s_1 + s_2 = -b_1/b_2$, finally gives

$$I = \frac{a_0 b_2 - a_2 b_0}{2 b_0 b_1 b_2}. \tag{9}$$

This result is quite fascinating as the a_1 term has totally dropped out giving a purely real result. This can be explained because $a_1 s$ is an *odd* function of s.

Another curious matter is that because of the conjugation on the denominator of the original integral, the Cauchy-Riemann equations are *not* satisfied. However, the method of residuals happens to nevertheless work because the integral is essentially real given that the a_1 term drops out. As a precaution, we integrated the real part of the integral the long hand way and found that it did indeed reduce to the same result provided by the method of residues. Due to the great length of this procedure, this was accomplished using the MAPLE math editor software.

The fact that the method of residues is found to work on a non-analytic integral, is apparently not discussed in the complex analysis literature. There maybe be scope for future work to formally define a class of such integrals.

Noise Analysis of 2-Stage RC Ladder

From nodal analysis of the circuit we find that,

$$v_1 = \frac{e_1(1 + sC_2R_2) - e_2 s C_2 R_1}{1 + s(C_1R_1 + C_2R_2 + C_2R_1) + s^2 C_1 C_2 R_1 R_2} \tag{10}$$

and

$$v_2 = \frac{e_1 + e_2(1 + sC_1R_1)}{1 + s(C_1R_1 + C_2R_2 + C_2R_1) + s^2 C_1 C_2 R_1 R_2}. \tag{11}$$

Using $e_1 = 2kTR_1$ and $e_2 = 2kTR_2$ and multiplying by complex conjugates, gives us the spectral densities,

$$S_{11} = 2kT \frac{R_1|1 + sC_2R_2|^2 + R_2|sC_2R_1|^2}{|1 + s(C_1R_1 + C_2R_2 + C_2R_1) + s^2 C_1 C_2 R_1 R_2|^2} \tag{12}$$

$$S_{22} = 2kT \frac{R_1 + R_2|1 + sC_1R_1|^2}{|1 + s(C_1R_1 + C_2R_2 + C_2R_1) + s^2 C_1 C_2 R_1 R_2|^2} \tag{13}$$

$$S_{12} = 2kT \frac{R_1(1 + s^2 C_1 C_2 R_1 R_2)}{|1 + s(C_1R_1 + C_2R_2 + C_2R_1) + s^2 C_1 C_2 R_1 R_2|^2}. \tag{14}$$

These spectral densities are now integrated using the general solution given in the last section. This yields the noise voltages in Volts squared per Hertz, and the integrals simply reduce to

$$\langle v_1^2 \rangle = \frac{kT}{C_1}, \quad \langle v_2^2 \rangle = \frac{kT}{C_2}, \quad \langle v_1 v_2 \rangle = 0$$

but if $R_2 \to 0$,

$$S_{11} = S_{22} = S_{21} = \frac{2kTR_1}{|1 + s(C_1R_1 + C_2R_1)|^2}$$

therefore,

$$\langle v_1^2 \rangle = \langle v_2^2 \rangle = \langle v_1 v_2 \rangle = \frac{kT}{(C_1 + C_2)}.$$

If $R_1 \to \infty$,

$$S_{11} = \frac{2kTR_2C_2^2}{|C_1 + C_2 + sC_1C_2R_2|^2}, \quad S_{22} = \frac{2kTR_2C_1^2}{|C_1 + C_2 + sC_1C_2R_2|^2},$$

$$S_{12} = \frac{-2kTR_2C_1C_2}{|C_1 + C_2 + sC_1C_2R_2|^2},$$

therefore,

$$\langle v_1^2 \rangle = \frac{kTC_2}{C_1(C_1 + C_2)}, \quad \langle v_2^2 \rangle = \frac{kTC_1}{C_2(C_1 + C_2)}, \quad \langle v_1 v_2 \rangle = -\frac{kT}{(C_1 + C_2)}.$$

References

1. P. Penfield, *Proc. IEEE*, **54**, 9 (1966).
2. D. Abbott *et al*, *IEEE Trans. Edu.* **39**, 1 (1996).
3. H. Nyquist, *Phys. Rev.* **32** (1928).

NOISE IN MAGNETIC MATERIALS

NOISE AND MAGNETIC DOMAIN STRUCTURE IN NiFe THIN FILMS

J. BRIAIRE, L.K.J. VANDAMME

Eindhoven University of Technology, Department of Electrical Engineering

5600 MB Eindhoven, The Netherlands

K.M. SCHEP, J.B. GIESBERS, M.A.M. GIJS

Philips Research Laboratories, Prof. Holstlaan 4,

5656 AA Eindhoven, The Netherlands

$Ni_{80}Fe_{20}$ films with a thickness of about 40 nm have been sputtered in a magnetic field on Si (100) substrates. Twin samples from the same substrate have been investigated: (i) with the easy magnetisation axis parallel to the current in the longitudinal direction and (ii) with the easy axis perpendicular to the current. The width of the samples was 30 µm. The resistance and noise as a function of the applied magnetic field perpendicular to the length have been investigated. At strong magnetic fields, the 1/f noise parameter is of the same order of magnitude as in nonmagnetic metal layers (Au, Cu, ...). For moderate fields with appreciable values of dR/dH we observe an additional contribution to the 1/f noise parameter which we associate with fluctuations in the magnetisation that modulates the resistance value.

1 Introduction

We have investigated anisotropic magneto resistance, AMR films. The AMR effect is based on the dependence of the resistance on the angle between current and magnetisation[1]. The change in noise power of AMR sensors can be explained qualitatively using simple models. In this paper, the noise of AMR films having a single domain state will be compared with films showing a multi domain pattern.

2 Experimental Procedure

The measured $Ni_{80}Fe_{20}$ films are grown by high-vacuum magnetron sputtering on a Si (100) substrate in an applied magnetic field in order to induce uniaxial magneto- crystalline anisotropy. The films are either 400 Å or 350 Å thick and are patterned into bars and form Wheatstone-bridges. The 30 µm wide samples have been chosen for investigations because they showed the most interesting domain structure. The long axis of the bars is orientated either parallel or perpendicular to the magnetic field applied during growing. We prefer to measure on bridge shaped samples because all spurious signals, common to all elements are largely cancelled out. The measurements were performed in a mu-metal shielded cage. Magnetic fields are generated using two pair of orthogonal Helmholtz coils. The current through the samples is generated with a battery in series with a relatively large resistor. The voltage noise over Wheatstone bridge is measured using a pair of battery fed ultra low-noise voltage preamplifiers (EG&G Brookdeal 5004). The outputs of the latter are connected to a HP3562A dynamic signal analyser which measures the sample noise using the cross-correlation technique. The noise is measured in the range between 1 Hz and 100 kHz. The equivalent noise resistance of the system is about 1 Ω above 10 Hz. The 1/f noise often obeys the empirical relation[2] where α is a dimensionless parameter for pure 1/f noise used to characterise its strength and N is the number of free carriers in the sample:

$$\frac{S_V(f)}{V^2} = \frac{\alpha}{Nf} \tag{1}$$

$S_V(f)$ is the spectral density of the voltage fluctuations induced by the resistance fluctuations, N is calculated for $Ni_{80}Fe_{20}$ by taking the number of electrons per unit volume as 5×10^{28} m^{-3} and f is the frequency in Hz. Tests showed that α is independent of the current density through each sample (at least upto 10^6 A/cm^2). The current densities used to measure the noise are between 10^5 and 6×10^5 A/cm^2.

3 Experimental Results

Bitter-fluid was used to show the magnetic domain structure of the films. Films with the easy axis of magnetisation along the long axis of the sample show a magnetic mono-domain structure. Films with an easy axis perpendicular to the long axis had a multi domain structure for a width larger than 20 μm. Only the 30 μm films had a stable pattern when no external fields are present. The observed noise spectra are proportional to f^{γ} with $0.8<\gamma<1.2$ in most cases. A weak constant magnetic field of 80 A/m was present parallel to the long axis during all measurements. This was done to force the magnetisation to rotate in one half plane.

3.1 Experimental results on single domain structures

Figure 1a shows AMR resistance change and figure 1b the measured 1/f noise in terms of the α parameter as a function of the magnetic field perpendicular to the long axis of the sample. The magnetoresistance curve of figure 1a is well described by[3]. The difference between the measured and calculated resistance is of the order of 1 %. The same parameters needed to model the resistance are also used to calculate the variance of the resistance. The calculated variance as a function of the perpendicular field is then compared with the experimentally observed 1/f noise. The similarity between the 1/f noise parameter α versus field and the variance or $(dR/dH)^2$ presented by a dotted line is surprisingly good although not perfect. In the model[3] the magnetisation of the complete sample is considered to be uniform, thereby overlooking almost all demagnetisation induced by edge effects. The α dependence on applied magnetic field in the AMR layer shows a dependence like

$$\alpha(H) = \alpha_\ell + K_A(dR/dH)^2 \tag{2}$$

with α_ℓ a typical value for non magnetic metal layers and the second term in eq. (2) gives the excess fluctuation in R by modulation due to spontaneous fluctuations of the magnetisation having a 1/f spectrum.

3.2 Experimental results on multi domain structures

Figure 2 shows an example of resistance and noise when the easy axis is perpendicular to the length axis of the sample. The absence of hysteresis in the magnetoresistance curve shows that there are no magnetisation jumps present but mainly rotation of the magnetisation. The large scattering in the 1/f noise, especially around zero field, is typical for the samples with their easy axis perpendicular to the length axis. For some samples we found high values (like in figure 2) and for some we found low values at zero external field, but all were very unstable. This points to domain wall activities. For comparison we have measured multilayers (not shown) which showed huge scattering in the transverse 1/f noise compared to longitudinal 1/f noise when no external field was present. This large scattering was not measured at high external fields. This also points to a non homogeneous 1/f noise source in the sample[4] due to domain wall activities.

4 Conclusions

Magnetic single layers showing the anisotropic magnetoresistance effect, have a strong 1/f noise where dR/dH shows a maximum. This is thought to originate from fluctuations in the magnetisation. For samples with a single domain structure this increase is more than a factor ten with respect to non-magnetic materials and for samples with a multi domain structure the increase is at least a factor hundred. The situation in which no domain walls are present was modelled and the calculated resistance fits the measured resistance very well. The intensity of the 1/f noise in the resistance and the variance of the resistance have the same dependence on the angle φ between current and magnetisation. However, the variance calculated from the spectrum (whatever limits of the 1/f region are used) is neglectible compared to the variance following from the dependence of magnetic energy on φ^3. This is the fundamental problem arising from our investigation. When domain walls are present, the 1/f noise strongly scatters. Experimental results on single domain structures are easily summarised by $\alpha(H) = \alpha_\ell + K_A(dR/dH)^2$.

References

1. T.R. McGuire and R.I. Potter, *IEEE Trans. Magn.* MAG-11, **1018** (1975).
2. F.N. Hooge, T.G.M. Kleinpenning, L.K.J. Vandamme, *Rep. Prog. Phys.* **44**, 479 (1981).

3. R.J.M. van de Veerdonck, P. Beliën, K.M. Schep. J.C.S. Kools, M.C. de Nooijer, M.A.M. Gijs, R. Coehoorn and W.J.M. de Jonge, to be published.
4. L.K.J. Vandamme and W.M.G. van Bokhoven, *Appl. Phys.* **14**, 205 (1977)

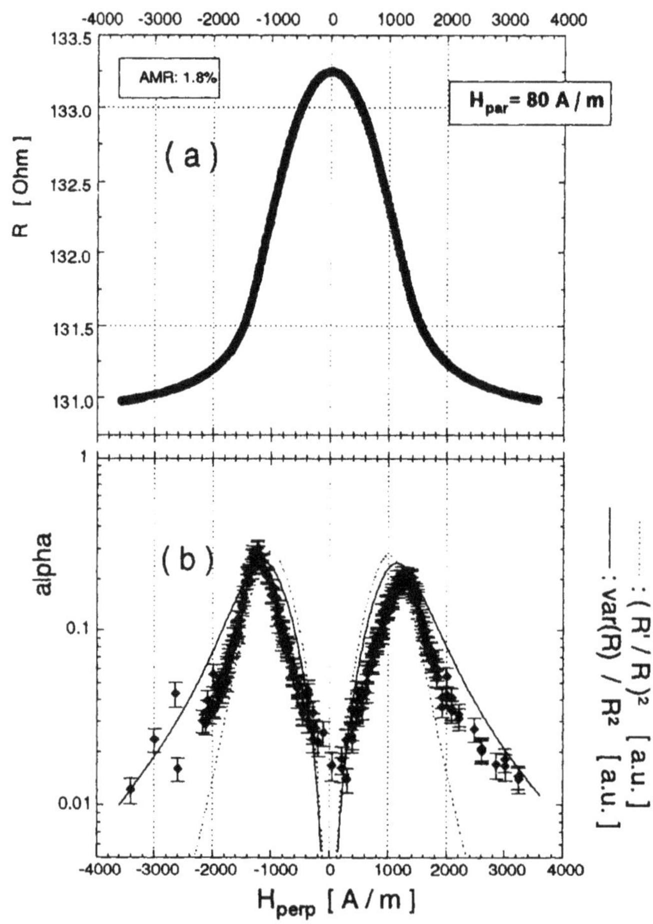

Figure 1: (a) Resistance and (b) 1/f noise as a function of the perpendicular magnetic field. The easy axis is parallel with the length of the sample and a constant field of 80 A/m is present in this direction. The 1/f noise is compared with the calculated variance (full line). The dimensions of an individual bridge resistance are: 360 μm x 30 μm x 40 nm. The dotted line represents $(dR/dH)^2$ indicating that $\alpha = \alpha_i + K_A\,(dR/dH)^2$ is also a good fit to experimentally obtained α-values.

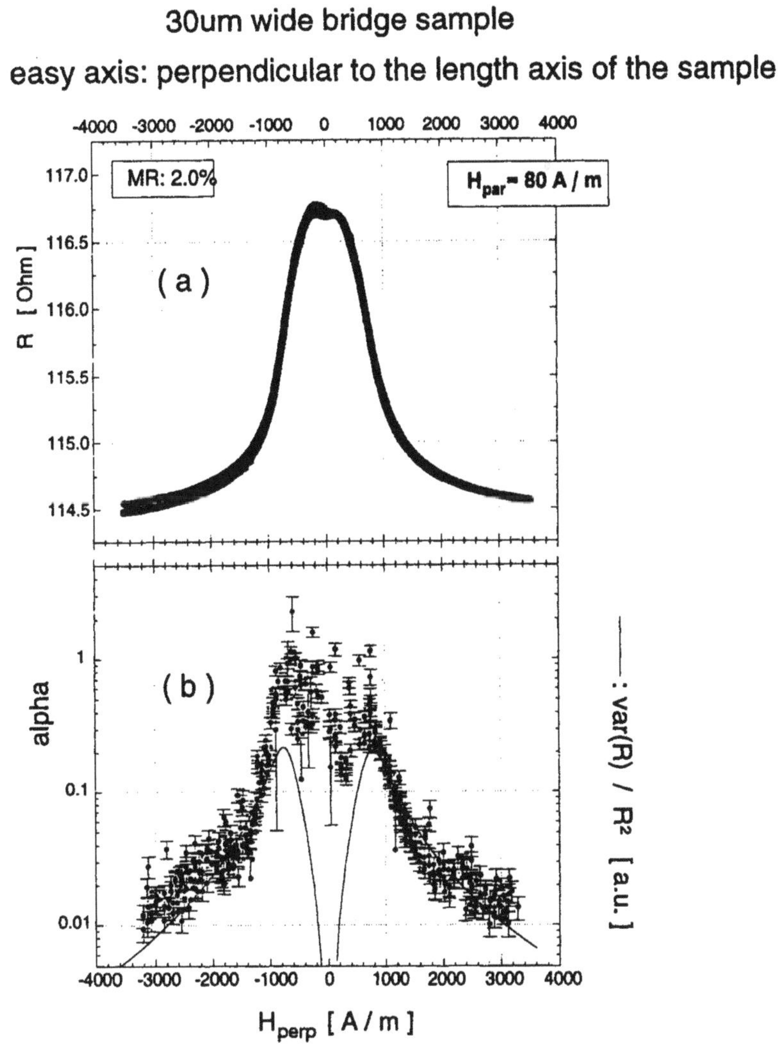

Figure 2: (a) Resistance and (b) 1/f noise as a function of the perpendicular magnetic field. The easy axis is perpendicular to the length of the sample and a constant field of 80 A/m is present in the direction parallel to the length of the sample. The 1/f noise power is compared with the calculated variance. The dimensions of a single resistance in the bridge structure are: 360 μm x 30 μm x 35 nm

PROBLEMS IN THE COMPREHENSION OF THE BARKHAUSEN NOISE

G. DURIN, G. BERTOTTI
Istituto Elettrotecnico Nazionale Galileo Ferraris, INFM and GSMN, C.so Massimo d'Azeglio 42, I - 10125, Torino, Italy

The characteristic features of the Barkhausen noise are presented in connection with the problem of self organisation in complex systems. The results experimentally verified and the main drawbacks of the existing theories are presented, togheter with the possible solutions of some intrigiung aspects.

1. Introduction

Since its discovery, the Barkhausen effect (BE) has been recognised as a fundamental tool for the investigation of magnetisation processes, and for non-destructive material testing. A detailed comprehension of the effect, together with an accurate description of experimental data, offers, in principle, the possibility to describe the huge variety of soft magnetic materials by a few macroscopic parameters, with considerable benefits in the applications. Beside this traditional field of studies, in the last years, with the increasing spread of research on non-equilibrium complex systems, broad renewed attention has been paid to BE, both from the theoretical and the experimental point of view, as a possible example a system showing self-organised criticality (SOC). Barkhausen noise, in fact, exhibits power law distributed avalanche instabilities and $1/f^{\beta}$ power spectra, and lacks of any inertial effect as shown, for instance, by sandpiles, earthquakes, etc. Up to now, however, the question has not been clearly solved. Following a stimulating and in a sense prophetical picture of R. Feymann[1,2], the central and unsolved question under debate is the following: is the 'sound' of Barkhausen noise that of sand grains falling over each other (i.e., the noise is a result of some system self-organisation) or of grains falling over something else (i.e., organisation is totally absent and the noise statistical properties simply reflect the disorder of the medium in which the system evolves)? An alternative and little more conservative way to pose the question may be: is it really necessary to invoke SOC in order to describe and predict the observed noise properties, or overdamped dynamics in *quenched-in* disorder is sufficient to justify the experimental results? Both points of view have estimators and detractors. Following the original work of Cote and Meisel[3], who, even with not too accurate results of avalanche distributions (see refs. 4,5 for discussion), claimed for a SOC evidence, some authors have simulated the Barkhausen signal using modified sandpile models[6] and related the scaling laws of avalanche distributions to power spectra using the SOC predictions [4,5]. On the other hand, Sethna *et al.*, studying zero-temperature random field Ising models[7], exclude any type of organisation and consider scaling laws as a simple consequence of quenched disorder of the medium. Otherwise, Urbach[8] *at al.* evaluated the presence of 'long range' demagnetising fields, not taken into account in the previous model, and concluded that they do give SOC while *local* demagnetising fields (mainly related to quenched disorder) do not. Finally, for O'Brien and Weissman[2], real self organisation requires that events on one temporal scale set the stage for events on another scale, and found no evidence of this feature in the existing theories and experimental results. Beside these different approaches, our traditional study is mainly focused on the possibility to describe BE statistics by a clear physical picture of domain wall (DW) motion[9], in terms of a (stochastic) dynamical equation; only as a second step, we investigate the presence of any type of organisation, as shown in the next paragraph.

2 Power spectra, avalanche distributions and SOC

Up to now, the more satisfactory explanation of the BE features is given by a stochastic model based on a Langevin-like description of DW dynamics in a randomly perturbed medium[9] (hereafter, ABBM). In this model, the field driving the DW is given by the difference between the external applied field H_a and an internal counterfield, made up of two components, the demagnetising field H_{dem} of magnetostatic origin due to sample geometry and the pinning field H_p, which takes into account all other sources of counterfields, like inclusions, defects, interactions with other DWs and so on. As the DW motion in metals is fully dominated by eddy currents, the DW velocity v is given by

$$v \propto H_a - H_{dem} - H_p \tag{1}$$

The DW dynamic equation is the time derivative of this equation where the applied field is a given function of time (e.g., $dH_a/dt=const$), and the pinning field is assumed to be a brownian process as a function of *space* (i.e. of DW position x). This interplay between space and time has far-reaching consequences: the amplitude distribution $P_o(v)$ of the DW velocity, which would be gaussian in the conventional Langevin approach, turns out to be a gamma distribution $P_o(v) = v^{c-1}\exp(-v)/\Gamma(c)$, where c is proportional to the applied field rate dH_a/dt. In addiction, the power spectrum amplitude is calculated to scale with dH_a/dt. These two results have been found to be a general property of BE in most magnetic materials[4,9,10], either single crystals, polycrystals, or amorphous alloys (see an example in figs. 1,2). This is a very surprising result still to be understood, also in connection with the fact that, on the contrary, the power spectrum data *never* fit with the $1/f^2$ shape predicted by the model, showing instead a $1/f^\beta$ spectrum with $1.5 \le \beta < 2$, in the simplest cases (fig. 1) and more complicated shapes in very complex materials such as some annealed amorphous alloys (see fig. 2 and ref. 10).

The hypothesis of a spatial random pinning field has been interpred[2,4] as the consequence of the structural disorder of the medium, and thus in the line of "grains falling over something else", in the sense that the DW moves on a *fixed* (structural) pinning field, excluding any type of cooperative effects required by SOC. Actually, the situation is not so simple. As sketched above, the correct interpretation of the pinning field is that of an equivalent counterfield including all the random effects[9] acting on the wall, consequence of structural disorder *and* other magnetic phenomena, such as flux propagation along the wall, interactions between DWs, an so forth. A spatial brownian pinning field only means that the effect of counterfields on the wall has independent increments, i.e. the wall does not retain any memory of the past interactions with the medium. Clearly, if this is the underlying physics of the magnetisation processes, no SOC occurs. Actually, the experimental data show that the hypothesis of a brownian pinning field, giving $1/f^2$ power spectra, is inadequate, and one has to introduce its natural extension to fractional brownian processes (fBP). A really subtle problem soon arises: how can we reconcile the request of a fBP (likely, in the space coordinate) with the proportionality of the power spectra with the applied field rate (and the related gamma distribution of signal amplitude)? In the ABBM model, it was assumed that $dx = v\, dt$ (the more natural assumption), so that the variance of the brownian process could be written as $<|dW|^2> \sim dx \sim v\, dt$, and so it was possible to consider and solve the associated Fokker-Planck equation[9]. In this case, we do not know how to handle a fBP with variance of the type $<|dW|^2> \sim dx^{2H}$, *unless* to consider $dx^{2H} = v\, dt^{2H}$ giving $<|dW|^2> \sim v\, dt^{2H}$, i.e. a fBP as a function of time with an *amplitude scaling with the DW velocity* v. In fact, this is the only assumption to give the spectra of the $1/f^\beta$ type, with $\beta = 2H+1$, proportional to the applied field rate[11], in the approximation $v \sim <v> = c \propto dH_a/dt$. This assumption has a very interesting physical meaning: because of $\beta<2$, and $H<0.5$, the pinning field has an antipersistent character, i.e. its increments are negatively correlated. Even if different explanations can be given (*structural* disorder as a fBP, to be explained), this is in agreement with the presence of local demagnetising counterfields which are active when the DW bows and tend to realign it, pushing (pulling) the wall at velocities lower (higher) than the average velocity c. This fact can be consider as a prerequisite for self-organisation, because states the presence of an internal mechanism which keeps the system not too far from its average dynamical state. On the other hand, this does not state any presence of precursor avalanches, a requisite considered essential[2]. The meaning of the proportionality of the amplitude with the velocity v is, on the other hand, not straightforward, and we cannot have any clear meaning in terms of known magnetisation processes.

Taking the first derivative of eq. 1 and considering a fBP pinning field, we get a type of dynamical equation which gives a power spectrum expression yet not too accurate to describe the experimental data, especially at frequencies lower than the maximum (fig.1). In fact, the calculation gives a sum $\alpha+\beta=2$ of the two exponents of the spectra before ($\sim \omega^{\alpha}$) and after ($\sim \omega^{-\beta}$) the maximum, while the data show different behaviours. One of the possible solutions[11] is the introduction of a fractional differential equation, as used in the simulation of turbulence[12], of the type

$$\frac{d^q}{dt^q}\left[v \exp(t/\tau_m)\right] = \sqrt{v}\frac{d^g N}{dt^g}\exp(t/\tau_m) \tag{2}$$

where τ_m is a the time constant related to the demagnetising field, assumed constant in the ABBM model, and here considered a function of the applied field rate, and N is the white noise. In the approximation $v \approx c$, it gives a power spectrum of the type

$$S(\omega) \sim \frac{\omega^{2g}}{(\omega^2 + \tau_m^2)^q} \tag{3}$$

so that, $\alpha=2g$ and $\beta=2(q-g)$, which fits with excellent accuracy the simplest spectra and describes its variability with the applied field rate with a single time constant τ_m. (fig. 1). The physical meaning of eq. (2) is otherwise obscure, as any further relation to SOC, and still needs to be analysed: it is worth noticing that the ABBM model is recovered in the case $q=1$ and $g=0$.

The analytical expression of $P_0(v)$ cannot be calculated as in ABBM, neither after the introduction of the fBP pinning field, nor from the eq. (2). Thus, it still remains unclear why the gamma distribution is so general and if the result of the ABBM model is a fortunate coincidence or its structure has a deeper connection with the magnetisation processes involved in the DW motion. Beside these considerations, the results on the amplitude distribution are also promising to test the relation between the avalanche distributions and the power spectra. We have calculated and experimentally verified[4] that the avalanche duration Δu and size Δx follow the power law distributions $P(\Delta u) \sim \Delta u^{-(2-c)}$ and $P(\Delta x) \sim \Delta x^{-(3/2-c/2)}$. Following Jensen *et al.*[13], the power spectrum can be calculated using a weighted distribution of the lifetimes of *independent* avalanches, getting $\beta=2$. Up to now, it is not clear whether this discrepancy is due to the assumption of independence or we should take into account the particular shape of the average avalanches, as recently proposed[5].

3 Final remark

The complex behaviour of BE and the difficulties in its comprehension has a direct physical explanation: as in many complex systems, we are considering a many body system (many interacting DWs) moving a random landscape *changing* during the evolution of the system. A similar behaviour have the piles of sand or rice, where avalanches change the structure of the pile. As said, many authors consider power laws and $1/f$ noise a direct evidence of self organisation, while others strongly refuse this statement, so that it is not universally clear which are the essential *ingredients* of SOC: intrinsic disorder, many body interactions, memory effects, precursors and aftershocks, or none of them. Without this clarification, in our opinion, the idea of SOC will remain as a general qualitative frame of interpretation rather than a valid physical theory to explain nature.

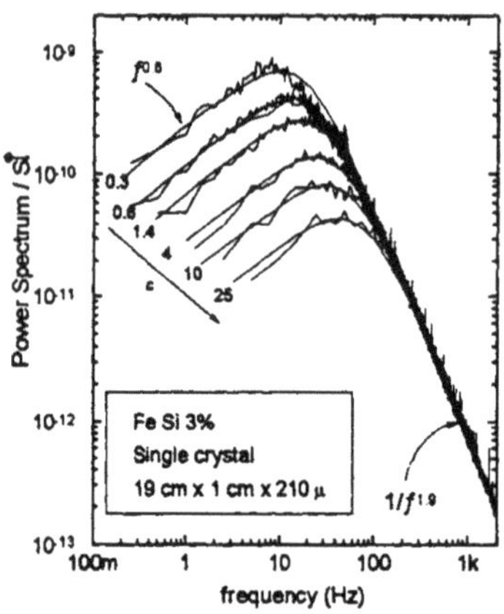

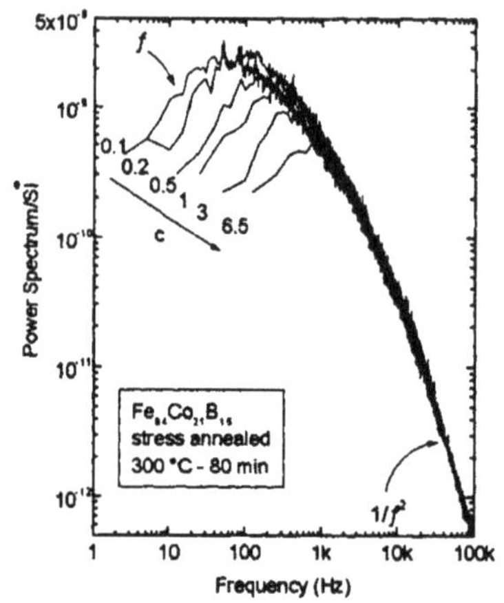

Fig. 1 Power spectra of a single crystal FeSi alloy at different average DW velocity $c \propto dH_a/dt$.
Fig. 2 Same of fig.1. for a $Fe_{64}Co_{21}B_{15}$ amorphous alloy. SI is the measured average flux rate $\propto c$.

References

1. R.P. Feymann *et al.* in *The Feymann Lectures on Physics* (Addison-Wesley, Reading, Feymann, MA 1964), Ch. 34, p. 9
2. O'Brien K.P. and M.B. Weissman, *Phys. Rev.* B **50**, 3446 (1994).
3. L. W. Meisel and P. J. Cote, *Phys. Rev.* B **46**, 10882 (1992).
4. G. Durin *et al.*, *Fractals* **3**, 351 (1995).
5. D. Spasojevic *et al.*, *Phys. Rev* E **54**, (1996), in press
6. O. Geoffroy and J. L. Porteseil, *J. Magn. Magn. Mater.* **133**, 1 (1994).
7. P Sethna *et al.*, *Phys. Rev. Lett.* **70** 3347 (1993).
8. J.S. Urbach *et al.*, *Phys. Rev. Lett.* **75**, 276 (1995).
9. B. Alessandro *et al.*, *J. Appl. Phys.* **68**, 2901 (1990); **68**, 2908 (1990).
10. G. Durin *et al.*, *J. Magn. Magn. Mater.* (1996), in press.
11. G. Durin and G. Bertotti, unpublished.
12. F.B. Tatom, *Fractals* **3**, 217 (1995)
13. H.J.Jensen *et al.*, *Phys. Rev.* B **40**, 7425 (1989).

PROBLEMS AT SIMULATING NOISE PROCESSES

IS THE IMPEDANCE FIELD APPROACH VALID FOR NOISE DESCRIPTION IN TRANSPORT QUANTUM SYSTEMS?

P. SHIKTOROV, V. GRUŽINSKIS, E. STARIKOV
Semiconductor Physics Institute, Goshtauto 11, LT2600 Vilnius, Lithuania

L. REGGIANI
Istituto Nazionale di Fisica della Materia, Dipartimento di Scienza dei Materiali, Università di Lecce, Via Arnesano, 73100 Lecce, Italy

L. VARANI
Centre d'Electronique et de Micro-optoelectronique de Montpellier (CNRS UMR 5507) Université Montpellier II, 34095 Montpellier, France

The extension of the impedance-field method to quantum systems is presented as yet an unsolved problem of relevant interest. A brief formulation of the problem for the simple case of one dimensional structures is given. Analogies and differences between classical and quantum systemts are illustrated in the framework of a Wigner-function approach.

1 Introduction

Since the seminal paper of Shockley et al [1], further extended in Refs. [2-4], the impedance-field method has been widely used for noise modeling in the framework of drift-diffusion [2,4], hydrodynamic [5] and classical kinetic approaches. In going from a classical to a quantum description of systems under conditions far from equilibrium these approaches have corresponding quantum analogues [6]. Therefore, the following problem arises in a natural way: *is it possible to use the usual concepts and formalism of the impedance-field method for electronic noise modeling of a quantum system of charged carriers?* In the framework of a quantum approach based on the Wigner function, the aim of this work is to present the main logic for constructing the impedance field of a one-dimentional structure and for determining the characteristics of the source of fluctuations. It should be stressed that both: the structure of the quantum equation for the Wigner function and the calculation of the average values of dynamical variables are analogues to those of the classical Boltzmann equation. Therefore, to appreciate similarities and differences between the quantum and classical cases, the most important points will be carried out in parallel for both approaches.

2 Main equations

Let us consider a nondegenerate ensemble of N carriers (electrons) which interact with a thermostat, i.e. phonon system. By taking for simplicity a one-dimensional system in phase space (p, x), the equation for both the classical and quantum distribution function, $f(p, x, t)$, can be written in the following form:

$$\frac{\partial f}{\partial t} + v\frac{\partial f}{\partial x} + F[f, U] = S[f] \tag{1}$$

Here v is the carrier group velocity and $S[f]$ the collisional term assumed of the same form in both formalisms:

$$S[f] = -f(p, x, t)\int w(p, p'; x)dp' + \int w(p', p; x)f(p', x, t)dp' \tag{2}$$

where $w(p, p')$ is the probability per unit time for a carrier to be scattered from state p into state p' at point x. The main difference between the quantum and classical equations concerns the form of the field term, $F[f, U]$, namely:

$$F_c[f, U] = eE(x)\frac{\partial f_c}{\partial p} \tag{3a}$$

$$F_q[f, U] = \frac{2\pi}{ih^2}$$

$$\int dp' \int dq\ exp[\frac{i(p - p')q}{\hbar}][U(x + \frac{q}{2}) - U(x - \frac{q}{2})]f_q(p', x) \tag{3b}$$

for the classical and quantum cases, respectively. Here $eE(x) = -\partial U(x)/\partial x$ is the electric force and $U = U_{bi} + U_{sc}$ the electrical potential associated with the built-in and self-consistent contributions. The long range Coulomb interaction is accounted for through the self-consistent potential determined by the Poisson equation:

$$\frac{\partial^2 U_{sc}}{\partial x^2} = -\frac{e}{\epsilon\epsilon_0}[N\int f(p, x, t)dp - N_d(x)] \tag{4}$$

where $N_d(x)$ is the profile of the donor concentration, $f(p, x, t)$ the distribution function normalized to unity in accordance with $\int_0^L dx \int dpf(p, x, t) = 1$, L being the structure length. Under constant-current operation mode, the electric field and, hence, the distribution of the self-consistent potential inside the system, satisfies the equation:

$$\epsilon\epsilon_0\frac{\partial E}{\partial t} + eN\int vf(p, x, t)dp = J \equiv const(x, t) \tag{5}$$

where J is the total current density flowing through the structure.

3 Impedance field formula

The impedance field formalism is based on linear-response theory when it is implicitly supposed that fluctuations generated by the system are linear with respect to the system. (A more detailed discussion of this issue will be given in the next section.) Let the set of equations (1), (4), (5) has a stationary solution labelled by subindex s. The correspondent set of equations which describe the linear fluctuations takes the form:

$$\frac{\partial \delta f}{\partial t} + v\frac{\partial \delta f}{\partial x} + F[\delta f, U_s] = S[\delta f] - F[f_s, \delta U] \tag{6}$$

$$\epsilon\epsilon_0 \frac{\partial \delta E(x,t)}{\partial t} + eN\int v\delta f(p,x,t)dp = -\delta j_c(x,t) \tag{7}$$

$$\delta U_{sc}(x,t) = \int_0^x \delta E(x',t)dx' \tag{8}$$

where $\delta j_c(x,t)$ is the random perturbing force which describes fluctuations of the conduction current in the system. The solution of Eqs. (6)-(8) can be represented in the form:

$$\delta E(x,t) = \int_{-\infty}^{t} dt' \int_0^L dx' G(t-t';x,x')\delta j_c(x',t') \tag{9}$$

where $G(t-t';x,x')$ is the Green-function of Eqs. (6)-(8), i.e. the transfer impedance in the time-domain representation which describes the evolution of the electric field fluctuation caused by the fluctuation $\delta j_c = \delta(x-x')\delta(t-t')$. The fluctuations of the voltage between the structure terminals are obtained by integrating Eq. (9) over all the structure:

$$\delta U_{sc}(t) = \int_0^L \int_{-\infty}^{t} \nabla z(x',t-t')\delta j_c(x',t')dt'dx' \tag{10}$$

where

$$\nabla z(x',t-t') = \int_0^L G(t-t';x,x')dx \tag{11}$$

is the impedance field describing additive contributions of conduction current fluctuations $\delta j_c(x',t')$ at points x' to the total value of $\delta U_{sc}(t)$.

From the standart calculation of the spectral density of a random quantity one obtains the following expression for the spectral density of voltage fluctuations between the strucure terminals:

$$|U_\omega^2| = \int_0^L dx' \int_0^L dx'' \nabla Z(x',\omega)\nabla Z^*(x'',\omega)S_j(\omega,x',x'') \tag{12}$$

$$S_j(\omega, x', x'') = 4\int_0^\infty \overline{\delta j_c(t, x')\delta j_c(t+\tau, x'')} cos(\omega\tau)d\tau \tag{13}$$

where $\nabla Z(x,\omega)$ is the Fourier transform of the impedance field given by Eq. (11), $\overline{\delta j_c(t, x')\delta j_c(t+\tau, x'')}$ is the autocorrelation function of conduction current fluctuations at points x' and x'' averaged over the whole time domain. In terms of the formalism considered here, which is identical for both the classical and quantum cases, the source of fluctuations is only indicated formally, thus its microscopic expression should be provided by other means. A fundamental source of fluctuations for the system here considered is connected with the energy exchange between the ensemble of carriers and the thermostat described by the collisional term in Eq. (1). Accordingly, without any additional assumption, the correlation characteristics of the fluctuations generated in the system must be completely determined by the structure of the collision term (Eq. (2)) and by the electron motion in the electric field given by the field term (Eqs. (3)). In the following section, we provide a logical base for such a description of fluctuations.

4 Fluctuations in classical case

The natural incorporation of fluctuations into a kinetic scheme is based on the ergodic hypothesis, when a stochastic Markov process continuous in time is constructed to describe the trajectory $\xi(t) \in (p, x)$ of a random walk of a trial particle in momentum and real spaces. The stochastic process must be constructed in such a way that, in the long-time limit, the time-average of an arbitrary function $\theta(p, x)$ along the trajectory $\xi(t)$ should coincide with the average over the steady-state distribution function, namely:

$$\lim_{t\to\infty} \frac{1}{t}\int_0^t \theta[\xi(t')]dt' = \int \theta(p, x)f(p, x)dpdx \tag{14}$$

Such a process can be constructed (e.g. by the Monte Carlo procedure) if the time dependence of the conditional probability of the process, $P(p, x, t|p_0, x_0, t_0)$, describing the probability to find the trial particle at point (p, x) and time t under the condition that at time t_0 the particle was in point (p_0, x_0), satisfies the time-dependent kinetic Eq. (1) and looses memory of its initial conditions at long times, i.e.

$$\lim_{t_0\to -\infty} P(p, x, t|p_0, x_0, t_0) = f_s(p, x) \tag{15}$$

The Green function $G(t - t_0; p, x|p_0, x_0)$, formally obtained for the classical kinetic equation, satisfies the same equation as $P(p, x, t|p_0, x_0, t_0)$ and, hence, can be interpreted as the conditional probability of a certain process. From the above, the autocorrelation function of the fluctuations of $\theta(p,x)$ can be represented as:

$$\overline{\delta\theta(t+\tau,x)\delta\theta(t,x')} = N\{\int dp \int dp'\theta(p,x)\theta(p',x')G(\tau;p,x|p',x')f_s(p',x')$$

$$- \int \theta(p,x)f_s(p,x)dp \int \theta(p',x')f_s(p',x')dp'\} \tag{16}$$

Even if the determination of the correlation function can be performed in a natural way, some hidden problems remain. Let us come back to the condition of linearity. The approach described above has implicitly assumed the linearity of the kinetic Eq. (1) with respect to the distribution function. From a mathematical point of view one is legitimate to introduce the Green-function $G(t-t_0;p,x|p_0,x_0)$. When the self-consistent electric field is accounted for by Eqs. (4) or (5), Eq. (1) becomes nonlinear. As a consequence, for these equations the Green-function concept looses its meaning. However, the concept of conditional probability remains valid even for nonlinear systems, as shown, for example, in simulations with the Monte Carlo procedure. This means, that the noise generated by a system is not necessarily linear with respect to the system. In other words, the amplitude of the noise at the output of the system is not directly proportional to the amplitude of its source. We conclude that, there appears the problem of nonlinear noise. Let us recall, that Eq. (9) implies linearity of fluctuations, i.e. the impedance field can describe fluctuations which are linear with respect to the system only. Therefore, when using the impedance field approach one should make an additional assumption on the noise source. The most natural one is that such a source is calculated from the kinetic equation with a frozen (i.e. nonfluctuating) electric field which coincides with the stationary value $E_s(x)$ obtained from the solution of the nonlinear task.

5 Fluctuations in quantum case

Within a quantum approach based on the Wigner-function formalism, an attempt to construct a logical scheme to determine the correlation function of fluctuations generated by the system meets serious difficulties at once. Even if the Wigner function can be normalized to unity in such a way that $\int_0^L \int f_q(p,x,t)dxdp = 1$, it cannot be interpreted as a probability since it is not positively defined everywhere. Nevertheless, from a formal point of view the approaches based on the classical and quantum kinetic equations are very similar. Therefore, let us discuss some analogies which follow from the classical approach and seems to be useful for the determination of the correlation functions of quantum fluctuations. Since the Wigner function is not positively defined, an attempt to interpret the retarted Green function $G_q(t-t';p,x,|p',x')$ of the Wigner-function equation as the conditional probability of some stochastic process fails. Nevertheless, let us suppose that $G_q(t-t';p,x,|p',x')$ is known and we formally have calculated some correlation functions in accordance with Eq. (16) where instead of the classical Green function and function $\theta(p,x)$ we have substituted their quantum analogues. The question arises: do these correlation functions describes fluctuations in the quantum case? Let us consider the interrelation between the quantum kinetic equation and its classical analogue. To this end, let us rewrite the quantum equation (1) in a form corresponding to the classical kinetic equation for the space nonhomogeneous case:

$$\frac{\partial f_q}{\partial t} + v\frac{\partial f_q}{\partial x} + eE(x)\frac{\partial f_q}{\partial p} - S[f_q] = \int \Psi(p-p',x) f_q(p',x,t)dp' \tag{17}$$

where

$$\Psi(p-p',x) = \frac{2\pi}{ih^2}$$

$$\int dq\; exp[\frac{i(p-p')q}{\hbar}][U(x+\frac{q}{2}) - U(x-\frac{q}{2})] - \frac{\partial U}{\partial x}\frac{\partial}{\partial p}\delta(p-p')$$

The r.h.s. of Eq. (17) describes the difference in the dynamical motion associated with quantum and classical systems. By using the Green function of the classical form of the kinetic equation (17), $G_c(t-t';p,x|p',x')$, a solution of Eq. (17) for the quantum case with arbitrary initial condition at $t=t_0$, $f_q(p,x,t_0) = f_0(p,x)$ can be written as:

$$f_q(p,x,t) = \int_{t_0}^{t} dt' \int dp'dx' G_c(t-t';p,x|p',x') \int \Psi(p'-p'',x') f_q(p'',x',t')dp''$$

$$+ \int dp'dx' G_c(t-t_0;p,x|p',x') f_0(p',x') \tag{18}$$

By applying to Eq. (18) an iteration procedure, a formal solution for the Green function of Eq. (17) can be written in the form

$$G_q(t-t_0;p,x|p_0,x_0) = \hat{M}_{c\to q}[p,x] G_c(t-t_0;p,x|p_0,x_0) \tag{19}$$

Here $\hat{M}_{c\to q}$ is a certain linear operator defined on a variety of classical trajectories described by the classical conditional probability. Let us consider some consequences given by such a representation. As we can see, the transition from the classical to the quantum kinetic can be represented as a linear transformation $\hat{M}_{c\to q}$ from the space of classical dynamic variables (p,x) to the corresponding of quantum variables. The stochastic part, which is responsible for the appearance of fluctuations in the system, is invariant for this transition. Therefore, to introduce fluctuations in the quantum case one can entirely retain the classical scheme based on the ergodic hypotheses. Then, one constructs a random Markov process with the conditional probability given by the Green-function of the classical analogue of the quantum system. In accordance with Eq. (14), it is also possible to perform a time-average along stochastic trajectories $\xi(t)$, if one succeed to construct the transformation $\hat{M}_{c\to q}$ on these trajectories. For quantum kinetics, the function $\theta[\xi]$ which is averaged in Eqs. (14) must be replaced by $\theta[\xi]\hat{M}_{c\to q}[\xi(t)]$. This means that Eq. (16) holds in the quantum case too.

Acknowledgments

This work has been performed within the European Laboratory for Electronic Noise (ELEN) and supported by the CEC through the contracts ERBCHRX-CT920047, PECO Project EAST ELEN ERBCIPDCT940020, NATO CN. NIG 951009.

References

1. W. Schokley *et al.*, Quantum theory of atoms, molecules and solid state, Ed. P.O. Lowdin, Academic Press (New York, 1966) p. 537
2. K.M. Van Vliet *et al.*, *J. Appl. Phys.* **46**, 1804 (1975).
3. J.P. Nougier *et al.*, J. Appl. Phys. **52**, 5683 (1981).
4. G. Ghione and F. Filicori, *IEEE Trans. CAD* **12**, 425 (1993).
5. P. Shiktorov *et al.*, *Appl. Phys. Lett.* **68**, 1516 (1996).
6. D.K. Ferry *et al.* Quantum Transport in Ultrasmall Devices, Plenum Press (New York, 1995).

IS CURRENT-NOISE OPERATION MORE PHYSICAL THAN VOLTAGE ONE?

L. VARANI, J.C. VAISSIERE, J.P. NOUGIER
Centre d'Electronique et de Micro-optoélectronique de Montpellier (CNRS UMR 5507), Université Montpellier II, F-34095 Montpellier, France

L. REGGIANI
Istituto Nazionale di Fisica della Materia, Dipartimento di Scienza dei Materiali, Università di Lecce, Via Arnesano, I-73100 Lecce, Italy

V. GRUŽINSKIS, E. STARIKOV, P. SHIKTOROV
Semiconductor Physics Institute, Goshtauto 11, LT-2600 Vilnius, Lithuania

Current- and voltage-noise operations are the two limiting modes that can be employed when studying electronic noise in semiconductors. Due to their mutual link through the differential impedance of the sample they are usually thought to be perfectly equivalent. We report some considerations pointing out the main differences, advantages and disadvantages, critical situations and bottlenecks emerging from a unaware use of these operation modes.

1 Introduction

Current and voltage operation modes represent two mutually exclusive ways to detect macroscopically electronic noise. The former is realized when a constant voltage is applied to the device and fluctuations of the total current as measured in the outside circuit are detected. In this case the quantity of interest is the spectral density of current fluctuation $S_I(f)$. The latter is realized when a constant total-current is forced to flow in the device and voltage fluctuations are measured at the terminals of the device. In this case the quantity of interest is the spectral density of voltage fluctuations $S_U(f)$. The two spectral densities are related as:

$$\frac{S_U(f)}{S_I(f)} = |Z(f)|^2 = \frac{1}{|Y(f)|^2} \tag{1}$$

where $Z(f)$ and $Y(f)$ are respectively the differential impedance and admittance. Using the generalized Nyquist relation, the two spectral densities can be written as:

$$S_I(f) = 4k_B T_n(f) Re[Y(f)] \tag{2}$$

and

$$S_U(f) = 4k_B T_n(f) Re[Z(f)] \tag{3}$$

where k_B is the Boltzmann's constant and $T_n(f)$ the noise temperature. Equation (1) states that it is possible to go from one representation to the other through the knowledge of the small-signal response of the system. Therefore, it would seem that the two operation modes are perfectly equivalent. In spite of these simple considerations, the actual situation presents some hidden difficulties that we aim to point out in this paper.

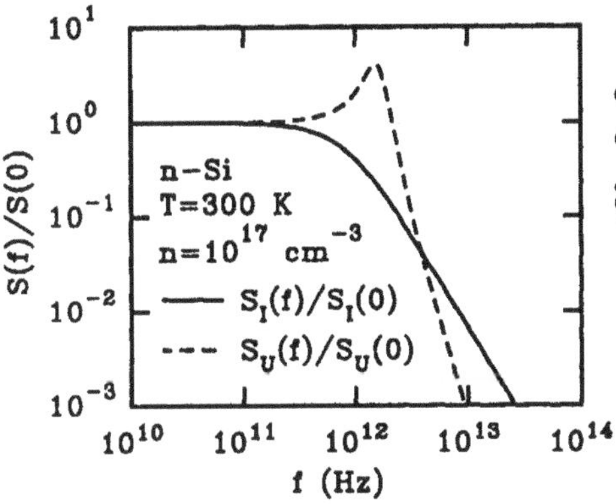

Figure 1: Spectral densities of current and voltage fluctuations at equilibrium for a homogeneous n-Si sample at $T = 300\ K$ with $n = 10^{17}\ cm^{-3}$ and $\mu = 1320\ cm^2V/s$.

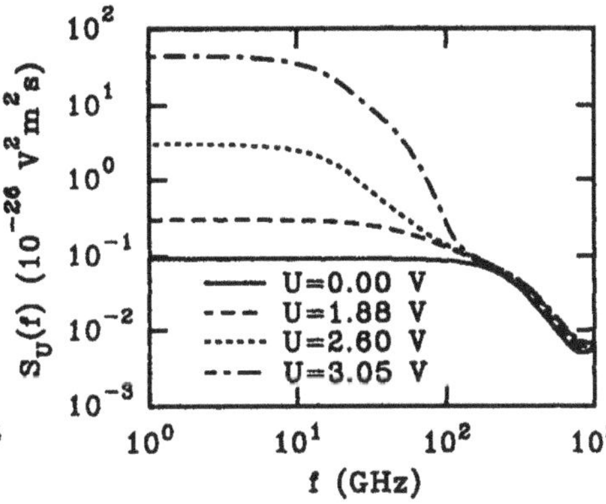

Figure 2: Spectral density of voltage fluctuations for a n^+nn^+ GaAs diode with 1.25-7.5-1.25 μm length and $n = 10^{15}$ and $n^+ = 2 \times 10^{16}\ cm^{-3}$ doping levels obtained from a hydrodynamic simulation for the reported applied voltages at $T = 300\ K$.

2 Are the noise spectra similar for both operation modes?

Let us consider a homogeneous semiconductor under ohmic conditions. In this simple situation the differential impedance can be calculated analytically and it is given by:

$$Z(\omega) = R\frac{1 + i\left[\omega\tau_m(1-\omega^2\tau_p^2) - \omega\tau_d\right]}{(1-\omega^2\tau_p^2)^2 + \omega^2\tau_d^2} \tag{4}$$

where R is the ohmic resistance, τ_m the momentum relaxation time associated with the time evolution of the mobility, τ_d the dielectric relaxation time and $\tau_p = \sqrt{\tau_m \tau_d}$ the plasma time. A trivial but not sufficiently stressed consequence of this relation is that, due to Eq. (1), the current and voltage spectral densities exhibit completely different frequency behaviors, as shown in Fig.

1 for a homogeneous n-Si sample at equilibrium. As a matter of fact, while the current fluctuations are associated with the momentum relaxation time, the voltage fluctuations are associated with the dielectric and the plasma time. Furthermore, when comparing two systems a common question which arises is: which of the systems is the most noisy? We believe that the answer to this question has no meaning unless the operation conditions are not precisely specified. These conditions usually depend on the specific application aimed by the system. An example of this situation is shown by the results obtained from hydrodynamic calculations (lines) in Figs. 2 and 3 which report the spectral densities of voltage and current fluctuations for a n^+nn^+ GaAs diode with doping levels of $n = 10^{15}$ cm^{-3} and $n^+ = 2 \times 10^{16}$ cm^{-3} and cathode, n-region and anode lengths, respectively of 1.25, 7.5 and 1.25 μm [1]. At low frequencies $S_U(f)$ is found to increase significantly with voltage while no significant change is observed for the corresponding $S_I(f)$.

3 Do both operations enable a spatial analysis of the noise?

When studying the noise in a inhomogeneous structure one of the most interesting point is to spatially locate the origin of the noise in the structure. If current-noise operation is used such a piece of information is lost since the only quantity which fluctuates is the total current flowing in the outside circuit. On the other hand, the voltage-noise operation is able to provide a spatial analysis of electronic noise since voltage fluctuations can be evaluated not only between the terminals of the device but also between two arbitrary points inside the structure thus providing a spatial map of the noise of the impedance-field formula. This is illustrated in Fig. 4 which reports the contribution of each point of an n^+nn^+ diode to the net noise, $s(x,0)$ and the local noise of the structure originated from noise sources at all positions $dS_U(x,0)/dx$. In this latter case a comparison between Monte Carlo (points) and hydrodynamic (dashed line) calculations proves the consistency of the curve reported.

4 What happens under strongly non-equilibrium conditions?

Both approaches are in principle equivalent for a description of electronic noise of a stable system at thermal equilibrium or even far from it, but under unstable conditions the situation can change significantly. Depending on the operation mode the system can be in different physical states [2]. When a constant current is forced to flow through the structure, the diode remains stable and the voltage drop between the terminals is imposed by the current. In contrast, under a constant voltage U_0 applied to the diode there exist regions of values of U_0 where periodic oscillations of the conduction current appear.

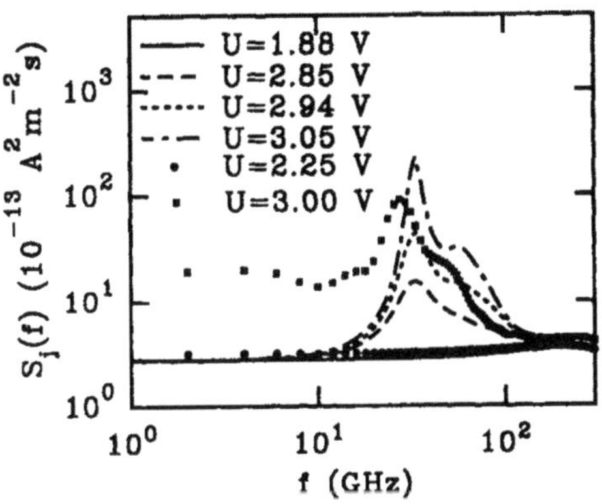

Figure 3: Spectral density of current fluctuations for the same structure of Fig. 2 obtained from a hydrodynamic simulation for the reported applied voltages (lines). Symbols refer to the results of a Monte Carlo simulation.

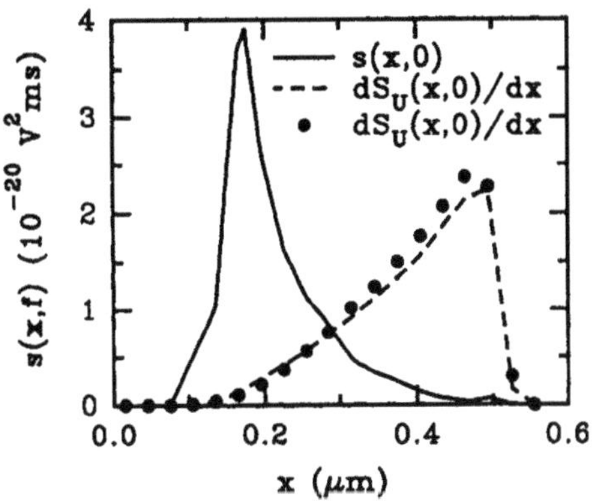

Figure 4: Space contributions to the voltage spectral density and space distribution of the local voltage noise calculated by HD and MC techniques at low frequencies for a submicron n^+nn^+ Si diode with 0.1-0.4-0.1 μm length and $n = 2 \times 10^{15}$ and $n^+ = 5 \times 10^{17}$ cm^{-3} doping levels biased at $U = 1.5$ V.

When a pronounced near-oscillatory macroscopic behavior is exhibited, the current spectral density tends to a δ-like behavior and in this case it is the spectral width of a line which characterizes the coherence of the periodical oscillations in the system[3].

Furthermore, we have observed the appearance of an additional source of noise near the threshold voltage when the system goes from the stationary state independent of time into the stable state periodical in time. Due to this reason one can observe a strong enhancement of the low-frequency noise under voltages slightly below the threshold. This is illustrated by symbols in Fig. 3 which have been calculated using a MC simulation under current-noise operation. The absence at low frequencies of this extra noise in the HD results (perhaps due to the presence of partition noise) remains an unsolved problem [4].

5 Are both operations "computationally" equivalent?

Finally, we want to comment a technical point concerning the theoretical calculation of noise spectra. To this end, nowadays several numerical techniques are available; among them the most popular are microscopic methods such as Monte Carlo, Scattered Packet, and Cellular Automata or phenomenological methods such as hydrodynamic and drift-diffusion methods. In the framework of microscopic methods it has been recently demonstrated that both operation modes can be treated at the same level with a comparable computational burden. On the other hand, when using phenomenological methods, which do not contain directly in themselves any information on fluctuations, only the well-known transfer impedance method is able to provide a technique for the calculation of the noise. This method is developed only under the voltage-noise operation and the information on the local source of noise must be taken elsewhere (actually, phenomenological methods are able to compute only the influence of the local source of noise on the device terminals, i.e. the impedance field).

Acknowledgments

This work has been performed within the European Laboratory for Electronic Noise (ELEN) and supported by the Commission of European Community through the contracts ERBCHRXCT920047 and PECO Project EAST ELEN ERBCIPDCT940020. Partial support from NATO networking infrastructure grant CN. NIG 951009 and CNRS PECO/CEI/2003 bilateral cooperation project are gratefully acknowledged.

References

1. L. Reggiani et al., Proceedings of ESSDERC (1996).
2. V. Gruzhinskis et al., J. Appl. Phys. **76**, 5260 (1994).
3. E. Starikov et al., Appl. Phys. Lett. **66**, 2361 (1995).
4. E. Starikov et al., these Proceedings.

IS THE DEVICE NOISE-TEMPERATURE SPECTRUM A "GOOD" PHYSICAL QUANTITY ?

E. STARIKOV, P. SHIKTOROV, V. GRUŽINSKIS
Semiconductor Physics Institute, Goshtauto 11, LT2600 Vilnius, Lithuania

L. REGGIANI
Istituto Nazionale di Fisica della Materia, Dipartimento di Scienza dei Materiali, Università di Lecce, Via Arnesano, 73100 Lecce, Italy

L. VARANI
Centre d'Electronique et de Micro-optoelectronique de Montpellier (CNRS UMR 5507) Université Montpellier II, 34095 Montpellier, France

The problem of defining the noise-temperature spectrum $T_n(f)$ of two-terminal semiconductor devices under far-from-equilibrium conditions is considered. Main attention is paid to the constraints of system stability and stationarity. Relevant features emerging when relaxing the above constraints are illustrated by numerical calculations for GaAs n^+nn^+ structures biased at increasing applied voltages.

1 Problem formulation

The noise-temperature concept is usually introduced to describe small fluctuations over stationary values of device characteristics [1,2] and, for the case of two-terminal structures, this has been widely investigated both experimentally [3-6] and theoretically [1-4,7-9]. Its knowledge is of general help to determine the equivalent sources of noise through Norton and Thevenin generators corresponding to constant-voltage and constant-current operations, respectively. The definition of the noise temperature is based on linear-response theory when the constraint of a stationary stable-state is rigorously defined. However, in both cases of experiments as well as of theoretical simulations, such as Monte Carlo (MC) procedures, a rigorous mathematical definition of a stationary stable-state cannot be used and it is usually replaced by the following phenomenological one. Since any measured or simulated quantity fluctuates, a stationarity of the system is understood as invariance in time (during a certain time interval τ) of some average physical-quantity, e.g. $Q(t)$. The noise is thus associated with the fluctuations around the stationary value, $\Delta Q(t)$. A simplified mathematical formulation of such a phenomenological procedure can be represented as follows:

$$\Delta Q(t) = Q(t) - \frac{1}{\tau}\int_{t-\tau/2}^{t+\tau/2} Q(t')dt' \qquad (1)$$

Near thermodynamic equilibrium the mathematical and phenomenological definitions of stationarity can be considered as equivalent for frequencies $f >> 1/\tau$. In this case, by neglecting quantum effects associated with black-body radiation [10], the noise-temperature for a two-terminal device, $T_n(f)$, satisfies a generalized Nyquist relation of two equivalent forms:

$$T_n(f) = \frac{S_U(f)}{4k_B Re[Z(f)]} = \frac{S_I(f)}{4k_B Re[Y(f)]} \qquad (2)$$

where $S_U(f)$ and $S_I(f)$ are the spectral densities of voltage and current fluctuations of the device, $Z(f)$ and $Y(f)$ are its small-signal impedance and admittance, respectively, k_B is the Boltzmann constant. The situation can be considerably complicated under conditions very-far from thermodynamic equilibrium. A significant example is a n^+nn^+ structure (diode) of GaAs when strong carrier-heating occurs because of an external voltage applied to the diode, especially when an N-shape of the current-voltage characteristic and associated Gunn-instabilities take place. Under constant-current operation, a Gunn-diode remains stable (i.e. it does not go to a generation regime) for any voltage applied between its terminals. In this case, the noise temperature definition through $S_U(f)$ given by Eq. (2) is valid outside the amplification band, i.e. where $Re[Z(f)] > 0$. Inside the amplification band, where $Re[Z(f)] < 0$, one can introduce the noise-measure, $M(f)$, a dimensionless quantity defined as [11,12]:

$$M(f) = \frac{S_U(f)}{4k_B T_0\{-Re[Z(f)]\}} \qquad (3)$$

where T_0 is the lattice temperature. The above quantity gives the intrinsic noise of an amplifyer with a shorted input. By assuming that the usual interrelations hold, namely:

$$S_U(f) = S_I(f)|Z(f)|^2, \quad S_I(f) = S_U(f)|Y(f)|^2, \quad Z(f)Y(f) = 1 \qquad (4)$$

one can expect to obtain the same results under constant-voltage operation too. However, in this latter case the diode can go into a generation regime, when a transition into another physical-state, periodically oscillating in time, takes place. Moreover, from one side the transition into the generation regime occurs usually sharply at some threshold voltage. (Such a transition can be easily controlled both theoretically and experimentally.) From another side, there exists a certain pre-generation region of applied voltages below the threshold value. Here, the diode is still stable, in accordance with the phenomenological definition of stationarity, but it already "feels" the possibility to settle in another physical-state and, indeed, some "virtual transition" between the two states occurs. For example, in the time analysis of the current such a transition is detected as spontaneous appearance of a series of quasi-harmonic oscillations which spontaneously damp. Such a situation can be monitored by the noise-temperature spectrum. It means, that the two definitions of $T_n(f)$ through $S_U(f)$ and $S_I(f)$ given by Eq. (2) can become nonequivalent and, for a generation regime, the concept of noise needs an essential revision. Therefore, an open problem arises: does the noise temperature and its spectrum remain a "good" physical quantity under these conditions?

2 Results and discussion

To better illustrate the open problem formulated above, we report and discuss some results of interest obtained for a n^+nn^+ structure of GaAs. Figure 1 shows $T_n(f)$ at increasing voltages for a $0.5 - 7.5 - 0.5\ \mu m$ GaAs diode with doping concentration $n = 10^{15}$ and $n^+ = 2 \times 10^{16}\ cm^{-3}$ calculated under constant current operation with a hydrodynamic (HD) approach based on the

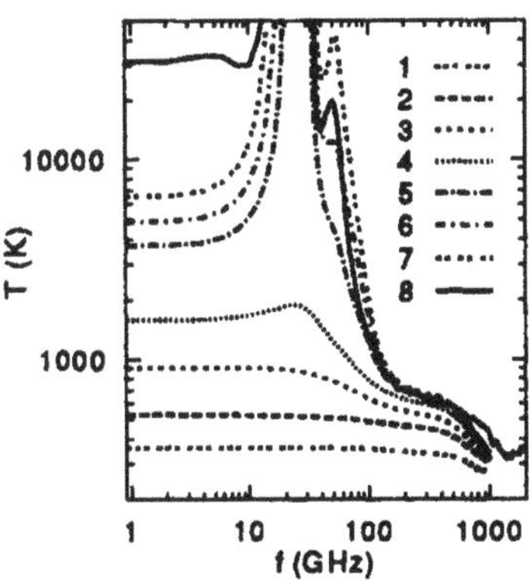

Figure 1: Noise temperature calculated for a GaAs diode by HD (curves 1 to 7 at $U = 0.9, 1.9, 2.2, 2.6, 2.85, 2.94, 3.05\ V$) and MC-HD (curve 8 at $U = 3.0\ V$) approaches under constant current and constant voltage operations, respectively.

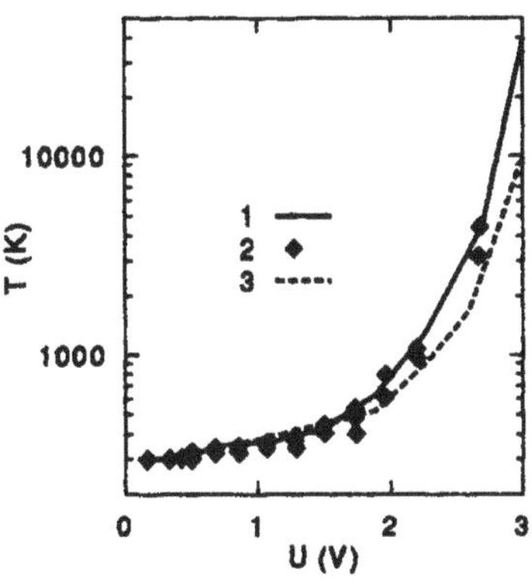

Figure 2: Voltage dependence of the noise temperature at $f = 10\ GHz$ for the GaAs diode of Fig. 1. Curves 1 and 2 refer to theory and experiment under constant voltage operation. Curve 3 is calculated under constant current operation.

impedance-field formalism [13]. It should be underlined that, apart from the generation band centered at $f = 25\ GHz$ where $T_n(f)$ takes negative values not reported in the figure, all positive values of $T_n(f)$ in Fig. 1 correspond to the maximum noise power which can be measured by an external detector. Thus, the noise temperature defined by Eq. (2) for positive values of $Re[Z(f)]$

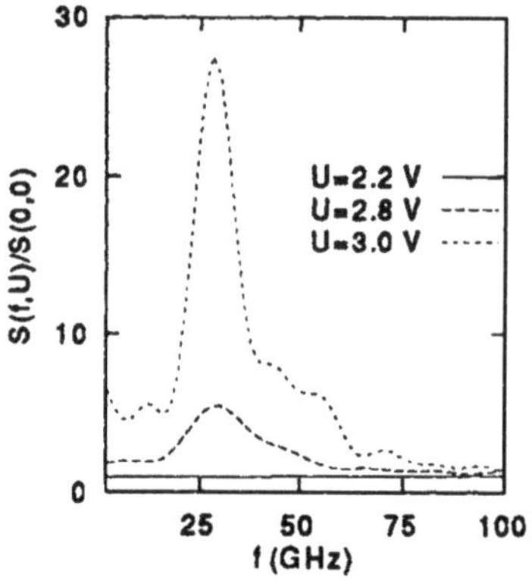

Figure 3: Spectral density of current fluctuations calculated at $T_0 = 300\ K$ for a n^+nn^+ GaAs diode of Fig. 1 by MC method under constant voltage operation.

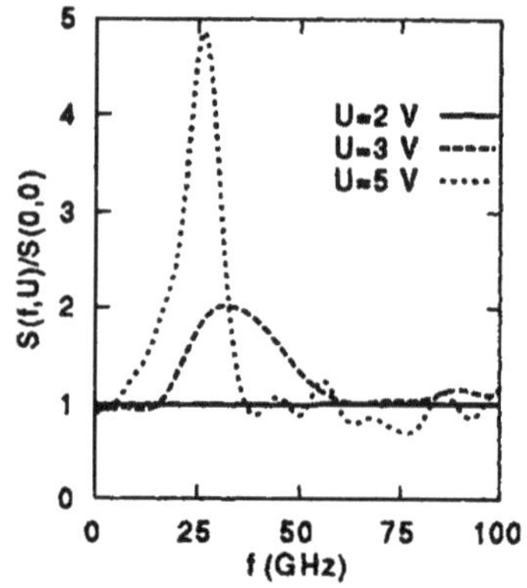

Figure 4: The same as in Fig. 3 but for a diode with 5-fold reduced concentration in n-region.

has a rigorous physical meaning. For comparison, curve 8 in Fig. 1 shows the noise temperature spectrum for the same diode calculated under constant voltage operation with a mixed MC-HD scheme, when MC and HD approaches are used separately to calculate $S_I(f)$ and $Re[Y(f)]$, respectively [14]. The general behavior of $T_n(f)$ is similar to that obtained under constant current operation. However, a significan difference is observed in the low-frequency range, where the value of $T_n(f)$ is much higher (over a factor of 5) than the correspondent calculated under constant-current operation. Moreover, when the applied voltage approaches the threshold value $U_{th} = 3.1\ V$ this difference tends to be further amplified. For voltages above this threshold value, $T_n(f)$ cannot be defined in the whole frequency range under constant-voltage operation due to the onset of self-generation at the transit-time frequency $f = 25\ GHz$. Anyway, for voltages below U_{th} the low-frequency noise temperature remains a good physical quantity and can be directly measured. This is illustrated in Fig. 2 where T_n at a frequency $f = 10\ GHz$ is reported as a function of the applied voltage. Curve 1 is calculated under constant-voltage operation with a mixed MC-HD scheme, curve 2 corresponds to experimental measurements performed on a similar diode [6], curve 3 reports the results of HD calculations carried out under constant-current operation. As can be seen from Fig. 2, the MC calculations and the experimental results obtained within similar conditions are in good agreement. We remark that the results obtained under constant-current operation (curve 3) exhibit systematic lower values of T_n when compared with those obtained under constant-voltage operation (curve 1), at $U = 3\ V$ the value of curve 3 being four times less than that of curve 1. We believe that this difference reflects the appearance of an extra-source of noise detectable under constant-voltage operation in the pre-threshold region of applied voltages. This extra-noise should be considered as a precursor signal that the diode is approaching the generation regime. Indeed, as a counterproof of this conjecture we notice its absence under constant-current operation when the diode remains always stable. The fact that an extra-noise is the precursor of a system instability is also illustrated in Figs. 3 and 4. Here, the spectral density of current fluctuations calculated directly with the MC approach is reported in Fig. 3 for the same diode of previous figures and in Fig. 4 for a similar diode but with a concentration of the n-region reduced by a factor of 5. In this way we have been able to avoid the self-generation for applied voltages above the threshold value under constant-voltage operation. For each diode $S_I(f, U)$ is normalized to its value at thermodynamic equilibrium and zero frequency $S_I(0,0)$. As follows by comparing Figs. 3 and 4, the extra-noise at low-frequency is entirely absent in the spectral density of the stable diode for both values of voltage near and above the threshold value for Gunn-effect. It should be stressed that, with a proper choice of the external resonant-circuit, also the diode of Fig. 4 can operate as an active device for $U > 3\ V$ and generate microwave power. To appreciate the physical implications of the extra-noise we recall that, by definition, the noise temperature is associated with the power extracted at the matched load and which can be measured in both operation modes. From an experimental point of view it means that the existance of the extra-noise can be easily verified. From a theoretical point of view it means that the two forms of the generalized Nyquist relation in Eqs. (2) are no longer equivalent in the pre-threshold voltage region (note, that the microscopic nature of this noise source is of no importance in this case). Furthermore, the inter-relations given by Eqs. (4) are obtained from linear-response theory based on the concept of a stationary state. On the one side, violation of these inter-relations can be connected with violation of stationary conditions in one of the operation modes (e.g. constant-voltage operation for Gunn-diodes). On the other side, it is implicitly supposed that the output noise results from a linear response, however, in this case one deals with a system operating under nonlinear response, and it is impossible to determine,

either experimentally or from numerical simulation, whether the noise is linear or not. In any case, by considering that:
(i) both in practice and numerical simulations through stochastic methods measured and calculated quantities fluctuate in time;
(ii) any experiment or simulation cannot take an infinite time but is always limited within a well defined time-interval;
there appear here the following unsolved problems which are common to all noise subjects:
(i) what shall we call as stationary part of a process (noise can be introduced after having provided an answer to this question only) ?
(ii) is it always possible to separate the stationary part of a process and the noise ?
(iii) is it possible to classify a noise as a result of linear or nonlinear response?
In our opinion, a solution of these problems would allow to consider some unsolved problems as solved ones and vice versa.

Acknowledgments

This work has been performed within the European Laboratory for Electronic Noise (ELEN) and supported by the CEC through the contracts ERBCHRXCT920047, PECO Project EAST ELEN ERBCIPDCT940020, NATO CN. NIG 951009, CNRS PECO/CEI/2003 bilateral cooperation project.

References

1. J.P. Nougier, *phys. stat. sol. (b)* **55**, K43 (1973).
2. J.P. Nougier, *IEEE Trans. Electr. Dev.* **41**, 1902 (1994).
3. J. Zimmermann *et al.*, *Appl. Phys. Lett.* **30**, 245 (1977).
4. J.P. Nougier *et al.*, *Solid State Electr.* **21**, 133 (1978).
5. D. Gasquet *et al.*, *Physica B* **134**, 264 (1985).
6. V. Bareikis *et al.*, *IEEE Trans. Electr. Dev.* **41**, 2050 (1994).
7. J. P. Nougier and M. Rolland, *Phys. Rev. B* **8**, 5728 (1973).
8. P. Hu and C.S. Ting, *Phys. Rev. B* **35**, 4162 (1987).
9. J.C. Adams and T.-W. Tang, *IEEE Electron. Dev. Lett.* **13**, 378 (1992).
10. L. Reggiani *et al.* these Proceedings (1996).
11. B.C. DeLoach, *IRE Trans. Electr. Dev.* **ED-9**, 366 (1962).
12. H.K. Gummel and J.L. Blue, *IEEE Trans. Electr. Dev.* **ED-14**, 569 (1967).
13. P. Shiktorov *et al.*, *Appl. Phys. Lett.* **68**, 1516 (1996).
14. E. Starikov *et al.*, *J. Appl. Phys.* **79**, 242 (1996).

PROBLEMS OF NOISE IN DEVICES

DIFFUSION COEFFICIENT TO CHARACTERIZE LOCAL NOISE SOURCES IN SUBMICRON DEVICES, IS IT THE GOOD MAGNITUDE?

T. GONZALEZ, J. MATEOS, D. PARDO
Departamento de Física Aplicada, Universidad de Salamanca, Plaza de la Merced s/n, 37008 Salamanca, Spain

V. GRUZINSKIS, E. STARIKOV, P. SHIKTOROV
Semiconductor Physics Institute, Gostauto 11, 2600 Vilnius, Lithuania

We present a theoretical analysis, performed by means of Monte Carlo and hydrodynamic calculations, on the applicability of the diffusion coefficient to characterize the local diffusion noise sources in submicron semiconductor devices. To this end the noise behaviour of a GaAs n^+nn^+ structure is studied.

1 Introduction

Traditional techniques dealing with the study of noise in electronic devices usually assume that there is no correlation between diffusion noise sources at two different positions.[1] However, it has been theoretically demonstrated that the velocity fluctuations remain strongly correlated over lengths shorter than the distance traveled by a carrier between two scattering events.[2,3] This means that while the diffusion coefficient (DC) is appropriate to describe the local noise sources in long devices (as compared with the correlation length), in the case of short and/or non-homogenous devices some doubts about its applicability appear, since the spatial correlation (SC) between local noise sources should be taken into account. By means of Monte Carlo (MC) calculations, the existence of SCs in GaAs has been confirmed[4] and the influence of these correlations on the diffusion noise in submicron devices has been analyzed for the case of an n^+nn^+ structure with an n region of 0.6 microns.[5] It has been found that the SCs do not correspond to those found in the homogenous case (specially far from equilibrium) and, consequently, the addition of all these correlations provides a value substantially different from the DC corresponding to the local electric field (or to the local mean energy). This result leads to conclude that the DC does not seem to be the good magnitude to characterize the local noise sources in short non-homogenous devices, and suggests that the noise should be analyzed by taking into account all the individual cross-correlations.

The objective of this work is to check this conjecture. With this aim, the spectral density of voltage fluctuations between the terminals of a GaAs n^+nn^+ structure has been calculated by using the impedance field (IF) method in the framework of a closed hydrodynamic (HD) approach.[6] Here, the noise sources are assumed to be uncorrelated and they are characterized by the DC corresponding to the local mean energy. The same calculation has been performed by using the MC technique, where all the SCs are naturally incorporated in the simulation.

2 Theoretical analysis

Within the IF formalism, the spectral density of voltage fluctuations between the terminals of a one-dimensional device of length L under constant-current operation is given by:

$$S_U(\omega) = \int_0^L \int_0^L \nabla Z(x,\omega) S_I(x,x',\omega) \nabla Z^*(x',\omega) dx dx' \tag{1}$$

where $\nabla Z(x,\omega)$ is the local IF and $S_I(x,x',\omega)$ is the noise source term associated to the positions x and x'. Usually it is assumed that the noise sources at two different points are uncorrelated and consequently, in the case of diffusion noise, $S_I(x,x',\omega)$ becomes:

$$S_I(x,x',\omega) = Aq^2 n(x) S_v(x,\omega) \delta(x-x') = 4Aq^2 n(x) D(x,\omega) \delta(x-x') \tag{2}$$

where A is the cross-sectional area of the device, q is the electron charge, $n(x)$ the local carrier concentration and $S_v(x,\omega)=4D(x,\omega)$ the local spectral density of velocity fluctuations, with $D(x,\omega)$ the local DC, used to characterize the local noise source. With Eq. 2, $S_U(\omega)$ becomes:

$$S_U(\omega) = Aq^2 \int_0^L n(x) |\nabla Z(x,\omega)|^2 S_v(x,\omega) dx \tag{3}$$

This final expression assumes that all the noise source associated to the position x is localized at x (uncorrelated with other positions) and it is characterized by the local DC. With the MC simulation, the calculation of $S_U(\omega)$[7] is made without introducing any assumption about the noise sources, since it intrinsically contains the microscopic fluctuations of the involved magnitudes. The comparison between the MC and IF-HD calculations of $S_U(\omega)$ allows to determine if the use of the DC to characterize the local noise sources is correct.

The evaluation of the SCs between the noise sources is also possible with the MC technique. The SCs are studied by decomposing the noise source related to a given cell n, represented by a magnitude $D_n(\omega)$, into the contributions coming from the correlations with near cells m, $D_{nm}(\omega)$,[4,5] so that:

$$D_n(\omega) = \sum_m D_{nm}(\omega) \tag{4}$$

When the field and carrier distributions are uniform over distances longer than the correlation length, $D_n(\omega)$ corresponds to the bulk longitudinal DC.

3 Results

Numerical simulations have been performed for a GaAs n^+nn^+ diode with the following parameters: doping levels of 10^{17} cm^{-3} in the n^+ regions and 5×10^{15} cm^{-3} in the n region; and cathode, n region and anode lengths of 0.3, 0.6 and 0.4 µm respectively. Abrupt homojunctions are assumed. The details of the MC and HD calculations can be found elsewhere.[5,6] The input parameters of the HD model were calculated from single-particle MC simulations.

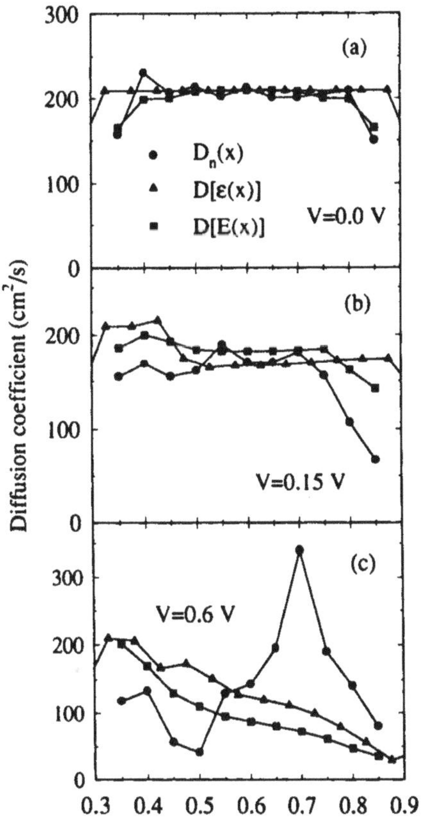

Figure 1: Low-frequency value of the diffusion coefficient corresponding to (•) addition of cross-correlations, (▲) local mean energy, (■) local electric field, as a function of the position in the n region of the n^+nn^+ structure for three applied voltages: (a) 0.0 V, (b) 0.15 V and (c) 0.6 V.

In Fig. 1 we compare the values of the DC corresponding to the local electric field and to the local mean energy with the values obtained for $D_n(0)$ by adding the contributions from the cross-correlations according to Eq. 4. The comparison is performed in the n region of the n^+nn^+ structure, which constitutes the main source of voltage noise, and for three different biasings. At equilibrium [Fig. 1(a)] the three sets of values show a good agreement due to the uniformity of the carrier velocity and energy inside the structure. In the case of a bias voltage of 0.15 V [Fig. 1(b)] the agreement is found only in the central part of the n region, where the carriers reach the steady state corresponding to the electric field. The effect of the non-stationary motion of the electrons is higher as the applied voltage is increased. For 0.6 V the steady state is not reached at any position inside the n region and the electric field is strongly non-uniform. All these factors increase the disagreement between $D_n(0)$ and the local DC (field- or energy-dependent) [Fig. 1(c)], specially in the near-anode region.

The previous results indicate that in the case of short non-homogenous

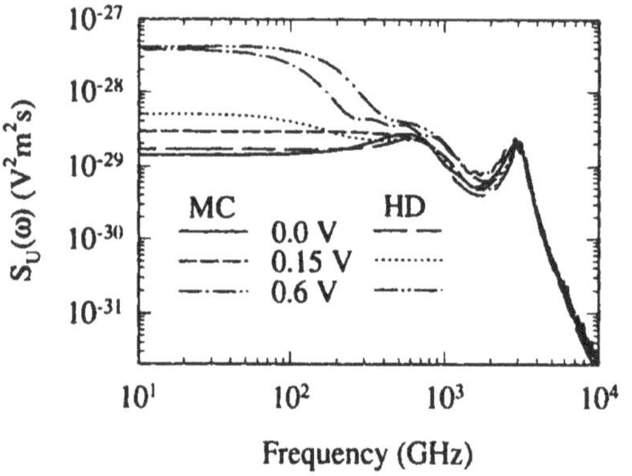

Figure 2: Frequency dependence of the spectral density of voltage fluctuations between the terminals of the n^+nn^+ structure for several voltages, calculated by Monte Carlo (MC) and hydrodynamic (HD) approaches.

devices, and specially far from equilibrium, the individual cross-correlations should be taken into account to analyze the noise and, therefore, the DC is not appropriate to characterize the local noise sources. To verify this conclusion, MC and IF-HD calculations of $S_U(\omega)$ in the GaAs n^+nn^+ structure have been performed. $D(x,\omega)$ corresponding to the local mean energy, and calculated from MC simulations, is used in the noise source term of Eq. 2 for the IF evaluation of $S_U(\omega)$. The comparison between MC and HD calculations for the three analyzed voltages is shown in Fig. 2. From the previous conjecture both results are expected to be rather different, however they are found to be reasonably similar. At equilibrium a very good agreement is found, which seems logical in view of the favourable comparison observed in Fig. 1(a) for the possible estimators of the noise sources. For 0.15 V a significant (but minor) difference appears at the lowest frequencies. And for 0.6 V, when the most important disagreement could be predicted from the values shown Fig. 1(c), again both results are quite similar except in the range 100-300 GHz. For frequencies above 500 GHz the agreement found between both calculations is excellent for the three voltages.

4 Problem discussion

Which is the origin of this unexpected agreement? Does this result mean that the DC remains the good magnitude to describe the local noise sources in submicron non-homogeneous devices? Is it due to the fact that the main influence of SCs takes place in the region of the device with the lowest contribution to the noise at the terminals? To illustrate this last possibility, Fig. 3 shows the

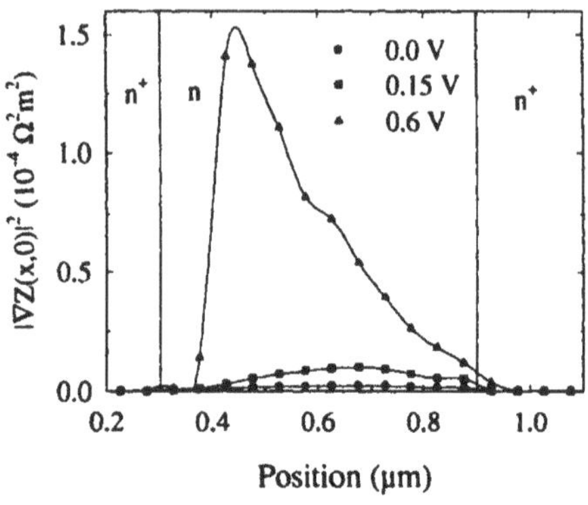

Figure 3: Profile of the impedance field squared at low-frequency along the n^+nn^+ structure for several voltages calculated by hydrodynamic approach.

profile of the IF squared at low frequency, $|\nabla Z(x,0)|^2$, along the n^+nn^+ structure. Following Eq. 3, this is the magnitude which weights the local noise sources to provide the contribution of each position to the total voltage noise at the terminals. For the three voltages it can be observed that $|\nabla Z(x,0)|^2$ takes the lowest values in the region where the most important differences between $D_n(0)$ and the local energy-dependent DC appear (Fig. 1), thus minimizing the influence of the SCs on the noise at the terminals. This is specially manifest for a biasing of 0.6 V, case in which $|\nabla Z(x,0)|^2$ is maximum in the near-cathode region, while the influence of the SCs is very important in the near-anode region.

5 Conclusions

Trying to prove the inadequacy of the DC to characterize local noise sources in submicron devices, we have performed a theoretical MC and IF-HD analysis of voltage noise in a submicron GaAs n^+nn^+ structure. Although the existence and influence of SCs in this structure are demonstrated, the results of the IF-HD approach using the local DC in the noise source term are in good agreement with the MC results. This fact would lead to conclude that the SCs can be ignored and that the DC remains the good magnitude for the noise analysis.

Acknowledgments

This work has been partially supported by the project TIC95-0652 from the Comisión Interministerial de Ciencia y Tecnología (CICYT).

References

1. K. M. van Vliet, A. Friedmann, R. J. J. Zijlstra, A. Gisolf and A. van der Ziel, *J. Appl. Phys.* **46**, 1804 (1975).
2. K. K. Thornber, *Solid-State Electron.* **17**, 95 (1973).
3. J. P. Nougier, J. C. Vaissiere and C. Gontrand, *Phys. Rev. Lett.* **51**, 513 (1983).
4. J. Mateos, T. González and D. Pardo, *J. Appl. Phys.* **77**, 1564 (1995).
5. J. Mateos, T. González and D. Pardo, *Appl. Phys. Lett.* **67**, 685 (1995).
6. V. Gruzinskis, E. Starikov and P. Shiktorov in *Proc. 13th Int. Conf. on Noise in Physical Systems and 1/f Fluctuations* (World Scientific, Singapore, 1995), p. 185.
7. T. González, D. Pardo, L. Varani and L. Reggiani, *Appl. Phys. Lett.* **63**, 84 (1993).

ARE NOISE SOURCES ASSOCIATED TO BASE AND COLLECTOR CURRENTS IN AlGaAs/GaAs HETEROJUNCTION BIPOLAR TRANSISTORS CORRELATED?

S. JARRIX, C. DELSENY, F. PASCAL, G. LECOY

Centre d'Electronique et de Micro-optoélectronique de Montpellier,
Laboratoire mixte de recherche CNRS-Université UMR5507, Place Bataillon,
U.M. II, 34095 Montpellier Cedex 5, France

In this paper, a correlation between base and collector currents of heterojunction bipolar transistors is studied versus bias and temperature.

1 Noise Measurement Procedure and Devices

Noise measurements were performed on heterojunction bipolar transistors (HBTs) in the 10 Hz - 100 kHz frequency range. Transistors are mounted in a common-emitter configuration. Voltage noise is measured simultaneously on the base and collector (cf. Fig. 1) of the device across bias resistances. The bias resistance on the base is always chosen with a much higher value than the dynamic input resistance of the intrinsic transistor r_d. Noise signals are fed into low-noise amplifiers and then into a spectral analyser. Through a FFT algorithm, Sv_b, Sv_c : voltage spectral densities associated to the base and collector; as well as Sv_bv_c : the cross-spectral density are measured.

Devices studied were non-self-aligned AlGaAs/GaAs heterojunction bipolar transistors (cf. Fig. 2) with different emitter surfaces. Base is C-doped with a doping of $4 \times 10^{19}\ cm^{-3}$.

2 Theoretical Expressions of Spectra

All spectral densities of all transistors exhibit excess noise composed of 1/f noise and generation-recombination (g-r) components[1,2]. The white noise is not reached in the frequency range used. Using a classical small-signal equivalent circuit, theoretical expressions of Sv_b, Sv_c[3] and Sv_bv_c have been established. Relations take into account the spectral densities Si_b, Si_c associated to base and collector current noise sources and the cross-correlation spectral density Si_bi_c. The hypothesis has been made that noise current sources i_b and i_c are partly correlated as previously suggested by Van der Ziel[4]. In the case where

$Rp >> r_d$, expressions are simplified and for the excess noise become :

$$\begin{pmatrix} Sv_b \\ Sv_c \\ RP\{Sv_b v_c\} \end{pmatrix} = M \begin{pmatrix} Si_b \\ Si_c \\ RP\{Si_b i_c\} \end{pmatrix} \tag{1}$$

with :

$$M = \begin{pmatrix} (r_\pi + r_e h_{fe})^2 & r_e^2 & 2(r_\pi + r_e h_{fe}) r_e \\ (R_L h_{fe})^2 & R_L^2 & 2R_L^2 h_{fe} \\ -R_L(r_\pi + r_e h_{fe}) h_{fe} & -r_e R_L & -2R_L r_e h_{fe} - R_L r_\pi \end{pmatrix}$$

$$IP\{Sv_b v_c\} = -R_L r_\pi \; IP\{Si_b i_c\} \tag{2}$$

The coherence function $\Gamma v_b v_c$ is defined by :

$$\Gamma v_b v_c = \frac{|Sv_b v_c|^2}{Sv_b \; Sv_c} \tag{3}$$

If $\Gamma v_b v_c = 0$, noise sources are entirely uncorrelated
If $\Gamma v_b v_c = 1$, noise sources are entirely correlated
If $0 < \Gamma v_b v_c < 1$, noise sources are partly correlated

3 Measurement of the Coherence Function and Temperature Study

An example of $\Gamma v_b v_c$ is given for HBTs with emitter surfaces of $S_1 = 28.5\mu m^2$ and $S_2 = 10\mu m^2$ on Fig. 3 and 4 respectively. For HBT with S_1, $\Gamma v_b v_c$ evolves between 0.9 and 1. The curve decreases regularly in the whole frequency range. This is the case for all HBTs studied with an emitter-surface over 10 μm^2. This phenomenon is also observed on classical silicon bipolar transistors. For HBT S_2, $\Gamma v_b v_c$ exhibits values comprised between 0.65 and 1. For this particular transistor, one can observe a minimum on the curves for a frequency f_m around 800 Hz.

The influence of temperature on this minimum has also been studied. To the frequency f_m, a time constant is associated by $\tau = 1/2\pi f_m$. An Arrhenius plot of ln τT^2 versus $1000/T$ could be drawn (Fig. 5). Slope of the line corresponds to an activation energy with a mean value of 180 meV. Could traps with this energy be responsible for the minimum of $\Gamma v_b v_c$?

4 Coherence function associated to currents

When $\Gamma v_b v_c$ approaches 1, we have shown the term associated to Si_b to be dominant[5] in the expressions of Sv_b, Sv_c, $Sv_b v_c$ in the frequency and bias range used. When values of $\Gamma v_b v_c$ diminish, the contribution of terms associated to Si_c and $Si_b i_c$ have to be considered. Using relations (1) and (2) we were able to calculate from the experimental results Si_b, Si_c, $Si_b i_c$ and $\Gamma i_b i_c$ for different bias currents. In the same way as for $\Gamma v_b v_c$, when $\Gamma i_b i_c$ is equal to 1 there is a total correlation between noise sources, in this case base and collector noise sources. An example of $\Gamma i_b i_c$ is given on Fig. 6 for HBT with S_2 at room temperature. Values of $\Gamma i_b i_c$ increase with bias. Also a maximum occurs at different frequencies f_M when bias evolves: $320\ Hz < f_M < 900\ Hz$ when $10\ \mu A < Ib < 80\ \mu A$.

5 Generation-Recombination Noise versus Temperature

To study some possible generation-recombination (g-r) process occurring at these frequencies, measurements have been performed versus temperature on HBT S_2. Spectral densities Si_b, Si_c and $Si_b i_c$ were extracted. An example of $|Si_b i_c|$, and of Si_b and Si_c at different temperatures is given on Fig. 7 and 8 respectively. On these figures, one can note that whatever the temperature the 1/f noise is clearly observed on Si_b only. Fig. 7 and 8 also show the presence of g-r components thermally activated on all spectra. The plateau increases when temperature decreases and cut-off frequencies shift towards lower values. Also, when temperature is lowered, more components appear. Finally, for a given temperature, curves Si_b and Si_c can be superimposed in the g-r frequency range.

6 Discussion-Conclusion

A possible explanation of the partial correlation observed between Si_b and Si_c is to consider a common current component between the base and collector current. This component would be linked to a recombination current depending on temperature through thermally activated traps. By considering the evolution of spectra with bias, these traps would be located in the base-collector space-charge-region[2, 6]. From a realistic point of view, this current component should not be a major one. Thus we are surprised by the high values measured for $\Gamma i_b i_c$. For example, at room temperature, $\Gamma i_b i_c \sim 0.95$ for $Ib = 80\mu A$ at $f = 1\ kHz$. At this frequency, the value of $\Gamma i_b i_c$ induces a nearly- total correlation between i_b and i_c. To our knowledge this phenomenon has never been put forward in the literature. This effect could be due to an increase in leakage current when scaling-down devices. All spectral densities measured on the transistors have shown excess noise which can be attributed to several noise sources. Therefore is it reasonable to think that correlation measurements may give more indications on the physical mechanisms producing this excess noise?

Figures

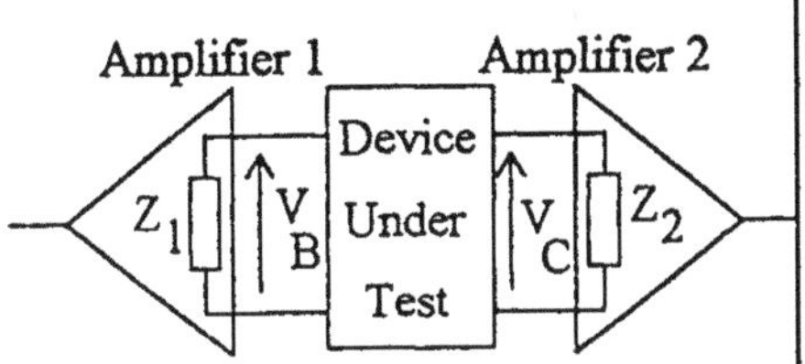

Fig. 1 : Experimental set-up

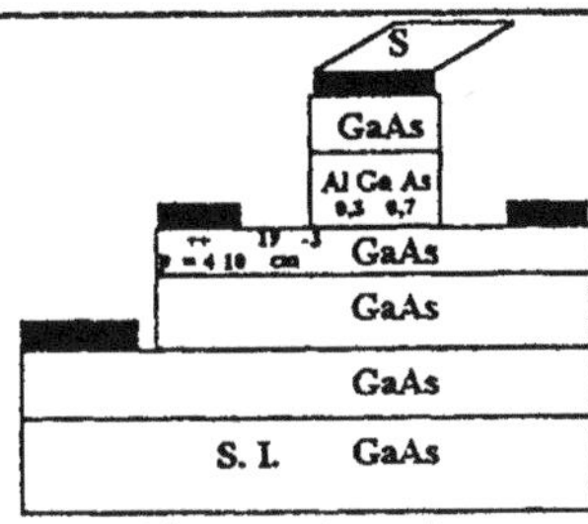

Fig. 2 : Device

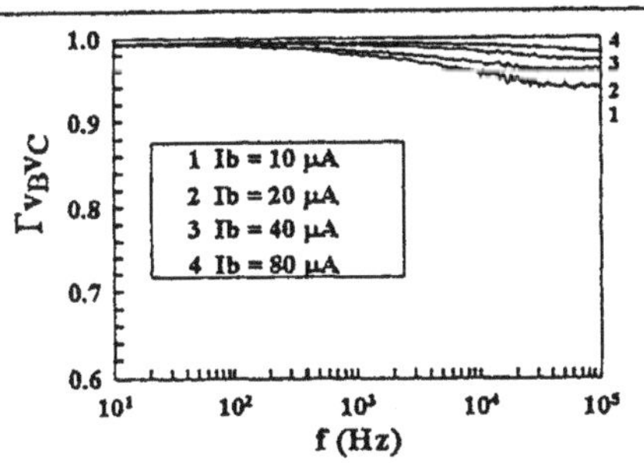

Fig. 3 : Voltage coherence function of HBT with $S_1 = 28.5\ \mu m^2$

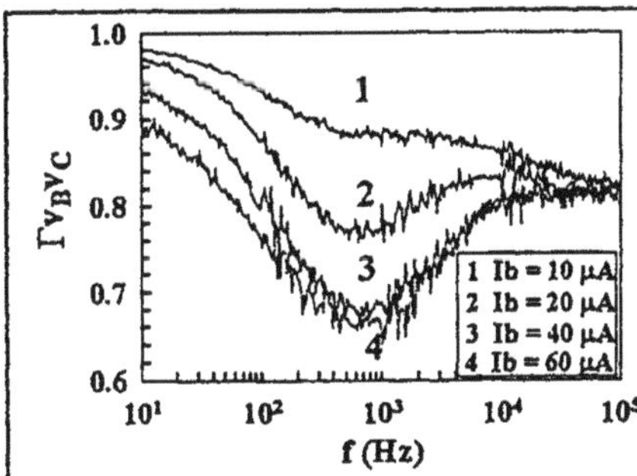

Fig. 4 : Voltage coherence function of BT with $S_2 = 10\ \mu m^2$

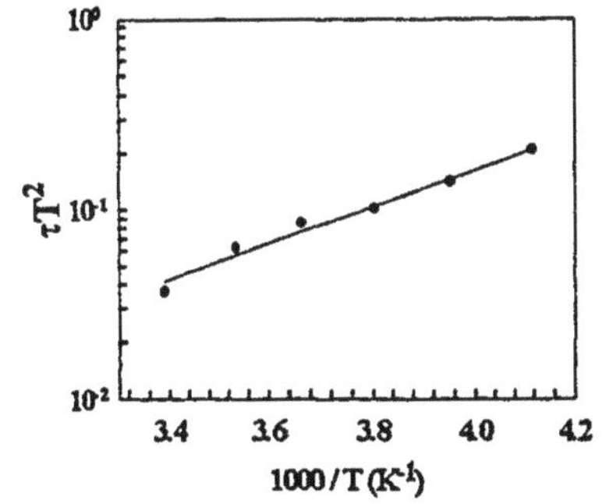

Fig. 5 : Arrhenius Plot at Ib = 40 µA of HBT with $S_2 = 10\ \mu m^2$

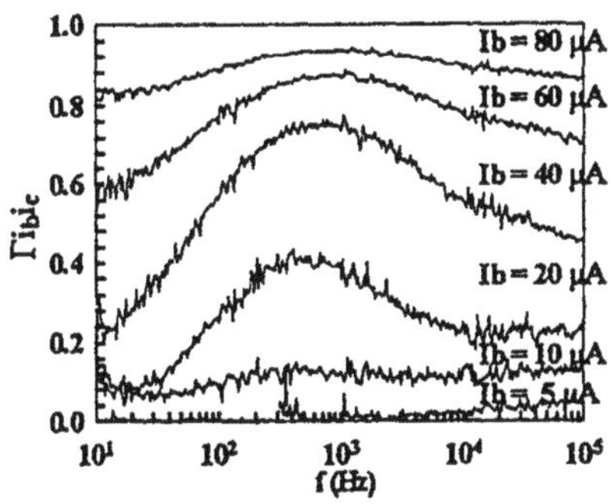

Fig. 6 : Current coherence function of HBT with $S_2 = 10\ \mu m^2$.

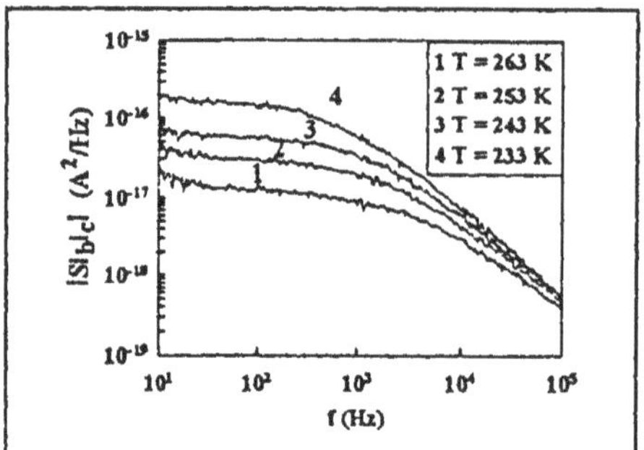

Fig. 7 : Cross-spectra of HBT with $S_2 = 10\ \mu m^2$ at Ib = 40 mA

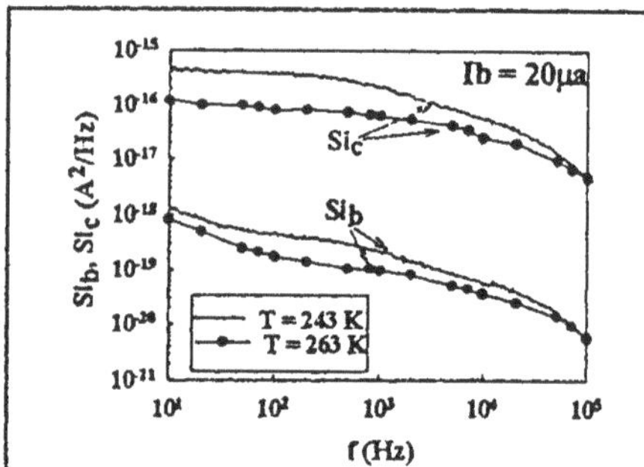

Fig. 8 : Base and collector current densities of BT with $S_2 = 10\ \mu m^2$

Acknowledgments

Authors sincerely thank the CNET Bagneux (French Telecommunications) for supplying HBTs and for the helpful discussions.

References

1. D. Costa and J. S. Harris Jr., IEEE Trans. on Electron Dev. 39, 2383 (1992).
2. C. Delseny, F. Pascal, S. Jarrix, G. Lecoy, IEEE Trans. on Electron Dev. 41, 2000 (1994).
3. T. G. M. Kleinpenning,IEEE Trans. on Electron Dev. 41, 1981 (1994).
4. A. van der Ziel, in *Noise in solid-state devices and circuits*, (John Wiley and sons Inc, Wiley Interscience Publication, 1986).
5. S. Jarrix, C. Delseny, F. Pascal, G. Lecoy, J. Appl. Phys., submitted.
6. S. Kuglar, J. Appl. Phys. 66, 219 (1992).

ISSUES IN MODELLING THE HIGH FREQUENCY NOISE PARAMETERS OF POLYSILICON EMITTER BIPOLAR TRANSISTORS

M. J. DEEN

School of Engineering Science, Simon Fraser University

Burnaby, British Columbia, Canada V5A 1S6

Analytical modelling of the high frequency noise parameters - minimum noise figure (NF_{MIN}), noise resistance (R_N), and optimal source resistance ($R_{S,OPT}$) and reactance ($X_{S,OPT}$), of polysilicon emitter bipolar transistors presents many challenges. This paper presents possible solutions for some of these issues, and also discusses some of the outstanding problems. In particular, we discuss how the four noise parameters are affected by the bonding pads. Finally, we present some of the outstanding issues that must be solved for improved noise modelling of devices with smaller emitter areas and as a function of higher current levels and varying temperatures.

Introduction

Modern bipolar and BiCMOS high speed circuits now widely use polysilicon emitter bipolar junction transistors (PE-BJTs) because of the compatibility of these transistors with the increasingly important self-aligned fabrication technologies. High unity gain frequencies (f_T's) coupled with high current gains, and high collector-emitter breakdown voltages (BV_{CEO}) make these PE-BJTs suitable candidates for analog applications; but high frequency analog circuits are very sensitive to the noise of the devices. However, accurate noise modelling must take into account all noise sources and electrical elements in the equivalent circuit model, including those from the pads, the base-collector capacitance, the collector substrate junction capacitance, current dependence of the base and emitter resistances etc.

In this paper, we present results illustrating the importance of the pad effects on the device noise and new expressions of the noise parameters that includes the emitter resistance. We also discuss some of the issues and problems related to better modelling of the high frequency noise parameters. Hopefully, this paper will stimulate researchers to investigate these issues so that improved noise modelling of the PE-BJTs as a function of both device and operating parameters will soon be realized.

Experiments

The transistors used in these studies were fabricated in a 0.8 μm BiCMOS technology and they had emitter areas ranging from 3.2 μm^2 to 144 μm^2 (five sets), with the emitter width being 0.8 μm. Noise figures were measured and modelled for frequencies from 0.5 GHz to 5 GHz, and for collector current densities from 0.016

mA/μm^2 to 0.31 mA/μm^2 for all devices. In addition, the high frequency s-parameters of both transistors and dummy devices were measured using an on-wafer microwave test system. From the s-parameters, the unity-gain frequency (f_T), base resistance (R_B) and emitter (R_E) resistance were determined. Measurements were made to determine the base-emitter junction capacitance (C_π) and d.c. current gain (β). The biasing range was chosen to cover the realistic circuit operating conditions, and the collector bias was kept constant at 5 V for all measurements. For the current levels used, β was between 70 and 100, and f_T between 8 and 10 GHz. Finally, devices with two sizes of pads 80x80 μm^2 to 80x50 μm^2 were used in to investigate the effects of bonding pads on the device noise parameters.

Results and Discussions

To investigate the effects of varying emitter areas on the high frequency noise, the 5 sets of devices with different emitter areas were studied over the frequency range from 0.5 to 5 GHz. These BJTs had nominally designed emitter areas of 3.2, 12.8, 48, 96 and 144 μm^2 and the biasing base current was varied from 0.031 mA/μm^2 to 0.31 mA/ μm^2 (5 values for all devices). The intrinsic device noise figure (NF) was modelled with a simplified T-equivalent circuit originally described in [2] and expanded in [2-7], from which the following expression was obtained.

$$NF = 1 + \frac{R_B + R_E}{R_S} + \frac{r_e}{2R_S} + \left(\frac{\alpha_0}{|\alpha|^2} - 1\right) \cdot \left(\frac{(R_S + R_B + R_E + r_e)^2 + X_S^2}{2 r_e R_S}\right) +$$

$$\frac{\alpha_0}{|\alpha|^2} \cdot \frac{r_e}{2R_S} \cdot \{(\omega C_\pi X_S)^2 - 2\omega C_{JE} X_S + (\omega C_\pi)^2 \cdot (R_S + R_B + R_E)^2\}.$$

From this NF expression, NF_{MIN} is then calculated to be

$$NF_{MIN} = \frac{\alpha_0}{|\alpha|^2} + \left(\frac{\alpha_0}{|\alpha|^2} - 1 + \frac{\alpha_0}{|\alpha|^2}\omega^2 C_\pi^2 r_e^2\right) \cdot (R_B + R_E + R_{S,OPT}),$$

with $R_{S,OPT}$, $X_{S,OPT}$ and R_N defined as

$$R_{S,OPT}^2 = (R_B + R_E)^2 - X_{S,OPT}^2 + r_e(2(R_B + R_E) + r_e)\left(1 - \frac{\alpha_0}{|\alpha|^2} + \omega^2 C_\pi^2 r_e^2\right)^{-1}$$

$$X_{S,OPT} = \omega C_\pi r_e^2 \cdot \left(1 - \frac{\alpha_0}{|\alpha|^2} + \omega^2 C_\pi^2 r_e^2\right)^{-1}$$

$$R_N = \left(\frac{\alpha_0}{|\alpha|^2} - 1 + \frac{\alpha_0}{|\alpha|^2}\omega^2 C_\pi^2 r_e^2\right) \cdot \left|R^2_{S,OPT} + X^2_{S,OPT}\right|$$

In these three expressions, r_e is the emitter dynamic resistance, α is the usual frequency dependent current gain with a d.c. value of α_0, C_π is the base-emitter capacitance, R_B is the base resistance, R_S and X_S is the source resistance and reactance, and ω is the angular frequency.

The pads were modelled as a series/parallel combination of impedances between the source and input of the device (see [3-5] for details), and from circuit manipulations, they were taken into consideration in modelling the noise figure through $R_{S,TH}$ and $X_{S,TH}$, the Thevenin equivalent of the source and pad impedances, and $E_{S,TH}$, the Thevenin equivalent of the source and pad noise sources. In the following plots, we show the effects of the pad impedances as a function of collector current. We also show the calculated results when the pad impedances were not considered. All results are for a frequency of 1 GHz, and these are typical of the results between 0.5 and 5 GHz.

Fig. 1 shows the variation of NF_{MIN} with I_C for all 5 devices. Note that when the pads' impedances are neglected, there is significant disagreement of up to 2 dB between calculations and experiments over a large current range for the 3 smaller devices, as clearly shown by comparing Figs. 1(a) to 1(b). However, inclusion of the pads impedances results in good agreement between experiments and calculations for all devices, and over the entire current range. Figs. 2 and 3 show the variation of $R_{S,OPT}$ and $X_{S,OPT}$ versus I_C. When the pad effects are not considered, there

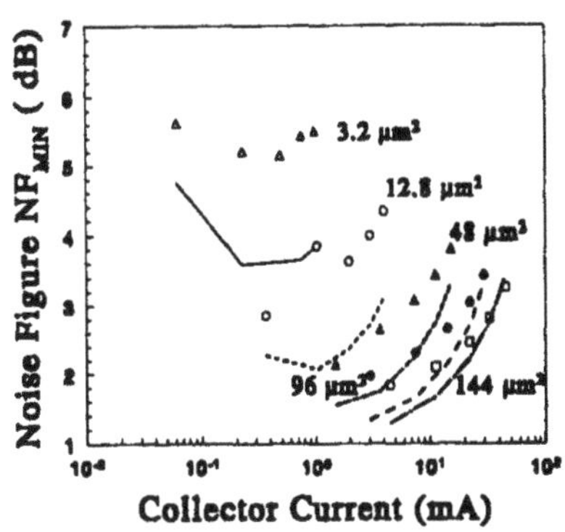

Fig. 1(a). NF_{MIN} versus I_C (uncorrected for pad effects). Note the significant disagreement between experiments and calculations.

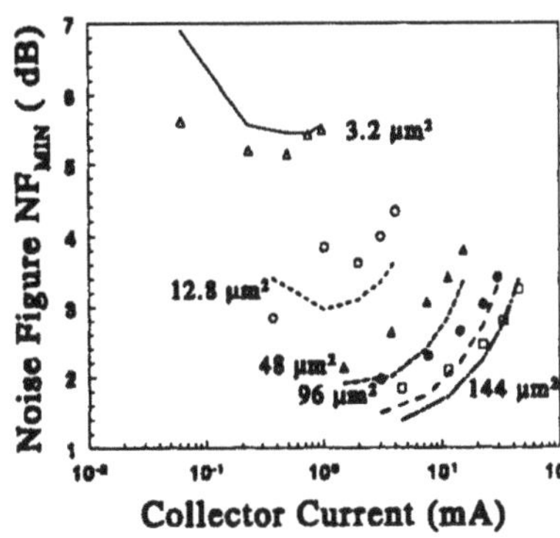

Fig. 1(b). NF_{MIN} versus I_C (corrected for pad effects). Note the improved agreement when the pad impedances are considered in the calculations.

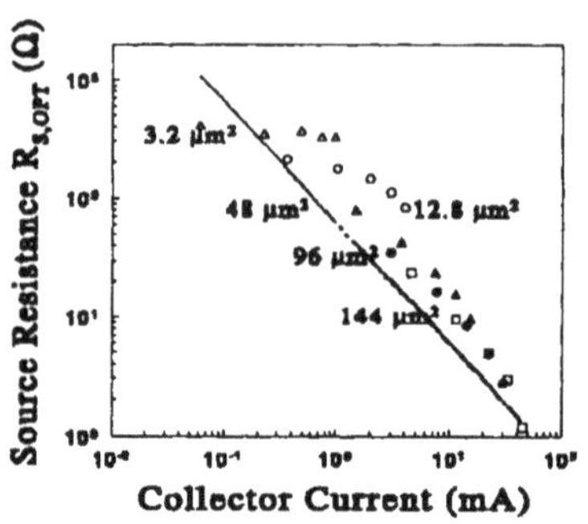

Fig. 2(a). $R_{S,OPT}$ versus I_C (uncorrected for pad effects). Note the significant disagreement between experiments and calculations.

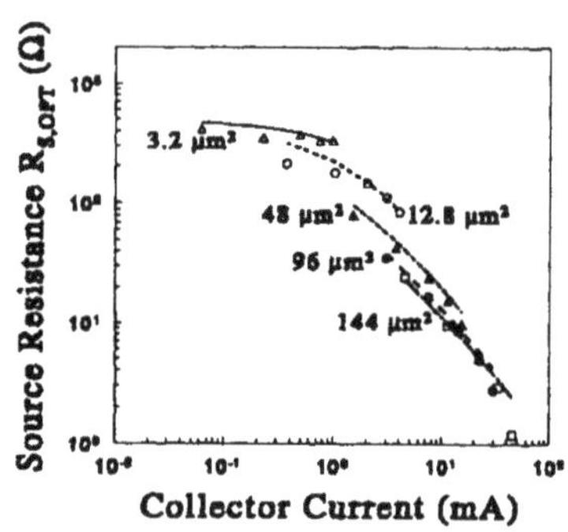

Fig. 2(b). $R_{S,OPT}$ versus I_C (corrected for pad effects). Note the improved agreement when the pad impedances are considered in the calculations

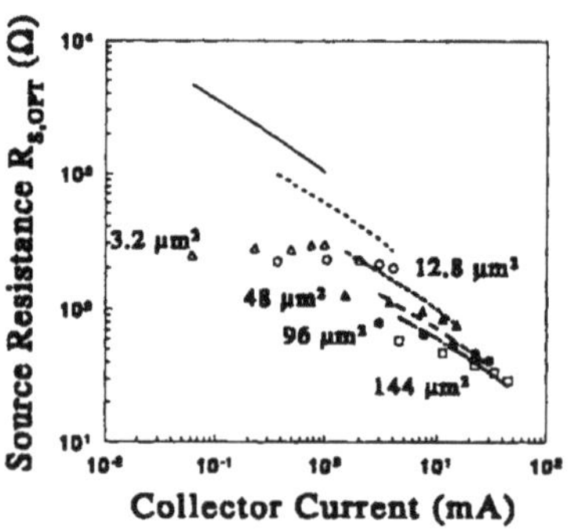

Fig. 3(a). $X_{S,OPT}$ versus I_C (uncorrected for pad effects). Note the significant disagreement between experiments and calculations.

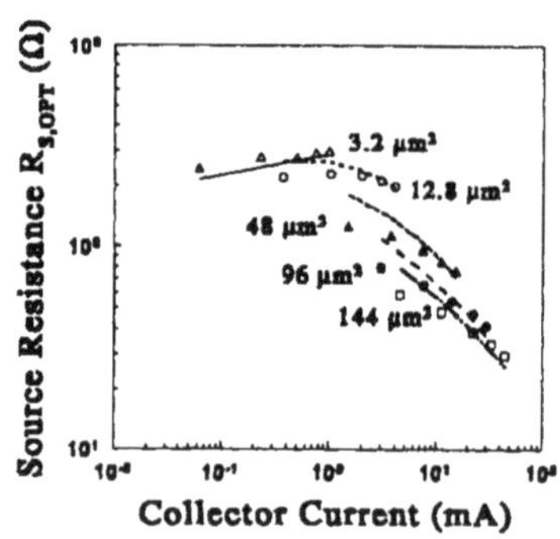

Fig. 3(b). $X_{S,OPT}$ versus I_C (corrected for pad effects). Note the improved agreement when the pad impedances are considered in the calculations

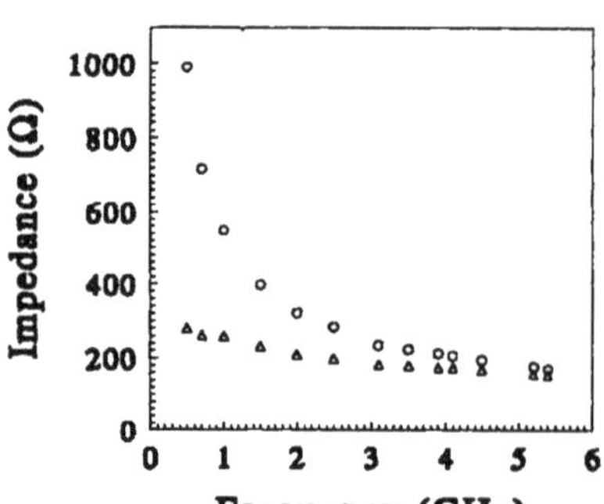

Fig. 4(a). Impedance versus frequency for a 80x80 μm^2 pad structure (larger signal pad structure).

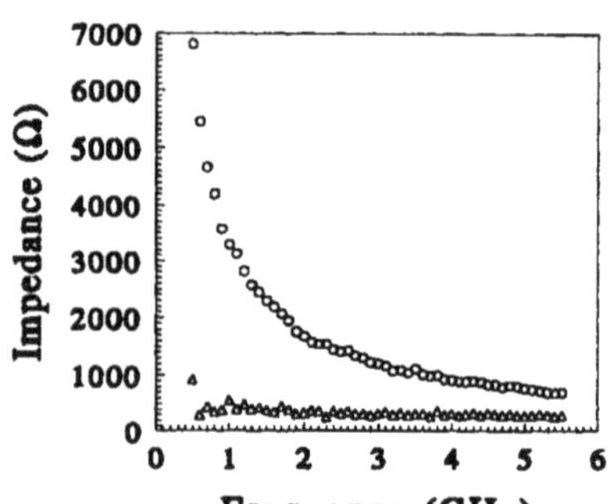

Fig. 4(b). Impedance versus frequency for a 80x50 μm^2 pad structure (smaller signal pad structure).

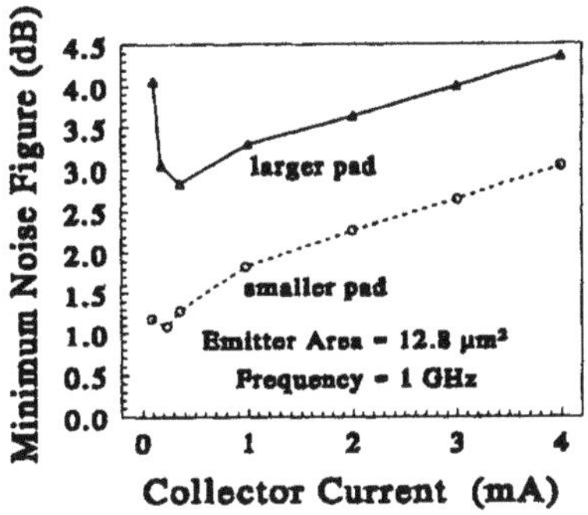

Fig. 5(a). NF_{MIN} versus I_C. Note that the smaller pad with larger impedance results in lower NF_{MIN} values for all collector currents.

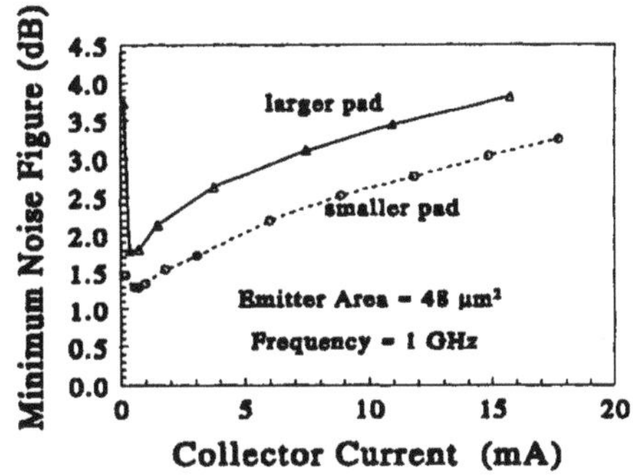

Fig. 5(b). NF_{MIN} versus I_C. Note that the smaller pad with larger impedance results in lower NF_{MIN} values for all collector currents.

is significant disagreement between experiments and calculations, especially for the smaller devices. With the pad impedances included, there is improved agreement between calculations and experimental data.

To illustrate the impact of the pad impedance on high frequency noise measurements, we measured the noise figure of two devices of emitter areas of 12.8 μm^2 and 48 μm^2 in which the signal pads were of different sizes - 80x80 μm^2 and 50x80 μm^2, and the pads' impedances. Figs. 4 and 5 show the results of the impedances of the pads and the minimum noise figure, respectively. These results demonstrate that the smaller pad with a larger impedance to ground results in the expected lower measured noise figure for both devices. This is because the larger pad impedance results in less input signal being lost to ground, thus resulting in NF_{MIN} values of the transistors being closer to their intrinsic values. Finally, these results, as well as those described above illustrate the need for proper accounting of the pads effects in modelling the noise figure, as well as proper design of their layout to minimize their influence on the measured minimum noise figure.

Effects to be investigated

In the model and results discussed above, several simplifying assumptions were made to derive simple analytical expressions for the noise parameters. These include the neglect of important elements in the PE-BJT equivalent circuit model [1-7] such as the base-collector capacitance which significantly affects the high frequency transistor performance; the collector-substrate junction capacitance; the variation of the base and emitter resistances with biasing current; the effects of high biasing currents (which results in higher gains and unity gain frequencies); perimeter effects which results in many parameters having two components, an area component and a perimeter component. In addition, the small signal properties of the transistor changes with temperature. To date, no systematic experimental or theoretical study has been conducted on the effects of varying high and low temperatures on the noise parameters on polysilicon emitter bipolar transistors.

However, for analytical modelling of the noise parameters, the resulting analytical model should not so complicated that physical insight into the noise properties are lost because of the complexity of the resulting expressions. Therefore, inclusions of the important effects described in the above paragraph in the analytical noise model should involve appropriate simplifications/limiting values to retain a form suitable for physical insight. The goal should be include the important electrical elements in the circuit description so that large signal, small signal and noise properties can be predicted and relevant trends such as scaling effects, temperature effects and high and low current effects accurately indicated through both calculations and appropriately simplified analytic expressions, as required.

Conclusions

In this paper, we demonstrated the effects on bonding pads on the noise parameters NF_{MIN}, $R_{S,OPT}$ and $X_{S,OPT}$ as a function of collector currents. The study also showed the effects of two different pad structures on NF_{MIN} values of two transistors. For test structure designers, the study also illustrate the need to minimize the pad areas and use metals that are well above the silicon substrate so that pad effects can be reduced. Finally, we described some of the important electrical elements in the transistor equivalent circuit model that have been neglected and the consequences. We also described the need to study the effects of varying operational conditions of biasing currents and temperatures, and the need to retain physical insight from the analytical model of the noise parameters.

References

1. R.J. Hawkins. Solid State Electronics, **20**, 191 (1977).
2. M.J. Deen and J.J. Ilowski, IEE Electronics Letters, **29**, 676 (1993).
3. M.J. Deen and J.J. Ilowski, *In* AIP Conference Proceedings 285 - Noise in Physical Systems and 1/f Fluctuations (ICNF '93), *Edited by* P.H. Handel and A.L. Chung, AIP Press, New York, 1993, pp. 216-219.
4. M.J. Deen and J. I. Ilowski, *In* Proceedings of the 13th International Conference on Noise in Physical Systems and 1/f Fluctuations (ICNF '95), *Edited by* V. Bareikis and R. Katilius, World Scientific Publishing, Singapore, 1995, pp. 458-461.
5. M.J. Deen, **Canadian Journal of Physics**, In Press, (1996).
6. M.J. Deen and J. Ilowski, **Canadian Journal of Physics**, In Press, (1996).
7. M.J. Deen, *in* **1996 Asia-Pacific Microwave Conference (APMC 96)**, New Delhi, 6 pages (17-20 December 1996).

PROBLEMS OF LOW-FREQUENCY NOISE IN DEPLETION MODE pMOSFETs UNDER INVERSION CONDITIONS

N.LUKYANCHIKOVA, M.PETRICHUK and N.GARBAR
Institute of Semiconductor Physics, 252650 Kiev, Ukraine

E.SIMOEN and C.CLAEYS
IMEC, Kapeldreef 75, B-3001, Belgium

The new results concerning the problem of low-frequency noise observed under inversion conditions in MOSFETs with a *p*-Si film are presented. It is shown that the source of high GR noise detected in DM *p*MOSFETs under inversion conditions manifests itself also in EM *n*MOSFETs. The characteristics of this noise are described. It is found that the correlation occurs between the low-frequency $1/f$ noise and the current in DM *p*MOSFETs under inversion conditions. The arguments are adduced in favour of the model where the fluctuations of the charge in centers located at some distance from the interface (below the inversion layer) can be responsible for the noise phenomena considered. However, many questions relating to this model are shown to be still open questions.

1 Introduction

An interface between a semiconductor and an oxide is known to be a powerful source of a low-frequency noise. The role of this noise can be depressed significantly in depletion mode (DM) MOSFETs under conditions where strong inversion occurs at the interface and the fluctuations of the interface potential are screened by the inversion channel. Such an effect was observed experimentally for $1/f$ noise[1] and generation-recombination (GR) noise[2] in the DM *n*MOSFETs (n^+nn^+MOS devices).

The situation appears to be quite different in the DM *p*MOSFETs (p^+pp^+MOS devices) where the low-frequency noise increases sharply as far as inversion conditions are approached and this noise does not disappear even in a strong inversion.

Such an anomalous behaviour was observed for the $1/f$ noise in bulk DM *p*MOSFETs[3] and SOI DM *p*MOSFETs prepared on ZMR substrates[4]. Recently the same behaviour has been revealed for the $1/f$ noise in SOS DM *p*MOSFETs and for the GR noise in the SOI DM *p*MOSFETs processed on SIMOX substrates[5,6]. More of this, the GR noise source of the same nature has been detected also in the enhancement mode (EM) *n*MOSFETs (n^+pn^+MOS devices) where the physical situation in the *p*-Si film under operation conditions is similar to that in the DM *p*MOSFETs but the current passes through the inversion layer itself. Those EM *n*MOSFETs were also SOI devices processed on SIMOX substrates.

New results on the noise phenomena considered are presented in this paper. It is shown that, though much of the behaviour of the noise is understandable, a lot of problems remains unsolved.

2 Results and Discussion

1. It has been found that the behaviour of the turn-over frequency f_0 and the plateau level $S_I(0)$ in GR noise spectra measured in the EM *n*MOSFETs is identical to that observed in the DM *p*MOSFETs under inversion conditions. This suggests that the same mechanism is responsible for the noise in both cases.

2. The properties of f_0 and $S_I(0)$ for this noise show specific features (Fig. 1) that are typical of the noise considered and differ significantly from the characteristics of the GR noise observed in the same devices in the absence of inversion.

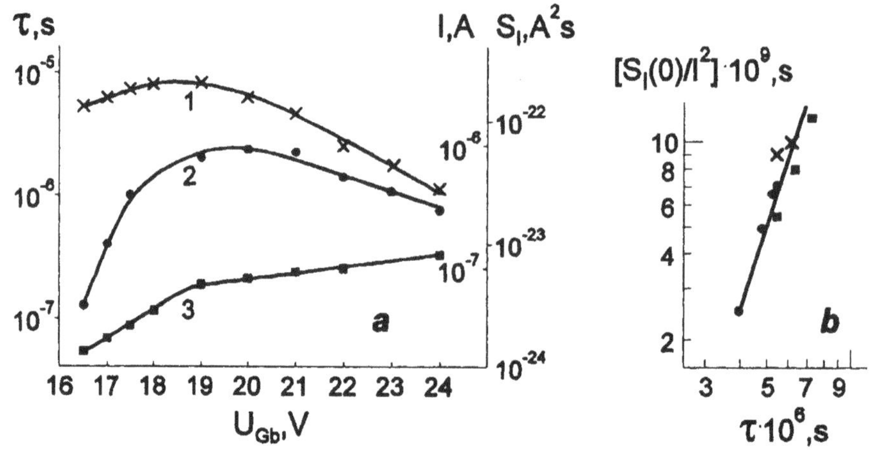

Figure 1: The dependences of τ (1), $S_I(0)$ (2) and I (3) on the back gate voltage at $U_{gf}=0$ (*a*) and the dependence of $S_I(0)/I^2$ on τ measured in the subthreshold regime (*b*) for the EM SOI *n*MOSFET

The increase of the noise with increasing gate voltage under weak inversion conditions is described by the relation:

$$S_I(0)/I^2 \sim \tau^3 \tag{1}$$

where $\tau = (2\pi f_0)^{-1} \sim \exp(\beta_\tau U_G)$ and the value of β_τ is much smaller than for the GR noise observed in the absence of inversion. For example, $\beta_{\tau f} = (4.7 \div 9)V^{-1}$ and $\beta_{\tau b} \approx 0.5V^{-1}$ for inversion in the front and back interface, respectively, while $\beta_{\tau f} \approx 25V^{-1}$ and $\beta_{\tau b} \approx 2.5V^{-1}$ for the front and back interface GR noises in the absence of inversion in *p*MOSFETs studied[6]. For the EM *n*MOSFETs investigated we obtained $\beta_{\tau b} = (0.26 \div 0.34)V^{-1}$ in the subthreshold regime.

Note that the dependence (1) is not typical for the GR noise. In the most cases the dependences $S_I(0)/I^2 \sim \tau^a$ where $a \leq 2$ are observed.

The behaviour of $S_I(0)$ and τ observed in strong inversion appears to be of two types: (i) both values do not change with the gate voltage; (ii) both values decrease and this decrease can be described by the following relations:

$$S_I(0)/I^2 \sim \tau \sim \exp\left(-b_{f,b} U_{G_{f,b}}\right) \tag{2}$$

In the DM *p*MOSFETs the decrease with $b = b_f = (1.1 \div 1.4)V^{-1}$ was observed for the front interface. In the EM *n*MOSFETs such decrease with $b = b_b = 0.5V^{-1}$ has been found for the back interface.

Note that this decrease of $S_I(0)$ in DM *p*MOSFETs might be attributed to the screening effect. However, the fact that the same decrease is observed in the EM *n*MOSFETs (where the interface noise influences the channel current without any screening) shows that this is not the case.

3. Considering that the inversion conditions near the interface (front or back) are responsible for the increase of the noise, the GR processes taking place on centers near the interface can be proposed as a possible source of the noise studied. However, because the screening effect does not suppress this noise even at very low frequencies (~ 1Hz), these centers can not be located at the interface itself. Then it remains to suppose that they are located in the depletion region below the inversion layer. Note that in this case the value of τ has to be independent of the gate voltage if the centers are distributed uniformly over the Si film thickness[7]. Then the fact that $\tau = \tau(U_G)$ suggests that the centers are located in some more or less narrow layer. The increase of τ in weak inversion can be explained by that the electron exchange takes place between the centers and the *v*-band and $\tau \sim 1/p$ where p is the hole concentration near the centers (the value of p decreases with increasing gate voltage). The decrease of τ in strong inversion can be due to increasing exchange between the centers and the *c*-band.

4. An interesting properties of the phenomenon considered has been revealed by measuring the $1/f$ noise in the DM *p*MOSFETs prepared on SOS substrates. These measurements were carried out on wafer level (i.e. without dicing the silicon wafer into chips) by using the dedicated home-made system with the special noise probes. This permits to study the distribution of the noise across the wafer area. The

results are shown in Fig.2. As is seen, the high scatter in the data for different samples is typical for the noise under inversion conditions whereas no

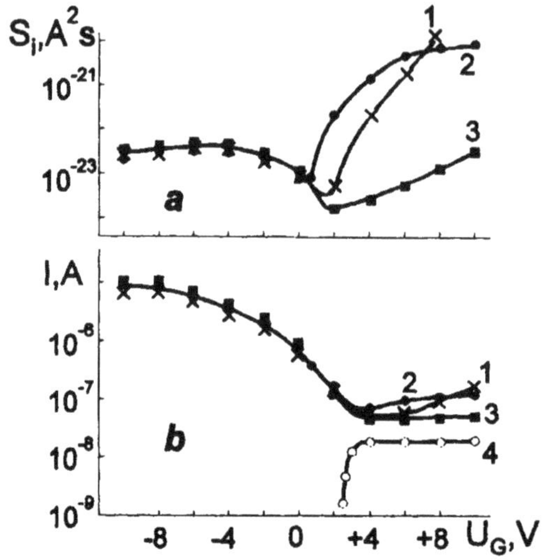

Figure 2: The dependences of the spectral density of $1/f$ noise (a) and of the current (b) on the gate voltage for three DM pMOSFETs (1,2 and 3) located at different regions of the SOS wafer and for one EM nMOSFET (4) located near the pMOSFET characteristics of which are shown by curves 2

dispersion of the experimental points is practically observed if the interface is depleted or even accumulated. What is more, the correlation is observed between the behaviour of the noise and of the current under strong inversion, namely: (i) the higher is the noise, the higher is the current; (ii) in those devices were the increase of the noise in strong inversion is too high the channel current appears to rise with the increasing gate voltage. In addition, when measuring the characteristics of the EM nMOSFETs prepared on the same SOS wafer where the DM pMOSFETs are situated, some non-trivial properties have been found in the dependences $I(U_G)$ measured at a small drain voltage in those nMOSFETs that are located near the pMOSFETs with too high $1/f$ noise in inversion. This is shown by curve 4 in Fig.2b where the saturation portion is observed in strong inversion ($U_G > 3.5$V) instead of a trivial linear increase of I with U_G.

3 Conclusions

It can be concluded that the noise phenomenon studied is observed in a wide variety of devices of different types, i.e. it is typical not for the devices of some special type but for the definite physical situation that takes place when the inversion occurs at the interface between the silicon of *p*-type and the oxide. This means that the effect considered is of a rather general nature. However, there is no full explanation of this effect till now.

It is shown that the physical reason for the phenomenon discussed may be the fluctuations of the charge at some centers in the depletion layer just below the inversion layer. However, in this case it is necessary to suppose that the concentration of these centers drops rapidly with increasing distance from the inversion layer. What is the nature of such centers? The other questions arised also and wait for their answer. They are the following. What physical situation may be responsible for the dependence (1)? Are the above $1/f$ noise and GR noise really of the same nature? Why, under conditions of high $1/f$ noise studied, the channel current in DM *p*MOSFETs can increase with increasing gate voltage and the channel current in EM *n*MOSFETs can become constant? Is not the reason of the well known high level $1/f$ noise in the EM *n*MOSFETs just that one which gives rise to the increase of such a noise under inversion conditions in the depletion mode *p*MOSFETs?

References

1. T.Elewa *et al*, *IEEE Trans. on Electron. Dev.* **ED-38**, 323, 1991
2. B.K.Jones and G.P.Taylor, *Solid-St. Electron.* **35(9)**, 1285, 1992
3. K.Amberiadis and A.Van der Ziel, *Solid-St. Electron.* **26(10)**, 1009, 1983
4. N.B.Lukyanchikova, N.P.Garbar and M.V.Petrichuk, In: Proc. 12th Int. Conf. on Noise in Physical Systems and $1/f$ Fluctuations, 1993, St.Louis, pp.374-377
5. N.B.Lukyanchikova *et al*, In: Proc. 7th Vilnius Conf. on Fluctuation Phenomena in Physical Systems, 1994, Palanga, pp.384-389
6. N.B.Lukyanchikova *et al*, In: Proc. 13th Int. Conf. on Noise in Physical Systems and $1/f$ Fluctuations, 1995, Palanga, pp.422-425
7. N.B.Lukyanchikova *Fluctuation Phenomena in Semiconductor Materials and Devices* (Radio i Svjaz', Moskow, 1990)

NOISE IN NONLINEAR DYNAMICAL SYSTEMS

SOME UNSOLVED PROBLEMS ON THE LEVEL CROSSING OF RANDOM PROCESSES

Tsutomu MUNAKATA
Tamagawa University, 6-1-1 Tamagawa-Gakuenn, Machida-shi, Tokyo,JAPAN

Some unsolved problems regarding to the level crossings of random processes will be presented. The works on the method of approximation to the probability densities of level crossing intervals are also reviewed.

1 Introduction

The level crossing problem in the mathematical theory of noise is to determine the distribution of the intervals between level crossing time points. This problem was firstly discussed by S.O.Rice [1,2] in the statistical communication theory, and it can be also found in other fields : oceanography, speech analysis, seismology, biological systems and Laser optics. In most of studies the level crossing of stationary Gaussian processes was discussed. But in application fields many studies are available for Rayleigh, Rice and K0 processes as the models of fading channels and speech signals[2,5,13,15,16].

An important quantity in the level crossing problem is the probability density or the variance of the level crossing interval length[3,6,8,9]. Up to now, none of them are known in analytical form. In field of communication theory, determination of the probability densities of the level crossing intervals was required, and some approximations are known[4,7,10,11,12]. These approximations are derived by applying certain assumptions on the statistical dependences between successive intervals.

On the other hand the problem was further extended to the case of two different levels or to the crossing with the barrior which changes as a time function. They are treated as a first passage time problems[13,14,15], and there are so many unsolved problems in this field.

In this paper some unsolved problems related to the crossings with constant level will be presented, and also the works on approximations or experimental approaches will be summarized.

2 Definition of the Problem

Let x(t) be a stationary random process. The process x(t) crosses the fixed level I upward or downward according to the time development. The time intervals between successive crossing points are called level crossing intervals. As it is shown in Fig.1, level crossing intervals are denoted as $\tau+$ or $\tau-$ depending on the start conditions, wheather the interval starts with upward crossing or downward crossing, resp.. In the same manner the crossing intervals between successive upward (downward) crossings are denoted as $\tau{+}{+}$, ($\tau{-}{-}$). The expected number of crossings in a unit time is called crossing rate, and it is written as NI. For the level crossing intervals, mean value, variance and probability density of them are the important quantities, but mostly they are not yet given analytically.

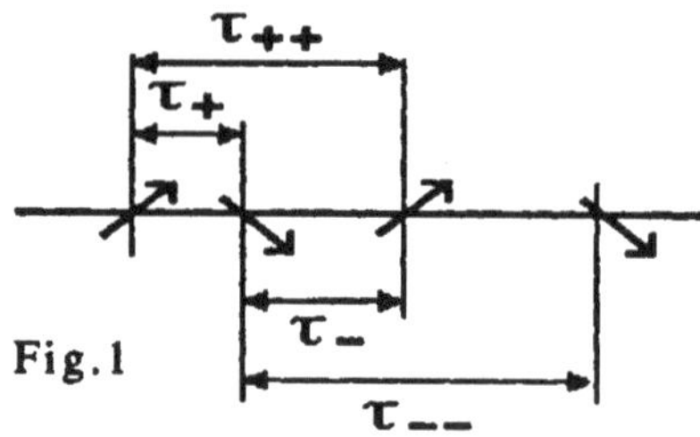

Fig.1

3 Solved and Unsolved Problems, and Known Results

3.1 Crossing rate NI. (solved)

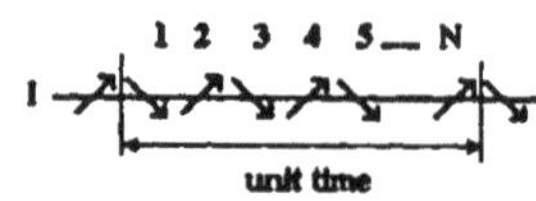

Fig.2

The expected number of crossing points, which are given by crossing the process with boundary upward or downword in a unit time, is called crossing rate, and it is written as NI. Crossing rate can be calculated from a joint probability density p(x,y) for the process x(t) and its derivatives y(t) as follows;

$$N_I = \int_{-\infty}^{\infty} |y|\, p(x=I,y)\,dy \tag{1}$$

For Gauusian processes this quantity is derived directly from the values of auto correlation function (ACF) of the processe and its derivatives at t=0, and it is given as;

$$N_I = 1/\pi\,(-\psi''(0)/\psi(0))^{1/2} \exp[-I^2/(2\psi(0))] \tag{2}$$

For many other types of random processes, crossing rates can be also calculated, and the following is an example for Rayleigh processes.

$$N_R = (2\pi)^{-1/2}(-\psi''(0))^{1/2}\, R/\psi(0) \exp[-R^2/(2\psi(0))] \tag{3}$$

3.2 Mean value of crossing intervals.

Case of single Level. (solved). For the crossings with single level, mean value of the crossing intervals can be calculated as the reciprocal of crossing rate. Let $\mu+$, $\mu-$, $\mu{+}{+}$ and $\mu{-}{-}$ be the mean values of the crossing intervals $\tau+$, $\tau-$, $\tau{+}{+}$ and $\tau{-}{-}$, resp., and they are given as

$$\mu_{+} = \langle\tau_{+}\rangle = P(I)\,\mu_{++} = P(I)\,2/N_I \tag{4}$$

$$\mu_{-} = \langle\tau_{-}\rangle = (1-P(I))\,\mu_{++} = (1-P(I))\,2/N_I \tag{5}$$

$$\mu_{++} = \langle\tau_{++}\rangle = \mu_{--} = \langle\tau_{--}\rangle = 2/N_I \tag{6}$$

where P(I) is the probability that the process x(t) stays above the level I.

Case of two levels. (unsolved) If we consider the intervals of crossings with two different levels, the problm becomes little bit complicated. Several types of intervals can be available by the combination of the crossing points on both levels. Here we will add the several new definitions for such types . As it is shown in Fig. 3, four different intervals are available by the combination of polarity of crossing points on both levels, and they are denoted as τ_{++},τ_{+-},τ_{-+} and τ_{--}. Notice that, in this case the process is permitted to have several times of crossing with level I_1 before it arrives to level I_2. This kind of crossing intervals are called "Usual first passage time".

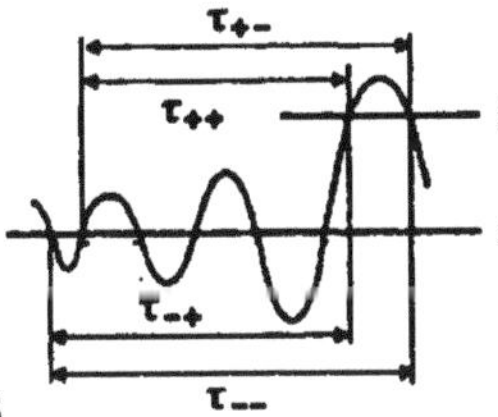

Fig.3

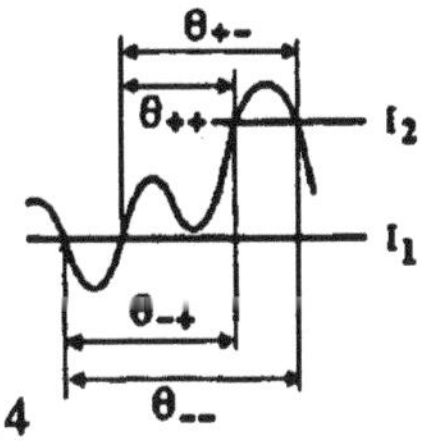

Fig.4

In the other hand, as it is shown in Fig. 4, if we take notice only of the shortest crossing intervals between both levels, other four different intervals can be defined by the combination of the polarity of the crossing points on each level. This kind of crossing intervals are called "Direct first passage time", and they are denoted by θ_{++}, θ_{+-}, θ_{-+} and θ_{--}. For these newly introduced intervals, mean values have not yet be calculated analytically.

3.3 Variance of the crossing intervals. (unsolved)

Let σ_+^2, σ_-^2, σ_{++}^2 and σ_{--}^2 be the variances of the crossing intervals τ_+,τ_-,τ_{++} and τ_{--}, resp., of single crossing level. Actually these quantities have not yet been known analytically. Also for the case of two levels, as a matter of course, none of the values for the variance of first passege times have been found analytically.

In the other hand, the properties of those quantities have been studied by experimental measurement. Such results were accumulated widely for many types of random processes. Some typical results are given in Fig.5,[8,9].

Fig.5

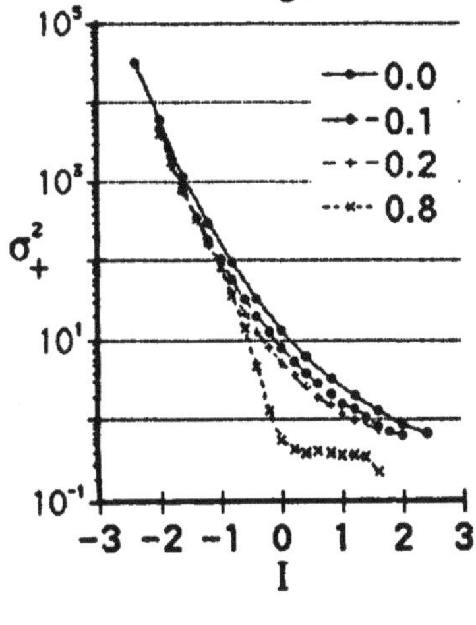

3.4 Correlations between successive crossing intervals. (unsolved)

As it is given in Fig. 6, the correlation between successive crossing intervals can be discussed[9], and the correlation between nth and n+i+1 th crossing intervals is defined as

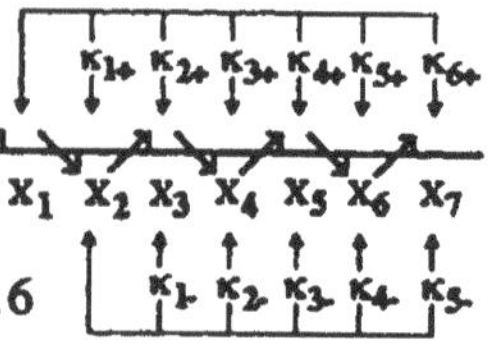

Fig.6

$$\kappa_{i+} = [E\{X_{n+}X_{n+i-}\} - \mu_+\mu_-]/(\sigma_+\sigma_-) \quad \text{for } i=1,3,5,\cdots \tag{7}$$

$$\kappa_{i+} = [E\{X_{n+}X_{n+i+}\} - \mu_+^2]/\sigma_+^2 \quad \text{for } i=2,4,\cdots \tag{8}$$

$$\kappa_{i-} = [E\{X_{n-}X_{n+i+}\} - \mu_+\mu_-]/(\sigma_+\sigma_-) \quad \text{for } i=1,3,5,\cdots \tag{9}$$

$$\kappa_{i-} = [E\{X_{n-}X_{n+i-}\} - \mu_-^2]/\sigma_-^2 \quad \text{for } i=2,4,\cdots \tag{10}$$

These quantities have not yet been known analytically. Difficulty of this problem is just same as those of variance. Since each value of κ_i is unknown, the following two relation have been deriveds between the sums of κ_i and the variance $\sigma+$ and $\sigma-$.

$$A/\beta = (1-a)^2\sigma_+^2\Big(1+2\sum_{i=1}^{\infty}\kappa_{2i+}\Big) + a^2\sigma_-^2\Big(1+2\sum_{i=1}^{\infty}\kappa_{2i-}\Big) - 4a(1-a)\,\sigma_+\sigma_-\sum_{i=0}^{\infty}\kappa_{2i+1+} \tag{11}$$

$$4(1+2B)/\beta^2 = \sigma_+^2\Big(1+2\sum_{i=1}^{\infty}\kappa_{2i+}\Big) + \sigma_-^2\Big(1+2\sum_{i=1}^{\infty}\kappa_{2i-}\Big) + 4\sigma_+\sigma_-\sum_{i=0}^{\infty}\kappa_{2i+1+} \tag{12}$$

where

$$A = \int_0^{\infty} [\, r(t,I) - (1-2a)^2 \,]\, dt \tag{13}$$

$$B = \int_0^{\infty} [\, W_+(t,I) - \beta/2 \,]\, dt \tag{14}$$

$\beta = N_I$, and $a = P(I)$: probability that the process is found upper the level I. $r(t,I)$ denotes the auto-correlation function of bivalued clipped random process given as 1 for $x(t) >= I$ and -1 for $x(t) < I$.

The properties of quantities κ_i s have been studied experimentally, and Fig. 7 show the typical result of the change of κ_i vs. level I for the Gaussian process with several bandwidth ($k=f_l/f_h$) of power spectrum,[8,9]. For the case of crossing with two levels, this kind of correlation problem has not yet been discussed.

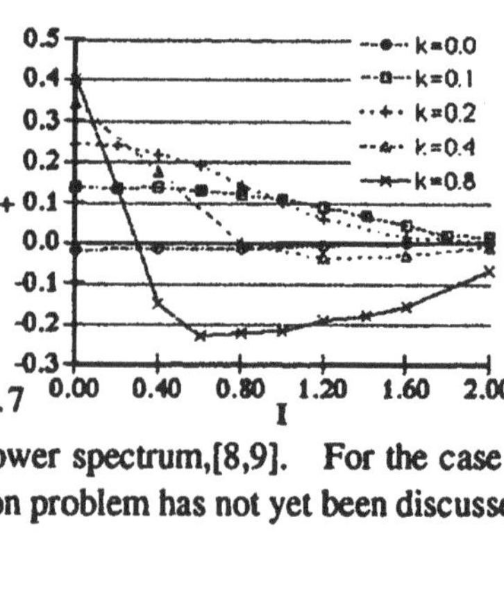

Fig.7

3.5 Probability densities of crossing intervals. (unsolved)

Probability densities of crossing intervals are the main things of this kind of research. In addition to the crossing intervals $\tau+$ and $\tau-$, the sum of successive intervals are introduced. The sum of n+1 successive crossing intervals is dnoted by τ_{n+} or τ_{n-}, here the sign + or - corresponds to the polarity of crossing at the biginning of the interval. The probability density of the intervals τ_{n+} and τ_{n-} are denoted as $P_{n+}(\tau,I)$ and $P_{n-}(\tau,I)$,resp.. In most studies $P_0(\tau,I)$ and $P_1(\tau,I)$ are discussed mainly rather than other $P_n(\tau,I)$, and the calculation of these probability densities are not yet be overcome except special cases. Probability density $P_0(\tau,I)$ can be expressed as the infinite series of multiple integrals, but it is enough difficult to calculate each multiple integrals. To avoid this difficulty, several methods of approximation were proposed, and they will be refered later.

In the other hand, on the results of experimental measurements, many interesting things were observed relating to the probability densities $P_0(\tau,I)$ or $P_n(\tau,I)$. One of the interesting things is a number of peaks on the curve. In many cases, if the value I of level is large, several number of peaks can be observed on the curve of probability densities $P_n(\tau,I)$, and it is called "Multi-Peak property". Typical results of multipeak are found in Fig.8. Another interesting thing is a decay property of the probability density. In the range of large τ, $P_n(\tau,I)$ is expected to decay exponentially, and the factor of decay is known by experiment, and some analytical approximations are known.

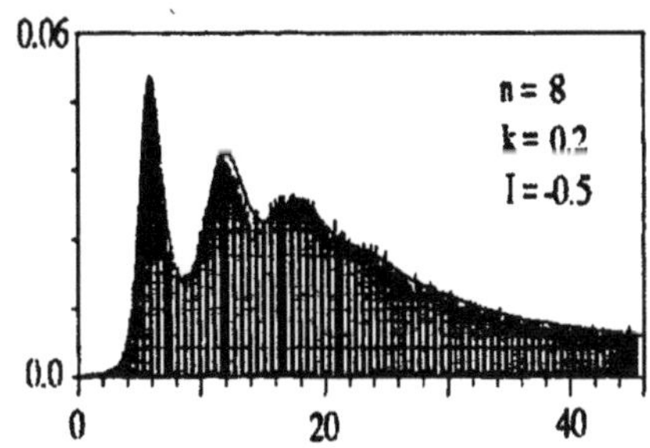

Fig. 8

4 Approximations of $P_0(\tau,I)$ AND $P_1(\tau,I)$

4.1 Rice function (S.O.Rice) [1,2]

Rice functions are defined as follows: $Q_+(\tau,I)$ and $W_+(\tau,I)$ are the conditional probability densities that the process crosses the level I downward or upward, resp., in infinitesimal interval $[t_1+\tau, t_1+\tau+dt]$, given an upward crossing with level I at t_1. Notice that, from the definition, the Rice functions describe only the crossings at time point t_1 and $t_1+\tau$, and the crossings happen in the interval τ are not considerd. From this reason Rice functions may correspond not to $P_0(\tau,I)$, or $P_1(\tau,I)$ but to the sum of $P_n(\tau,I)$. Following relations are known between these quantities.

$$Q_+(\tau,I) = \sum_{l=0}^{\infty} P_{2l,+}(\tau,I) \quad (15) \qquad W_+(\tau,I) = \sum_{l=0}^{\infty} P_{2l+1,+}(\tau,I) \quad (16)$$

Rice functions are able to approximate $P_0(t,I)$ and $P_1(t,I)$ only for small τ.

4. 2 Quasi independent model. (McFadden) [4]

Quasi-independent assumption means that each of the crossing intervals τ+ and τ- is statistically independent of the sum of following 2n+2 level crossing intervals, (n=0,1,2,...). Under quasi independent assumption, $P_0(\tau,I)$ and $P_1(\tau,I)$ are given as the solution of following integral equations;

$$P_0(\tau,I) = Q_+(\tau,I) - P_0(\tau,I) * W_+(\tau,I) \tag{17}$$

$$P_1(\tau,I) = W_+(\tau,I) - P_1(\tau,I) * W_+(\tau,I) \tag{18}$$

This approximation has following difficulties;

1) In many cases the value of solution goes minus in some portion of time range, and it is not acceptable for the approximation of probability density.

2) Results of experimental measurements for the correlations between intervals are contradictory to this assumption.

As an Extension of this kind of approach, geschwaecherte (weakened) quasi independent model (Wolf and Brehm)[7] is available, and more precise approximations can be obtained. This method employs still complicated triple integral for joint probability densities, and it is not so easy to calculate for non-Gaussian processes.

4.3 Multi state model, 4 state and 6 state models. (Munakata) [10,11,12]

As it is illustrated in Fig.9, the 4-state model is written by the 4 signal states of the process defined as follows:

Given an upward crossing of level I at t_1, then the process in

Z_1 remains above the level I at least until the time $t_1+\tau$;

Z_2 is found below the level I at the time $t_1+\tau$ after just one subsequent downward crossing;

Z_3 is found above the level I at time $t_1+\tau$ after at least one subsequent upward crossing during the time interval τ;

Z_4 is found below the level I at time $t_1+\tau$ after at least two subsequent downward crossings during the time interval τ.

The process, which is found in state Z_1 by having un upward crossing at t_1, proceeds to other states like Z_1-> Z_2 -> Z_3 -> Z_4 -> Z_3 -> Z_4 -> ... by counting new crossing. The change of each state probability is written by following differential equations;

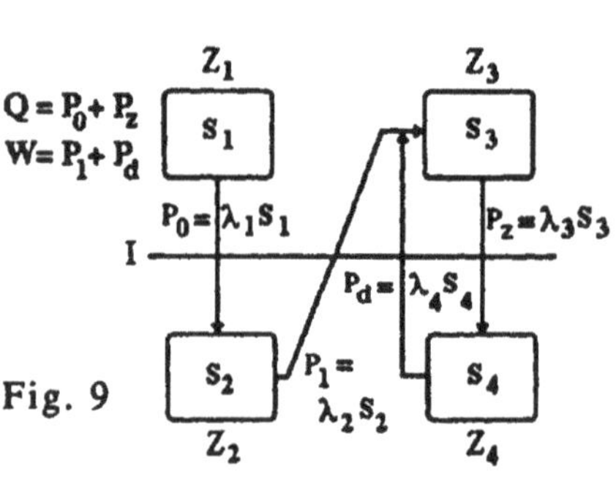

Fig. 9

$$
\begin{aligned}
dS_1(t)/dt &= -\lambda_1 S_1(t) = -P_{0+}(t) \\
dS_2(t)/dt &= \lambda_1 S_1(t) - \lambda_2 S_2(t) = P_{0+}(t) - P_{1+}(t) \\
dS_3(t)/dt &= \lambda_2 S_2(t) + \lambda_4 S_4(t) - \lambda_3 S_3(t) \\
&= P_{1+}(t) - P_z(t) + P_d(t) = W_+(t) - Q_+(t) + P_{0+}(t) \\
dS_4(t)/dt &= \lambda_3 S_3(t) - \lambda_4 S_4(t) \\
&= P_z(t) - P_d(t) = Q_+(t) - P_{0+}(t) - W_+(t) + P_{1+}(t)
\end{aligned}
\tag{19}
$$

By assuming

$$\lambda_1 = \lambda_3 = g \qquad \lambda_2 = \lambda_4 = h \tag{20}$$

the quantities g and h can be calculated as

$$g = Q_+(t)/P_+(t) \qquad h = W_+(t)/(1 - P_+(t)) \tag{21}$$

where

$$P_+(t) = S_1(t) + S_3(t) = 1 - \int_0^t [Q_+(t) - W_+(t)]\, dt \tag{22}$$

For the calculation of these quantities only the Rice functions are required. Then S1(t) and S2(t) (or P0+(t) and P1+(t)) can be given by solving following differential equations;

$$dS_1(t)/dt = -g\, S_1(t) = -P_{0+}(t) \tag{23}$$

$$dS_2(t)/dt = g\, S_1(t) - h S_2(t) \tag{24}$$

and they are solved for P0+(t) and P1+(t) as

$$P_{0+}(t) = g(t) \exp[-G(t)] \tag{25}$$

and

$$P_{1+}(t) = h(t) \exp[-H(t)] \int_0^t g(t) \exp[-G(t) + H(t)]\, dt \tag{26}$$

where

$$G(t) = \int_0^t g(t)\, dt \qquad H(t) = \int_0^t h(t)\, dt \tag{27}$$

For large value of lebel |l|, this approximation is enough useful, but for small value of lebel |l|, low precision may be expected. It was required to make some modification to this model.

The 6 state model was proposed as a modification of 4 state model. As it is shown in Fig.10, 6-state model were derived by introducing two relaxation states Z31 and Z41 into the state Z3 and Z4,resp., of 4 state model. After having arrived to Z31 or Z41, the process remains there for a certain time T1 and T2,resp., before it proceeds to Z32 or Z42. Generally such relaxation time must be a random variable, but for simplisity, they are assumed to be constant. State probability S3 and S4 are written as

S3 = S31 + S32 and S4 = S41 + S42, and the relaxation probabilities can be written as

$$S_{31}(t) = \int_{t-T_1}^{t} W_+(t)\,dt \qquad (28)$$

$$S_{41}(t) = \int_{t-T_2}^{t} [Q_+(t) - P_{0+}(t)]\,dt$$

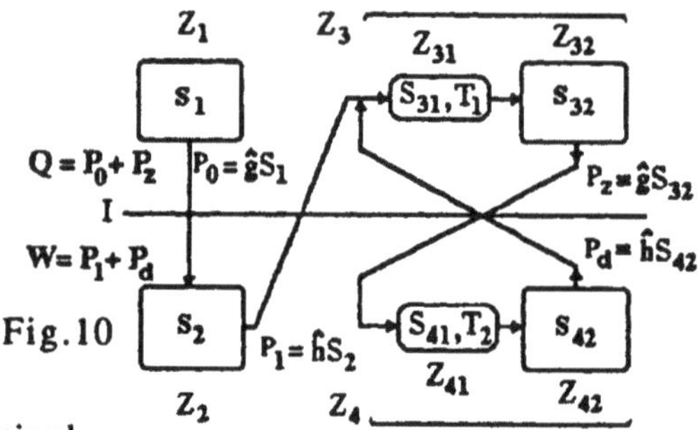

Fig.10

By replacing S3 by S32 and S4 by S42 on right hand of eq.(19), following relations are obtained,

$$P_{0+}(t) = S_1(t)\,\hat{g} = S_1(t)Q_+(t)/[S_1(t)+S_{32}(t)]$$
$$= S_1(t)Q_+(t)/[P_+(t)-S_{31}(t)] \qquad (30)$$

$$P_{1+}(t) = S_2(t)\,\hat{h} = S_2(t)\,W_+(t)\,/[S_2(t)+S_{42}(t)]$$
$$= S_2(t)\,W_+(t)\,/[1-P_+(t)-S_{41}(t)] \qquad (31)$$

Only by replacing g by $\hat{g}$ in eq.(23) and h by $\hat{h}$ in eq.(24), solutions of 6 state model are obtaind as

$$P^{(6)}_{0+}(t;T_1) = \hat{g}(t;T_1)\exp[-\hat{G}(t;T_1)] \qquad (32)$$

$$P^{(6)}_{1+}(t;T_1,T_2) = \hat{h}(t;T_1,T_2)\exp[-\hat{H}(t;T_1,T_2)]\cdot \qquad (33)$$
$$\cdot\int_0^t \hat{g}(t;T_1)\exp[-\hat{G}(t;T_1)+\hat{H}(t;T_1,T_2)]\,dt$$

where

$$\hat{G}(t;T_1) = \int_0^t \hat{g}(t;T_1)\,dt \qquad (34)$$

and

$$\hat{H}(t;T_1,T_2) = \int_0^t \hat{h}(t;T_1,T_2)\,dt \qquad (35)$$

Since these solutions are still function of the relaxation times T_1 and T_2, one must found the metod to determine these parameters. For this purpose, by adjustong T_1 and T_2 the first moment of $P^{(6)}_{0+}(t;T1)$ and $P^{(6)}_{1+}(t;T1,T2)$ is set to be equal to the value m_0 or m_1, the first moment of $P_{0+}(t)$ and $P_{1+}(t)$, which can be given as the mean value of crossing intervals. The first moments of of the solutions are given as

$$m_0(T_1) = \int_0^\infty t\,P^{(6)}_{0+}(t;T_1)\,dt = \int_0^\infty \exp[-\hat{G}(t;T_1)]\,dt \qquad (36)$$

and

$$m_1(T_1,T_2) = \int_0^\infty t\,P^{(6)}_{1+}(t;T_1,T_2)\,dt = m_0(T_1) + \qquad (37)$$
$$+\int_0^\infty \exp[-\hat{H}(t;T_1,T_2)]\cdot\int_0^t P^{(6)}_{0+}(u;T_1)\exp[\hat{H}(u;T_1,T_2)]\,du\,dt$$

Only Rice functions Q , W and mean values μ_+, μ_- of the crossing intervals are required for the calculations. From this reason it is very easy to apply the model for non-Gaussian processes. Fairly good approximatios are obtained for many cases.

5 Conclusion

Some unsolved problems on the level crossing of random processes are summarized. These problems are simple to state and easy to understand, but they are still unsolved after half a century. In many cases the only reliable information that we have about these problems are experimental results. We need further attention to this field and more intensive investigations.

References

1. S.O.Rice, The Mathematical Analysis of Random Noise, B.S.T.J., 24,1945, 46-156
2. S.O.Rice, Distribution of The Duration of Fades in Radio Transmissions, B.S.T.J.,37,1958,581-635
3. M.S.Longuett-Higgins, The Distribution of Intervals Between Zeros of Stationary Random Function, Phil.Transact.Roy.Soc.,A254,1962,557-599
4. J.A.McFadden,The Axis-Crossing Intervals of Random Functions II, IRE Trans. Inform. Theory, IT-4 1958,14-24
5. A.J.Rainal, Axis-Crossing Intervals of Rayleigh Processes, B.S.T.J.,44,1965, 1219-1224
6. T.Mimaki, Zero-crossing intervals of Gaussian processes, J.Appl. Phys., 44, 1973, p.477
7. D.Wolf and H.Brehm, Die Verteilungsdichte der Zeitintervalle Zwischen Null-durchgaengen bei Gaussian Stochastischen Signalen, AEU,27,1973, 477-489
8. T.Mimaki and T.Munakata, Experimental Results on the Level-Crossing Intervals Of Gaussian Processes, Correspondence,IEEE.Trans. on Inf. Theory, IT-24 1978, 515-520
9. T.Mimaki, T.Munakata and D.Wolf, On the Correlation between Level-Crossing Time Interval Length of Random Process, The Trans of the IECE of Japan, Sept. 1979, Vol. E62 No. 9, 583-589
10. T.Munakata and D.Wolf, Neue Theoretische Loesungen fuer das Problem der Pegel-Ueber-schreitungen bei Gaussschen Prozessen, 4.Aachener Kolloquium "Theorie und Anwendung " RWTH Aachen, 1981,151-154
11. T.Munakata and D.Wolf, On the distribution of the level-crossing time intervals of random processes, Proc. 7th Int. Conf.on Noise in Physical Systems, North-Holl. and Publ. Co., Editor M.Savelli, Montpellier, 1983, 49-52
12. T.Munakata, Mehr Zustaende Modelle zur Beschreibung des Pegelkreuzungs-verhaltens Stationaerer Stochastischer Prozesse, Inaugural Dissertation, Fachbereiches Physik, J.W.Goethe Universitaet Frankfurt am Main,1986
13. A.J.Rainal, First ans Second Passage Times of Rayleigh Processes, IEEE Trans Inform. Theory vol IT-33,419-425, May 1985

14. H.Perez, T.Kawabata and T.Mimaki, First Passage Time Intervals of Gaussian Processes,JJAP,26,No.8,1987,1378-1383
15. T.Munakata and T.Mimaki, First Passage Time Intervals of Rayleigh Processes, AEU Band 43 Heft 5 Sept./Okt. 1989 , 278-283
16. N.Youssef, T.Munakata and T.Mimaki, Experimental Results on the Axis-Crossing of the Phase of Sine Wave Plus Noise, Proc. 12th Symposium on Information Theory and Its Applications (SITA89) Inuyama,Dec. 1989
17. N.Youssef, T.Munakata and M.Takeda, Rice probability functions for level-crossing intervals of speckle intensity fields, Optics Communications 123 (1996) 55-62.
18. L.Gammaitoni, F.Marchesoni, E.Menichella-Saetta and S.Santucci, Zero-crossing distributions for smoothed random signals , Physics Letters A 158 (1991) 9-13, North-Holland
19. J.Abrahams, A survey of recent progress on level-crossing problems for random processes, in Communications and Networks, A Survey of Recent Advance, I.F.Blake and H.V.Poor, eds. (Springer-Verlag, New York, 1986)
20. I.F.Blake and W.C.Lindsey, Level-crossing problems for random processes, IEEE Trans. Inf. Theory IT-19, 295-315 (1973)

STOCHASTIC RESONANCE AT PHASE NOISE

K. LOERINCZ, G. BALAZSI[$], Z. GINGL, [+]L.B. KISS
JATE University, Dept. of Experimental Physics, Dóm tér 9, Szeged, H-6720 Hungary
[$] *Babes-Bolyai University of Cluj-Napoca, Kogalniceau street 1., Cluj-Napoca, 3400 Romania*
[+]*Uppsala University, Angstrom Lab, Box 534, Uppsala, S75121 Sweden*

A surprising new stochastic resonance phenomenon is reported. The particular set-up is a level-crossing detector with a supra-threshold sinusoidal excitation, and the noise is an additive band-limited white noise in the phase angle of the sinusoidal time-function. We observe stochastic resonance phenomenon in the first harmonic at the output power spectral density of the system. It means that, if the phase noise is not zero, there is an optimal strength of that phase noise where the signal to noise ratio reaches a local maximum. At the moment, there is no theory to explain this phenomenon.

1 Introduction

Stochastic resonance (SR) is a widely investigated phenomenon of statistical and solid state physics. SR can occurs in special nonlinear systems, when one can identify a periodic and noisy input excitation. SR means that the signal-to-noise ratio (SNR) at the output has a maximum as a function of the intensity of the input noise [1]. There are several real and many artificial systems, which can produce stochastic resonance [1], for example SQUIDS, noisy neural networks, lasers, level-crossing detectors [2-6], Schmitt-triggers and a lot of different mathematical models. First, SR was observed and explained in dynamical, two-state systems, however, later it was shown, that the real meaning of SR is different: SR is basically a level-crossing dynamical problem of the noisy signal [5]. Recently a promising possibility of increasing the output SNR over the input SNR in level-crossing detectors has been reported [5,6], highlighting the application possibilities of SR.

Previously, input signals used for stochastic resonators were usually additive, sometimes multiplicative, with the input signal. Here, we introduce a new type of stochastic resonance based on random phase modulation of the periodic input signal.

2 Model

The principle of the new stochastic resonator set-up can technically be visualised by a phase modulator followed by a level-crossing detector (LCD) [2], see Figure 1. The input of the phase modulator is a sine wave with frequency f_0 and the modulating input is fed by a band-limited white noise serving as the phase modulating quantity. For simplicity, the amplitude distribution of the phase noise has been chosen to be uniform. The phase modulator adds the noise to the phase of the input sine wave which can be expressed in the following way:

$$y(t) = A \sin[2\pi f_0 t + w(t)] \tag{1}$$

where w(t) represents the phase noise.

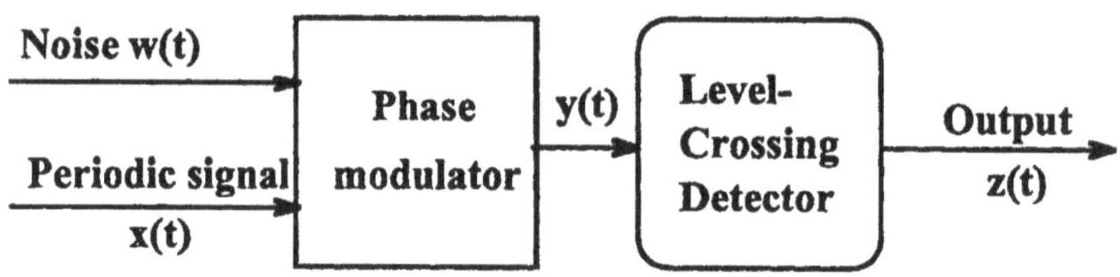

Figure 1: Visualisation of the set-up for stochastic resonance at phase-noise.

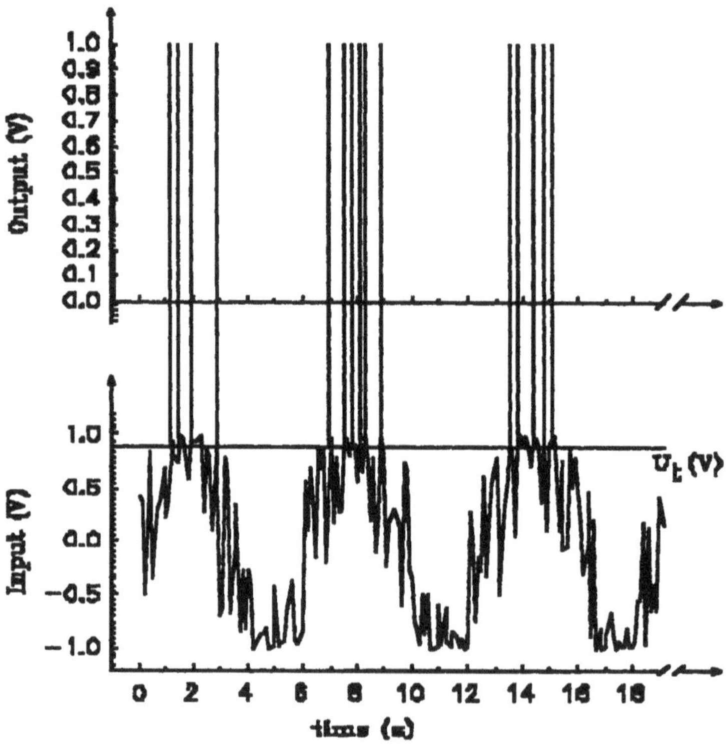

Figure 2: An example for the input and output amplitudes

The phase modulated sine wave gets into an LCD to produce the output signal. The LCD emits a pulse, whenever its input value is crossing the threshold value U_t upwards, and the LCD output is zero otherwise. Figure 2 shows a typical input and the corresponding output signal of the system.

3 Results

We used numerical simulations to investigate the behaviour of the phase modulator based SR. A (high-performance) uniform random number generator was used to represent the input noise. These numbers were used to generate samples of length of 16000. The power spectrum was calculated using FFT and by averaging 1000 samples. An example for the output spectra can be observed on Figure 3.

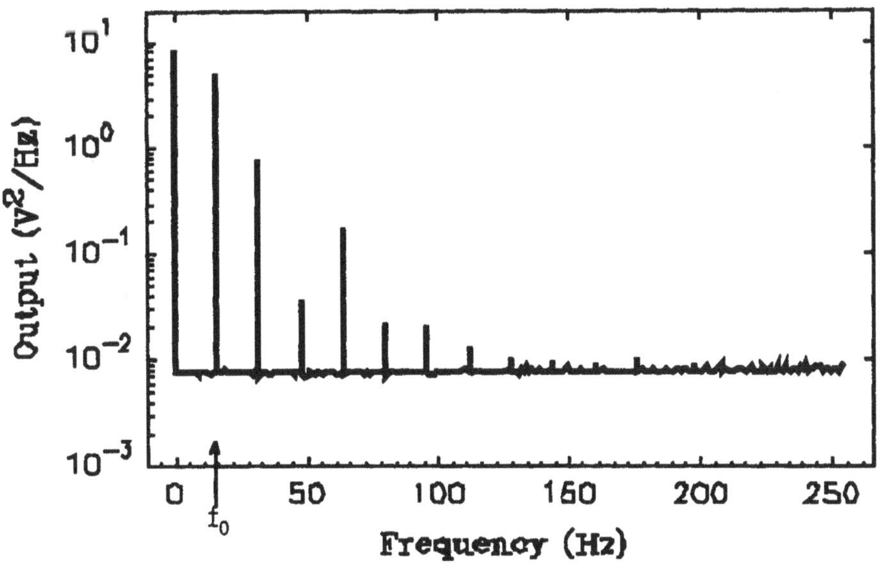

Figure 3:. An example for the output power density spectrum

The amplitude of the sine wave was fixed to unity, and the threshold and noise amplitude were varied. The output signal strength and the SNR values were calculated and plotted as the function of the threshold of the LCD and RMS value of the input noise. This dependence is illustrated on Figure 4. The existence of strong stochastic resonance is very clear from the behaviour of both quantities.

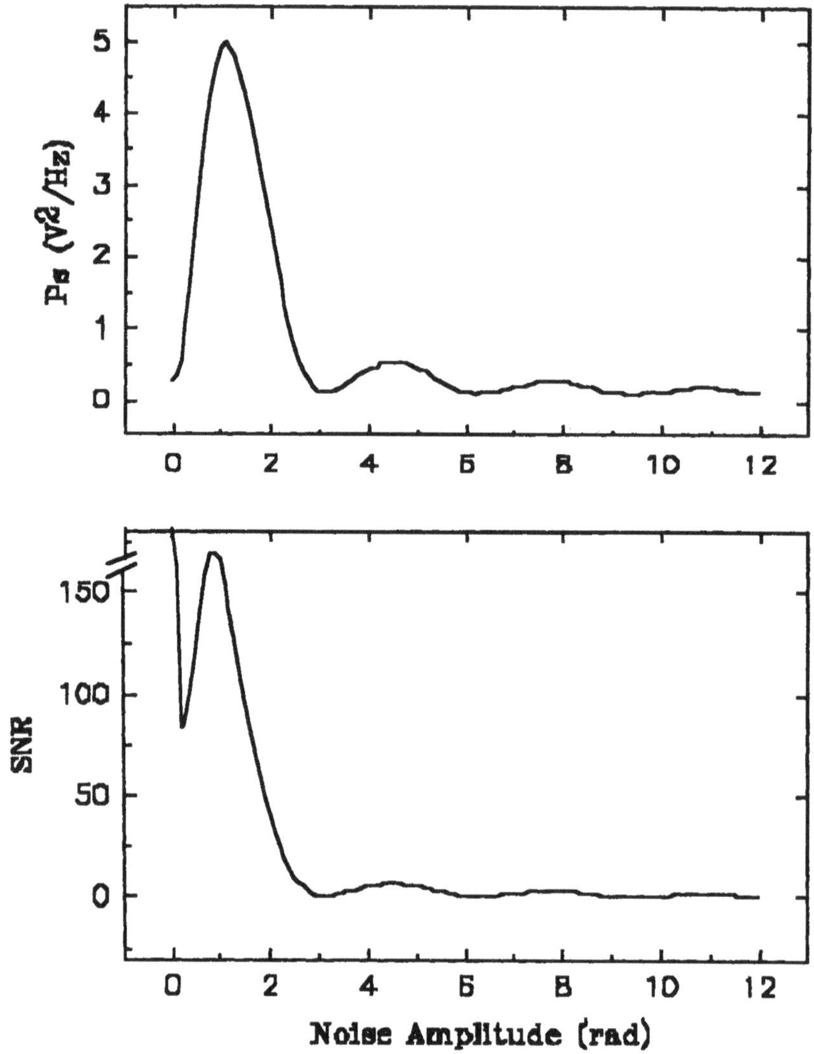

Figure 4: The output mean-square signal amplitude P_s and SNR versus the phase noise amplitude

It is important to note that to get any signal at the output, the system has to work in the *supra-threshold* limit:

$$-A < U_t < A \quad , \tag{2}$$

otherwise the input signal never crosses the threshold. On Figure 5, we can see the output spectrum when the threshold level varies between its limits. This behaviour is completely different from the previously considered SR systems additive noise.

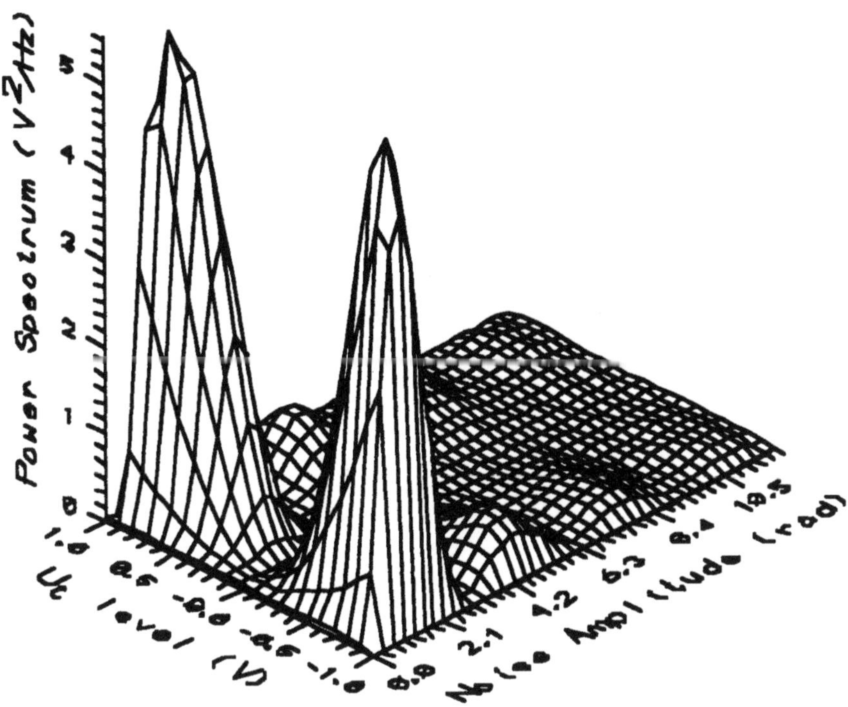

Figure 5: Power Spectrum at the signal frequency with various threshold levels and phase noise

4 Unsolved problems/questions

a. Theory. This is the really unsolved problem here. There is no theoretical explanation of these new results at the moment.

Some open questions:

b. The same effect with different noises and different signals.

c. Importance of this effect at practical applications. LCD systems are often used in the information technology and the occurrence of phase noise is rather general there, too. Whenever the threshold level of the LCD is not zero, stochastic resonance can probably be used to optimise the signal transfer.

Acknowledgements

The research has been supported by OTKA (Hungary) under grants T016342, F014309 and T014290 and MKM (Hungary).

References

1. Proc. of NATO Adv. Research Wksp. *on Stochastic Resonance in Physics and Biology* , eds. F. Moss, A.R. Bulsara and M. F. Shlesinger, *J.Stat.Phys.* **70** (1993)
2. Z. Gingl, L.B. Kiss and F. Moss, *Europhys.Lett.* **29**, 191 (1995);
3. Z. Gingl, L.B. Kiss and F. Moss, in Int. Wksp. on *Fluctuations in Physics and Biology: Stochastic Resonance, Signal Processing and Related Phenomena*, eds, A.R. Bulsara, S. Chillemi, L.B. Kiss, P.V.E. McClintock, R. Mannella, F. Marchesoni, K. Nicolis and K. Wiesenfeld, eds., *Nuovo Cimento* D **17**, 795 (1995)
4. P. Jung, *Phys.Rev.* E **50**,.2513 (1994)
5. L.B. Kiss, in *Chaotic, Fractal, and Nonlinear Signal Processing*, ed. R. Katz, American Institute of Physics Press 1996, p. 382
6. K. Loerincz, Z. Gingl and L.B. Kiss, *Phys.Lett.* A, in press (1996)

STOCHASTIC RESONANCE IN A REACTION-DIFFUSION SYSTEM: ENHANCEMENT DUE TO COUPLING

HORACIO S. WIO AND FERNANDO CASTELPOGGI
Centro Atómico Bariloche (CNEA) and Instituto Balseiro (UNC)
8400 S.C. de Bariloche, Argentina.

Abstract

Stochastic resonance in an extended system is studied in a simple version of a reaction–diffusion (RD) model. The known form of the nonequilibrium potential is exploited to obtain the probability for the decay of the metastable extended states and expressions for the correlation function and the signal-to-noise ratio (SNR). It is shown that the SNR increases with the diffusivity that plays the role of a coupling parameter.

1 Model and Nonequilibrium potential

One of the most fascinating cooperative effects arising out of the interplay between deterministic and random dynamics in a nonlinear system is the phenomenon of *stochastic resonance* (SR).[1] The particular features of this phenomenon in the case of extended systems are still under study, with particularly interesting recent results of numerical simulations of arrays of coupled nonlinear oscillators[2] where it was shown that the coupling enhances the response.

We present here a brief analysis of this phenomenon in a spatially extended system by exploiting our previous results obtained using the notion of the *nonequilibrium potential*[3] in a simple RD model. We shall focus on a one–dimensional, one–component model of an electrothermal instability,[4] that corresponds to an approximation to the continuous limit of the coupled system studied by Lindner *et al.*.[2] For this model we have studied the effect of boundary conditions (b.c.) in pattern selection, the *global stability* of the nonhomogeneous structures, and the critical like behaviour due to the coalescence of two patterns as a control parameter is varied.[4–6]

The particular form of the model that we work with is[4]

$$\partial_t \phi = D\partial^2_{xx}\phi - \phi + \theta(\phi - \phi_c), \qquad (1)$$

in the bounded domain $x \in [-L, L]$ and with Dirichlet b.c. at both ends, i.e. $\phi(\pm L, t) = 0$. Clearly, we have the trivial solution $\phi_0(x) = 0$, which is linearly stable and exists for the whole range of parameters. The piecewise linear approximation of the reaction term, mimicking a cubic polynomial, allows us to find analytical expressions for the spatially symmetric solutions of Eq.(1). In particular we find only one stable nonhomogeneous structure, $\phi_s(x)$, that presents a central excited zone where $\phi_s(x) > \phi_c$. Besides that, we find another similar unstable structure, $\phi_u(x)$, with a smaller central excited zone. This pattern corresponds to the saddle separating both attractors $\phi_0(x)$ and $\phi_s(x)$. There are other unstable nonhomogenous solutions, but playing no role in this problem. [4–6]

The indicated patterns are extrema of the nonequilibrium potential or Lyapunov functional (LF) of our system that reads [5,6]

$$\mathcal{F}[\phi,\phi_c] = \int_{-L}^{+L} \left\{ - \int_0^{\phi} (-\phi + \theta[\phi - \phi_c])\, d\phi . + \frac{D}{2}\left(\frac{\partial \phi}{\partial x}\right)^2 \right\} dx. \quad (2)$$

The functional $\mathcal{F}$ is such that $\frac{\partial \phi}{\partial t} = -\frac{\delta \mathcal{F}}{\delta \phi}$ and, as a good LF, fulfills the condition $\dot{\mathcal{F}} = -\int_{-\infty}^{+\infty} \left(\frac{\delta \mathcal{F}}{\delta \phi}\right)^2 dx \leq 0$. It offers the possibility to study the global stability of the patterns and the changes due to variations of model parameters. [5,6]

In Fig. 1 we depict the LF $\mathcal{F}[\phi,\phi_c]$ evaluated at the stationary patterns ϕ_0 ($\mathcal{F}[\phi_0] = 0$), $\phi_s(x)$ ($\mathcal{F}^s = \mathcal{F}[\phi_s]$) and $\phi_u(x)$ ($\mathcal{F}^u = \mathcal{F}[\phi_u]$), for a system size $L = 1$, as a function of ϕ_c and for two values of D. In the bistable zone, the upper branch of each curve is the LF for $\phi_u(x)$, where $\mathcal{F}$ attains an extremum (as a matter of fact it is a saddle of the nonequilibrium potential). On the lower branch, for $\phi_s(x)$, and also for $\phi_0(x)$, the LF has local minima. For each value of D the curves exist up to a certain critical value of ϕ_c at which both branches collapse. It is interesting to note that, since the LF for $\phi_u(x)$ is always positive and, for $\phi_s(x)$, $\mathcal{F}^s$ is positive for some values of ϕ_c and also $\mathcal{F}^s \to -\infty$ as $\phi_c \to 0$, $\mathcal{F}^s$ vanishes for an intermediate value of $\phi_c = \phi_c^*$, where $\phi_s(x)$ and $\phi_0(x)$ exchange their relative stability.

In order to account for the effect of fluctuations, we need to include in the time-evolution equation of our model (Eq.(1)) a fluctuation term $\xi(x,t)$, modeled as an additive noise source, yielding a stochastic partial differential equation for the random field $\phi(x,t)$. We assume that $\xi(x,t)$ is a Gaussian white noise with zero mean value and a correlation function given by $\langle \xi(x,t)\, \xi(x',t') \rangle = 2\, \gamma\, \delta(t-t')\, \delta(x-x')$, where γ denotes the noise strength.

We now exploit an scheme that allows us to describe the decay of extended metastable states, [7] yielding the following Kramers' like result for the first-passage-time $\langle \tau \rangle$:

$$\langle \tau \rangle = \tau_0 \exp\left\{ \frac{\Delta \mathcal{F}[\phi,\phi_c]}{\gamma} \right\}, \quad (3)$$

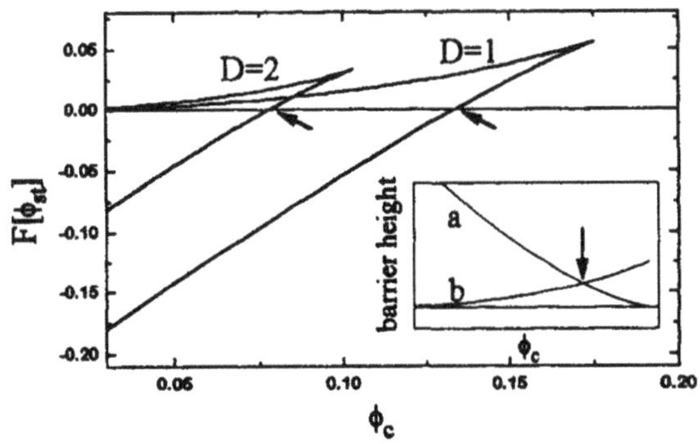

Figure 1: Nonequilibrium potential $\mathcal{F}$, for the stationary patterns, as a function of ϕ_c, for $L = 1$. The bottom curve corresponds to $\phi_s(y)$ and the top one to $\phi_u(y)$. The points ϕ_c^*, are indicated. The insert shows the behaviour of the barrier height as a function of ϕ_c.

where $\Delta\mathcal{F}[\phi,\phi_c] = \mathcal{F}[\phi_{unst}(y),\phi_c] - \mathcal{F}[\phi_{meta}(y),\phi_c]$. The prefactor τ_0 is determined by the curvature of $\mathcal{F}[\phi,\phi_c]$ at its extrema (minima) and is typically several orders of magnitude smaller than the average time $\langle\tau\rangle$. In the insert of Fig. 1 we show the form of $\Delta\mathcal{F}[\phi_0,\phi_c]$ (line (b)) and $\Delta\mathcal{F}[\phi_s,\phi_c]$ (line(a)), as a function of ϕ_c. It also corresponds to the behavior of $\ln(\langle\tau\rangle/\tau_0)$.

2 Stochastic resonance and the effect of coupling

We now assume that, due to an external harmonic variation, the parameter ϕ_c has an oscillatory part $\phi_c(t) = \phi_c^* + \delta\phi_c\cos(\Omega t + \varphi)$. For the spatially extended problem, we need to evaluate the space–time correlation function $\langle\phi(y,t)\phi(y',t')\rangle$. To do this we will use a simplified point of view, based on the two state approach of McNamara and Wiesenfeld, [8] that allows us to apply almost directly some of their results. To proceed with the calculation of the correlation function we need to evaluate the transition probabilities between our two states ϕ_0 and ϕ_s, $W_\pm = \tau_0^{-1}\exp\left(-\Delta\mathcal{F}[\phi,\phi_c]/\gamma\right)$, where $\Delta\mathcal{F}[\phi,\phi_c] \approx \Delta\mathcal{F}[\phi,\phi_c^*] + \delta\phi_c[\frac{\partial\Delta\mathcal{F}[\phi,\phi_c]}{\partial\phi_c}]_{\phi_c^*}\cos(\Omega t+\varphi)$. This yields for the transition probabilities $W_\pm \approx \frac{1}{2}\Big(\alpha_0 \mp \alpha_1\frac{\delta\phi_c}{\gamma}\cos(\Omega t+\varphi)\Big)$, $\alpha_0 \approx \exp(-\Delta\mathcal{F}[\phi,\phi_c^*]/\gamma)$ and $\alpha_1 \approx \alpha_0\frac{d\Delta\mathcal{F}}{d\phi_c}|_{\phi_c^*}$. With this identification, and using the fact that $\phi_0 = 0$, only one term remains. Hence, after averaging over the random phase φ, we end up with an expression similar to their correlation function but in which the position of their minima, $\pm c$, is replaced by $c^2 = \phi_u(x)^2$.

To obtain the generalized susceptibility $S(\kappa,\omega)$, we need to perform the Fourier transform of the correlation function in time as well as in space. Due

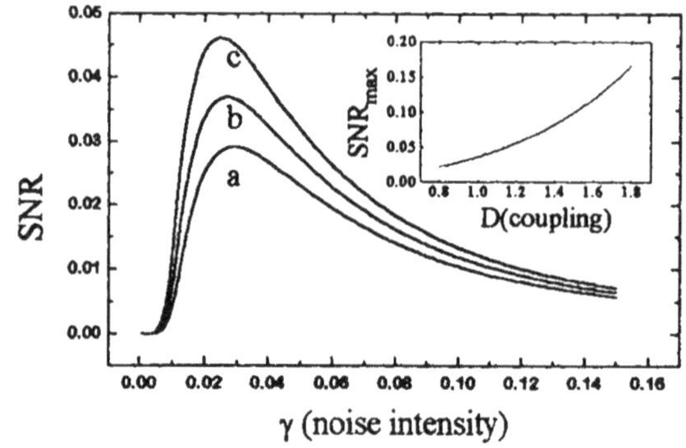

Figure 2: SNR as a function of the noise intensity γ (Eq.(17)), for (a) $D = 0.9$, (b) $= 1.0$, (c) $= 1.1$. We fixed $\phi_c = \phi_c^*$, $L = 1$, $\delta\phi_c = 0.01$ and $\Omega = 0.01$. The insert shows the maximum of SNR as a function of D.

to the fact that the space and time dependences of the correlation function factorize, $S(\kappa,\omega)$ factorizes too, and it is enough to analyze its time dependence. The Fourier transform of this time dependence yields a function analogous to the usual power spectrum function $S(\omega)$. [8] Finally, the result for the SNR is

$$SNR \sim (\Lambda\lambda\gamma^{-1})^2 \exp\Big(-2\Delta\mathcal{F}[\phi,\phi_c^*]/\gamma\Big), \tag{4}$$

where λ is an estimation of the potential curvature at the potential minima (for instance given by the linear stability eigenvalue), and $\Lambda \sim \frac{d\Delta\mathcal{F}}{d\phi_c}|_{\phi_c^*}\delta\phi_c$. Eq.(4) is analogous to what has been obtained in zero dimensional systems, but where $\Delta\mathcal{F}[\phi,\phi_c^*]$ contains all the information regarding the spatially extended character of the system.

In Fig. 2 we show the dependence of the present SNR on γ, for typical values of the parameters (same as in Fig. 1), and different values of D. It is seen that the response increases for increasing values of D. The insert shows the dependence of the maximum of the SNR as a function of D (that plays here the role of the coupling parameter). This is in agreement with recent numerical results for a system of coupled nonlinear oscillators. [2]

It is worth remarking here that the present calculation breaks down for large values of D. This is due to the fact that, for increasing D, the curves in Fig. 1 shift to the left while the barrier separating the attractors tends to zero. It is also worth noting that, besides the approximation involved in the Kramers' like expression in Eq.(3) and the two level approximation used for the evaluation of the correlation function, all the previous results (form of the patterns, nonequilibrium potential) are analytically exact. However, in a more careful analysis of the problem, as indicated by the present rough calculation, we can expect different strengths for the SR phenomena for different wave lengths, as the dependence of the generalized susceptibility $S(\kappa,\omega)$ on κ and ω –that will not necessarily factorize– will also imply that $SNR \sim SNR(\kappa,\omega)$.

The present form of analysis will be extended to activator–inhibitor or multicomponent models that, besides applications in chemical and biological systems, are related to spatio–temporal synchronization problems. [1,2] Also, in those models we have that a non-local coupling exists, in contrast with the nearest neighbour coupling presented here. [9]

ACKNOWLEDGEMENTS

The authors thanks D. Zanette and R. Deza for useful discussions and V. Grunfeld for a critical reading of the manuscript. Partial support from CONICET, Argentina, and from Fundación Antorchas, are also acknowledged.

1. (a) F. Moss, in *Some Problems in Statistical Physics*, edited by G. Weiss (SIAM, Philadelphia, 1992); (b) *Proc. NATO Advanced Research Workshop on Stochastic Resonance in Physics and Biology*, F. Moss, A. Bulsara and M. Shlesinger, editors, J. Stat. Phys. **70** No. 1/2 (1993); (c) *Proc. 2nd. Int. Workshop on Fluctuations in Physics and Biology: Stochastic Resonance, Signal Processing and Related Phenomena*, R.Mannella and P.V.E. McClintock, editors, Il Nuovo Cim. **D 17** (1995).
2. J.F. Lindner, B.K. Meadows, W.L. Ditto, M.E. Inchiosa and A. Bulsara, Phys. Rev. Lett. **75**, 3 (1995); J.F. Lindner, B.K. Meadows, W.L. Ditto, M.E. Inchiosa and A. Bulsara, Phys. Rev. E **53**, 2081 (1996).
3. R.Graham, in *Instabilities and Nonequilibrium Structures*, Eds. E.Tirapegui and D.Villaroel (D.Reidel, Dordrecht, 1987); R. Graham and T. Tel, Phys. Rev. A **42**, 4661 (1990).
4. C.L. Schat and H.S. Wio, Physica A **180**, 295 (1992).
5. G. Izús, R. Deza, O. Ramírez, H.S. Wio, D. Zanette and C. Borzi, Phys. Rev. E **52**, 129 (1995); G. Izús, H.S.Wio, J. Reyes de Rueda, O. Ramírez and R. Deza, Int.J.Mod.Physics B. (1996) to appear.
6. D. Zanette, H.S. Wio and R. Deza, Phys. Rev. E **53**, 353 (1996); F. Castelpoggi, H.S. Wio and D. Zanette, *Critical slowing–down of spatially nonhomogeneous patterns ...*, submitted to Phys. Rev. E (1996).
7. A. Foster and A.S. Mikhailov, Phys. Lett. A **126**, 459 (1988); S.P. Fedotov, Phys. Lett. A **176**, 220 (1993).
8. B. McNamara and K. Wiesenfeld, Phys. Rev. A **39**, 4854 (1989).
9. G. Drazer and H.S. Wio, *Nonequilibrium potential for an activator–inhibitor model with fast inhibition*, submitted to Physica A (1996).

WHAT MINIMAL NOISE IS NECESSARY FOR GENERATION OF TRANSPORT IN PERIODIC STRUCTURES ?

J. LUCZKA
Department of Theoretical Physics, Silesian University, 40-007 Katowice, Poland

P. HÄNGGI
Department of Physics, University of Augsburg, Memminger Str.6, D-86135 Augsburg, Germany

T. CZERNIK
Institute of Physics, Silesian University, 40-007 Katowice, Poland

Abstract

In spatially periodic structures, nonequilibrium fluctuations of specific statistics are able to generate non-zero current (Brownian ratchets). An unsolved problem is: What minimal statistics for the fluctuations is required for inducing finite transport in ratchet-type systems? In particular, is there a chance that a *symmetric* and *delta-correlated* additive noise does in fact yield directed motion?

In generic cases, random fluctuations (and other irregular forces) are reckoned to act destructively on processes, starting from physical through biological up to sociological ones. On the other hand, constructive influence of uncontrollable perturbations can be observed in nature. Examples are activation processes, stochastic resonance phenomena, Brownian ratchets, etc. In the latter, transport (non-zero current) in spatially periodic structures can be generated by random fluctuations of *zero* average values, without any field gradients and external bias forces.[1] The interest of such a transport mechanism is considerable: In biology (protein motors: transport of vesicles and ogranelles, locomotion, segregation of chromosomes),[2] material sciences (separation or pumping of particles), electronics (nano- and micro-technologies) and physics.

Periodic structures possess or do not possess a reflection symmetry. It means that systems can be described in terms of a spatially periodic potential $V(x) = V(x + L)$ with period L. For systems with a reflection symmetry, there is a constant C such that $V(C - x) = V(C + x)$. Moreover, fluctuations that act in systems can be symmetric or asymmetric. Symmetric noise $\xi(t)$ is characterized by the fact that all its odd numbered cumulant averages are identically vanishing; in contrast, asymmetric noise of zero mean can possess nonvanishing odd-numbered higher order cumulants.

From previous investigations, it has been known that in periodic structures:
(a) Transport can be induced by correlated symmetric noises in systems with a broken spatial symmetry (i.e., when the spatial potential is asymmetric).[3,4]
(b) Transport can be genereted by correlated asymmetric fluctuations in systems without broken spatial symmetry (i.e., when the spatial potential is symmetric).[5]
(c) Transport can be induced by uncorrelated (or δ-correlated) asymmetric shot noise in systems without broken spatial symmetry.[6]
(d) Transport cannot be generated by thermal fluctuations (symmetric Gaussian white noise).

An open fundamental question thus reads: What minimal statistics of noise should be sufficient for inducing a macroscopic current in periodic structures? To be more precise, let us formulate the problem in the form of overdamped motion of Brownian particles (of unit masses) in spatially periodic potential $V(x)$, namely,

$$\dot{x} = f(x) + \Gamma(t) + \xi(t), \tag{1}$$

where $f(x) = -dV(x)/dx$. The process $\Gamma(t)$ represents thermal fluctuations: It is Gaussian delta-correlated noise of zero mean and of strength $D_T \equiv k_B T/\gamma$ with T and γ denoting the temperature of the system and the friction coefficient, respectively. The process $\xi(t)$ is a "driving force" and models another, nonequilibrum source of fluctuations.

Let us recast the question as follows: Is it possible to generate non-zero current by *symmetric and δ-correlated* fluctuations $\xi(t)$? We think the question interesting since usually it is believed that symmetric white noise cannot give rise to directed transport. According to Ref. 3: "... all that is needed to generate motion and forces in the Brownian domain is loss of symmetry and substantially long time correlations". In Ref. 4 it is stated: "... if (in our notation) $\xi(t)$ is another symmetric white noise process, the stationary state corresponds to a thermal equilibrium state satisfying the condition of detailed balance, in which case no net current is possible for any shape of the potential". So, taking the above statements virtually, the answer to the question stated above would be in the negative.

It is indeed a challenge to find white noise sources which are able to convert random walk into directed motion. Hence, the first problem is to construct models of such δ-correlated fluctuations $\xi(t)$, which, by virtue of the statement in (d), should be *non-equilibrium and non-Gaussian*. There are several candidates for such stochastic processes as e. g. symmetric Poissonian white noises (white shot noises or Poisson point processes),[7] *composite* noises, in particular, multi-state diffusion processes [8] (in each state, the system is subject to diffusion with various diffusion coefficients and randomly jumps between states) or the so-called randomly interrupted (or flashing) Gaussian white noise: [9] Jumps between the Brownian diffusional state (a Feynman ratchet carrying zero current) and a deterministic flow (also carrying zero current) are steered by a symmetric two-state Markov process.

If $\xi(t)$ is Poissonian white shot noise, the output process $x(t)$ defined by (1) is a Markovian stochastic process. A master equation for the probability distribution $P(x,t)$ of it is a partial integro-differential equation [7] with the integral kernel $\rho(z)$ being a probability density of weights of the δ-impulses of shot noise. If statistical cumulants of odd order of the noise source are all equal to zero, and even numbered ones are non-zero, then the shot noise is symmetric. This is the case when $\rho(z) = \rho(-z)$. In dependence of a form of the distribution $\rho(z)$, one can construct a wide class of white shot noises. Such noises can be realized in electronic and physical systems under a variety of experimental conditions. These are visible in systems where events (such as emission of any excitations) occur with an average spacing greater than the characteristic time duration of each event. For $\xi(t)$ being composite noise, the resulting process $x(t)$ is *non-Markovian*, the treatment of which is much more complicated. For example, let $\xi(t)$ be two-state diffusion noise defined as

$$\xi(t) = \frac{1}{2}[1+\eta(t)]\Lambda_1(t) + \frac{1}{2}[1-\eta(t)]\Lambda_2(t) \tag{2}$$

where $\Lambda_i(t)$ $(i = 1,2)$ are independent δ-correlated Gaussian white noises of strengths D_1 and D_2, respectively. The process $\eta(t) = \{-1, 1\}$ is a dichotomous Markovian process, which switches back and forth between two states $1 \leftrightarrow -1$ with the rate ν. It is zero averaged and exponentially correlated proces with the correlation time $\tau_0 = (2\nu)^{-1}$. The two-state diffusion noise $\xi(t)$ is symmetric white noise of zero mean. However, the output process $x(t)$ in (1) is non-Markovian: it jumps with Poissonian statistics between dynamics $\dot{x} = f(x) + \Gamma(t) + \Lambda_1(t)$ and dynamics $\dot{x} = f(x) + \Gamma(t) + \Lambda_2(t)$. This unusual two-state noise can experimentally be realized.[10]

Finally, the open problem is to obtain the stationary, *nonzero* probability current of the combined, output process $x(t)$; if so, the answer to the open problem is in the positive. The insight which the problem presented provides is clear: White and symmetric noise might - contrary to common intuition - produce a net directed current.

Acknowledgments

J. Ł. and T. Cz. thank Komitet Badań Naukowych for support through the Grant 2 P03B 079 11; P.H. acknowledges the support by the Deutsche Forschungsgemeinschaft (Az. Ha1517/13-1). J. Ł. thanks Prof. Jean Brini for discussion on experimental realizations of two-state noise.

1. P. Hänggi and R. Bartussek, *Brownian Rectifiers: How to Convert Brownian Motion into Directed Transport*, in: *Nonlinear Physics of Complex Systems - Current Status and Future Trends*, J. Parisi, S. C., Müller and W. Zimmermann (eds.) Springer Series 'Lecture Notes in Physics' (Springer, Berlin, 1996) Vol. **476**, p. 294; C. R. Doering, *Nuovo Cimento* D **17**, 685 (1995).
2. J. Łuczka, *Cell. Mol. Biol. Lett.* **1**, 311 (1996).
3. M. O. Magnasco, *Phys. Rev. Lett.* **71**, 1477 (1993);
4. C. R. Doering, W. Horsthemke, and J. Riordan, *Phys. Rev. Lett.* **72**, 2984 (1994).
5. J. Kula, T. Czernik and J. Łuczka, *Phys. Lett. A* **214**, 14 (1996).
6. J. Łuczka, R. Bartussek and P. Hänggi, *Europhys. Lett.* **31**, 431 (1995); P. Hänggi, R. Bartussek, P. Talkner and J Łuczka, *Europhys. Lett.* **35**, 315 (1996).
7. P. Hänggi, *Z. Phys. B* **31**, 407 (1978).
8. V. Balakrishnan, C. Van den Broeck and P. Hänggi, *Phys. Rev. A* **38**, 4213 (1988).
9. J. Łuczka, M. Niemiec and P. Hänggi, *Phys. Rev. E* **52**, 5810 (1995).
10. J. Brini, P. Chenevier and P. D'onofrio, *RTS Variance Fluctuations in Ultrasmall MOSFET's at* $V_D = 0$ (to be published).

A Global Dynamical Modeling Scheme Using Transformed Variables in the Presence of Noise

Claudia S. M. Lainscsek, Ferdinand Schürrer

Institut für Theoretische Physik; Technische Universität Graz
Petersgasse 16; A-8010 Graz; Austria

James B. Kadtke

Institute for Pure and Applied Physical Sciences
MS-0360, University of California at San Diego
9500 Gilman Drive; La Jolla, CA 92093; USA

Abstract

The information contained in a scalar time series and its numerical derivatives is used to construct a global model for the underlying dynamical system, using a model transformation presented previously. Here, we present some modifications of this method to mitigate the effects of additive noise, test this method by reconstructing global models for known chaotic dynamical systems, and compare the dynamical properties of the re-constructed and original systems.

1 Global Modeling by Means of a Transformed System

The construction of a global dynamical model from measured time series is an important challenge in different fields of natural science. Very often the underlying physical model of complex systems is not known (e.g., convection cells on the surface of the sun). Unlike simple models of data evolution, global dynamical models are constructed to capture fundamental physical properties of the underlying system.

On the other hand, in real physical situations, usually only one or a few observables can be measured as a function of time with satisfactory precision. In such cases, if the underlying system of differential equations is strongly coupled, it is sometimes possible to reconstruct the global dynamical properties (e.g., dimension, Liapunov exponents, topological information) from scalar time series.

If all state variables of a dynamical system can be measured, the construction of a global model based on such data leads to a system of ODEs which can describe the original system very well. In the case of a more realistic situation, where only one time series is available, the global model can often be unsatisfactory, due to the problems of model choice and parsimony. In this paper, we show how to use a single variable time series instead of Taken's trajectory matrix, and construct a global model in terms of a transformed system. Since in practice, no a priori knowledge of the original dynamical system is available, we compare the original and model properties by integrating the model numerically and comparing to the solutions of the original system. We find that these systems are usually in good accordance with each other, and they are often comparable to the results of constructing a model from the whole set of state variables. Unfortunately, at present this method may only be defined for systems of dimension three or less.

To describe this method, we start from a discrete scalar time series of length N of some measured physical observable, say $(x_1, x_2, \ldots, x_i, x_{i+1}, \ldots, x_N)$, whose underlying generating system is governed by a dynamical system of the general form

$$\begin{aligned} \dot{x} &= a_0 + a_1 x + a_2 y + a_3 z + a_4 xy + a_5 xz + a_6 yz + a_7 x^2 + a_8 y^2 + a_9 z^2 , \\ \dot{y} &= b_0 + b_1 x + b_2 y + b_3 z + b_4 xy + b_5 xz + b_6 yz + b_7 x^2 + b_8 y^2 + b_9 z^2 , \\ \dot{z} &= c_0 + c_1 x + c_2 y + c_3 z + c_4 xy + c_5 xz + c_6 yz + c_7 x^2 + c_8 y^2 + c_9 z^2 , \end{aligned} \tag{1}$$

where we here restrict ourselves to systems with 3 effective degrees-of-freedom. We first try to rewrite system (1) in the form of

$$X = x , \dot{X} = Y , \dot{Y} = Z , \dot{Z} = f(X, Y, Z) . \tag{2}$$

It can be shown that this is not possible in general (Lainscsek, et al.[1]). However, two examples of systems that can be transformed in general are

$$\begin{aligned} \dot{x} &= a_0 + a_1 x + a_2 y + a_7 x^2 , \\ \dot{y} &= b_0 + b_1 x + b_2 y + b_4 xy + b_5 xz + b_7 x^2 + b_8 y^2 , \\ \dot{z} &= c_0 + c_1 x + c_2 y + c_3 z + c_4 xy + c_5 xz + c_6 yz + c_7 x^2 + c_8 y^2 + c_9 z^2 , \end{aligned} \tag{3}$$

and

$$\begin{aligned} \dot{x} &= a_0 + a_1 x + a_2 y + a_7 x^2 , \\ \dot{y} &= b_0 + b_1 x + b_2 y + b_3 z + b_4 xy + b_7 x^2 + b_8 y^2 , \\ \dot{z} &= c_0 + c_1 x + c_2 y + c_3 z + c_4 xy + c_5 xz + c_6 yz + c_7 x^2 + c_8 y^2 + c_9 z^2 . \end{aligned} \tag{4}$$

The function $f(X, Y, Z)$ in the exact case can be estimated by constructing a matrix equation:

$$\mathbf{f}(X_i, Y_i, Z_i) = \mathbf{P}_i \boldsymbol{\alpha}_i \tag{5}$$

with

$$\begin{aligned}\mathbf{P}_i = \Big(&1, \frac{1}{X_i}, X_i, X_i^2, X_i^3, X_i^4, X_i^5, X_i^6, X_i^7, Y_i, \frac{Y_i}{X_i}, X_iY_i, X_i^2Y_i, \\ &X_i^3Y_i, X_i^4Y_i, X_i^5Y_i, Y_i^2, \frac{Y_i^2}{X_i}, X_iY_i^2, X_i^2Y_i^2, X_i^3Y_i^2, Y_i^3, \frac{Y_i^3}{X_i}, X_iY_i^3, \\ &\frac{Y_i^4}{X_i}, Z_i, \frac{Z_i}{X_i}, X_iZ_i, X_i^2Z_i, X_i^3Z_i, Y_iZ_i, \frac{Y_iZ_i}{X_i}, X_iY_iZ_i, \frac{Y_i^2Z_i}{X_i}, \frac{Z_i^2}{X_i}\Big)\end{aligned} \tag{6}$$

for Eqs. (3) and

$$\begin{aligned}\mathbf{P}_i = \Big(&1, X_i, X_i^2, X_i^3, X_i^4, X_i^5, X_i^6, X_i^7, X_i^8, Y_i, X_i\,Y_i, X_i^2\,Y_i, X_i^3\,Y_i, X_i^4\,Y_i, \\ &X_i^5\,Y_i, X_i^6\,Y_i, Y_i^2, X_i\,Y_i^2, X_i^2\,Y_i^2, X_i^3\,Y_i^2, X_i^4\,Y_i^2, Y_i^3, X_i\,Y_i^3, X_i^2\,Y_i^3, \\ &Y_i^4, Z_i, X_i\,Z_i, X_i^2\,Z_i, X_i^3\,Z_i, X_i^4\,Z_i, Y_i\,Z_i, X_i\,Y_i\,Z_i, X_i^2\,Y_i\,Z_i, \\ &Y_i^2\,Z_i, Z_i^2\Big)\end{aligned} \tag{7}$$

for Eqs. (4). The terms in $\mathbf{P}_i$ are obtained by transforming Eqs. (3) or Eqs. (4) into Eqs. (2). The over-determined system (5) for the unknowns α_i, (N equations for 35 unknowns) can be solved in a least-square sense (e.g. by using the method of Singular Value Decomposition).

Using this method we are able to build a dynamical model in the manner of Gouesbet[2]. However, we show that we may mitigate noise effects by calculating numerical derivatives from time-delayed estimates.

In so doing, we finally obtain a transformed model which can be re-transformed to the original system (3) or (4) in the case of an exact transformation. In such a case a system of 35 equations for 21 unknowns a_i, b_i, c_i from Eq. (3) have to be solved:

$$\begin{aligned}\alpha_0 = {} & a_2b_0c_3 - a_0b_2c_3 + \frac{a_0^2b_8c_3}{a_2} - a_0b_0c_6 + \frac{a_0^2b_2c_6}{a_2} - \frac{a_0^3b_8c_6}{a_2^2} - \frac{2a_2b_0b_1c_9}{b_5} + \\ & \frac{2a_1b_0b_2c_9}{b_5} + \frac{2a_0b_1b_2c_9}{b_5} - \frac{2a_0a_1b_2^2c_9}{a_2b_5} + \frac{2a_0b_0b_4c_9}{b_5} - \frac{2a_0^2b_2b_4c_9}{a_2b_5} - \frac{4a_0a_1b_0b_8c_9}{a_2b_5} - \\ & \frac{2a_0^2b_1b_8c_9}{a_2b_5} + \frac{6a_0^2a_1b_2b_8c_9}{a_2^2b_5} + \frac{2a_0^3b_4b_8c_9}{a_2^2b_5} - \frac{4a_0^3a_1b_8^2c_9}{a_2^3b_5} \\ \alpha_1 = {} & -\frac{a_2b_0^2c_9}{b_5} + \frac{2a_0b_0b_2c_9}{b_5} - \frac{a_0^2b_2^2c_9}{a_2b_5} - \frac{2a_0^2b_0b_8c_9}{a_2b_5} + \frac{2a_0^3b_2b_8c_9}{a_2^2b_5} - \frac{a_0^4b_8^2c_9}{a_2^3b_5} \\ \alpha_2 = {} & \ldots \\ & \vdots\end{aligned}$$

This is again done by using a least-square procedure. A similar system has to be solved for Eqs. (4).

If one has a time series where the original system is unknown, the transformed model in the case of the "wrong" ansatz (e.g. ansatz (6) instead of (7)) could give satisfactory results in the sense of invariant dynamical properties. But when trying to re-transform such a transformed model, one obtains no solution. Therefore, this method can be used to distinguish between different classes of original systems (e.g. Eqs. (3) or Eqs. (4)) and it also can be used to find the minimum set of ODE's for a given time series.

The estimated solution is not unique because the 35 equations for the 21 unknowns are under-determined; it can be shown that some of the equations are linearly dependent.

2 Global Modeling of Simulated Systems

2.1 Lorenz System Without Noise

We can apply the above method to a simple example. The Lorenz equations[3,4] can be written as

$$\dot{x} = -\sigma x + \sigma y,\ \dot{y} = \mathcal{R}x - y - xz,\ \dot{z} = -bz + xy, \tag{8}$$

where $\mathcal{R}$ is the bifurcation parameter and σ and b are constants.

The numerical experiments (see Lainscsek, et al.[1]) of computing a transformed model and comparing the invariant properties with that of the original time series indicate good agreement between the global properties of the vector field of the reconstructed dynamical system with that of the original system. This agreement is possible because the form of model (rational polynomial)[4] allows recovery of the essential topology of the vector field. In addition, the SVD-based hybridization results in a more robust estimation method.

The re-transformed model is given by:

$$\begin{aligned}
\dot{x} &= 11.1468 - 10.6650x - 15.5306y - 0.0012x^2 \\
\dot{y} &= 0.3021 - 14.9591x - 0.3742y + 0.0256xy \\
&\quad +0.5652xz - 0.0600x^2 + 0.0024y^2 \\
\dot{z} &= -7.4430 + 0.9620x - 1.8031y - 2.9667z \\
&\quad -1.7684xy - 0.0219xz + 0.0009yz - 0.0644x^2 \\
&\quad -0.0056y^2 + 0.0052z^2
\end{aligned} \tag{9}$$

This system is compared with the original system by integrating Eqs. (10) (Fig. 1). The projections to the x, y-, y, z- and x, z-plane in Fig. 1 show the same

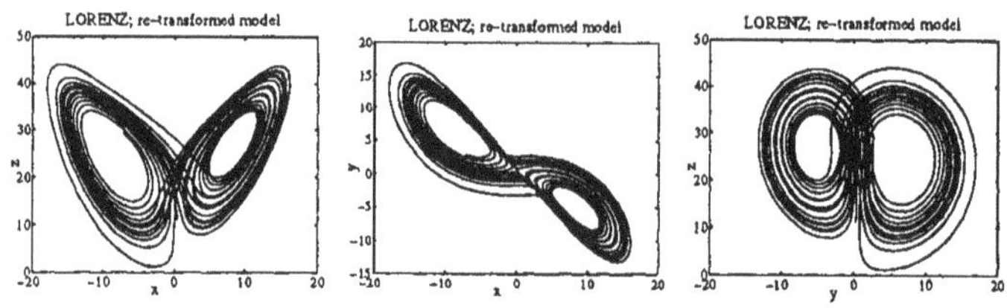

Figure 1: Time evolution and projections of the attractor of the re-transformed model

structure as the projections of the original Lorenz system to these planes. Also the unstable fixed points have the same values as in the original Lorenz System (8). These results are generally better than those obtained from methods which do not produce a rational polynomial form.

2.2 Lorenz System With Noise

We now consider the effects of additive observational noise on our modeling procedure. The same procedure as in the former section is applied to the same time series x of the Lorenz system with additional noise ξ. We use the signal-to-noise-ratio (SNR), SNR $= 10 \log_{10} \frac{\sigma^2(x)}{\sigma^2(\xi)}$, which was set to $\text{SNR}_1 = 30$; $\text{SNR}_2 = 20$; $\text{SNR}_3 = 10$. With these noise-levels a model in the form of Eq. (2) was computed and compared to the original time series by integrating the model system (see Fig. 2). Comparing the embedded transformed Lorenz

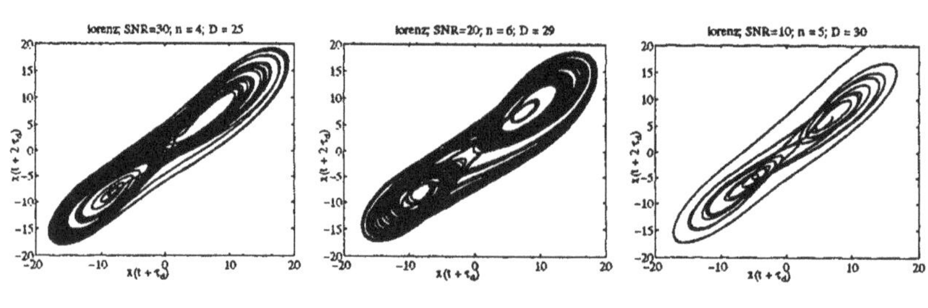

(a) SNR = 30 (b) SNR = 20 (c) SNR = 10

Figure 2: Integration and embedding of Lorenz models (Eq. (2)) with different noise levels

models (see Fig. 2), one can see the following: at a SNR of $\mathrm{SNR}_1 = 30$, the embedded attractor is similar to the original attractor. With $\mathrm{SNR}_2 = 20$, the model is even better, but with $\mathrm{SNR}_3 = 10$ the model becomes worse. This behaviour is symptomatic of the least-squares solution, and can also be mitigated using singular value editing.

The same conclusion can be drawn, if one compares the correlation time in the re-constructed phase space between the original and model time series (seperation of the two trajectories compared to the amplitudes has to be under 10 %). This times are for $\mathrm{SNR}_1 = 30 : t_{corr} = 2.03$ (about three cycles); $\mathrm{SNR}_2 = 20 : t_{corr} = 2.35$; $\mathrm{SNR}_3 = 10 : t_{corr} = 0.04$.

3 Conclusion

For dynamical systems defined by certain classes of differential equations, methods for global modeling have been developed when only one single variable time series is available. These methods are based on a transformation of the dynamical system to variables given by the time series and its derivatives and the re-transformation of the estimated model to the original system. Its application on simulated, noise-free time series (tested for the Lorenz system) yields good results. In the case of noisy data we modify the method using Takens variables and SVD, and find generally good results.

References

1. C.S.M. Lainscsek, F. Schürrer, and J.B. Kadtke. A general form for global dynamical data models for three-dimensional systems. *Int. J. of Bif. and Chaos*, to appear.
2. G. Gouesbet and J. Maquet. Construction of phenomenological models from numerical scalar time series. *Physica D*, 58:202, 1992.
3. E.N. Lorenz. Deterministic nonperiodic flow. *J. Atmos. Sci.*, 20:130 – 141, 1963.
4. C. Sparrow. *The Lorenz Equations : Bifurcations, Chaos and Strange Attractors*. Applied Mathematical Sciences , Volume 41. Springer - Verlag, 1982.

TRANSITION TO A FRACTAL ATTRACTOR VIA ON-OFF INTERMITTENCY IN A MODEL WITH DICHOTOMOUS NOISE

S.P. KUZNETSOV, P.V. KUPTSOV

Institute of Radioengineering and Electronics,
410019, Saratov, Russia

We consider dynamics of the discrete time model system driven by the dichotomous noise. The system has the fractal attractor and we show that the scenario of its appearance is an on-off intermittency — recently reported sort of intermittency which is typical for systems with multiplicative random parameter.

Phase transitions in systems with external noise is a subject of many investigations. Recently in this context a paradigm of on-off intermittency has been suggested [1,2]. The intermittency of this new kind is typically demonstrated by systems with multiplicative random parameter. This phenomenon may occurs both in presence of external noise and as a result of interaction between subsystems of a complex chaotic system.

Irwin, Farser, Kapral in Ref. [3] and Fraser, Kapral in Ref. [4] consider the linear system with discrete time forced by dichotomous noise. (Responsible for this noise variable switches with some probability between two possible values at every time step.) Such system may has a fractal attractor with Cantor's structure (Cantor's attractor) at some conditions. Authors of referred articles show that the probability distribution generated by the system consists of self similar series of peaks and troughs when system's control parameter lies in some range. This behavior authors call as a resonance regime. We will call this one as IFK–resonances.

We study in this report the model system with discrete time which demonstrates on-off intermittency and IFK–resonances as a scenario of transition to a Cantor's attractor. In absence of noise the system is a nonlinear map with pitch-fork bifurcation.

$$x_{n+1} = \xi_n \frac{x_n}{\sqrt{1 + x_n^2}}, \tag{1}$$

where ξ_n is a noisy parameter, $\xi_n = a$ or b with probability 1/2. The positive values a and b we will use as control parameters. The dynamics of the system (1) can not cause the change of x sign and the equation is symmetric under operation $x \to (-x)$. Hence we can consider only a nonnegative values of x.

The attractor of the system (1) is disposed inside the interval $[X_a,\ X_b]$:

$$X_a = X(a),\ \ X_b = X(b),$$

where

$$X(\xi) = \begin{cases} 0, & \text{when } \xi \le 1 \\ \sqrt{\xi^2 - 1}, & \text{when } \xi > 1. \end{cases} \tag{2}$$

The system comes to the boundary points X_a and X_b when $n \to \infty$ if the sequence ξ_n consists of the one repeating symbol a or b respectively.

It is important that the system (1) can be reorganized to a linear form by the substitution:

$$x_n = C_n^{-1/2}, \tag{3}$$

For new variable C_n one gets:

$$C_{n+1} = \xi_n^{-2}\,(C_n + 1) \tag{4}$$

Therefore we can use the results of Refs.[3,4] when analyzing the system (1).

In changing the parameters a and b one can observe four specific kinds of behavior of the system (1): a) Cantor's attractor; b) regime with IFK-resonances; c) on-off intermittency; d) zero point attractor. To illustrate them we plot in Fig. 1 the time series of dynamical variable x for all of these cases. In Fig. 2 the parameter plane $(a,\ b)$ is presented with four respective domains a) — d).

The small insertions in Fig. 2 we used to show the structure of an attractor of the system (1). Two dimensional phase portraits are plotted with x through the horizontal axis and with an artificial variable y on the vertical axis. The last one is generated by the map:

$$y_{n+1} = \begin{cases} y_n/2, & \text{when } \xi_n = a \\ (1 + y_n)/2, & \text{when } \xi_n = b \end{cases} \tag{5}$$

The same realizations of ξ_n drive both map (1) and map (5). The map (5) is chosen for its attractor to fill everywhere the unit interval with uniform distribution. An original attractor of system (1) is a projection of this portrait to the horizontal axis.

Now let us consider the regimes of the system (1) in different domains of the plane $(a,\ b)$ (Fig. 2).

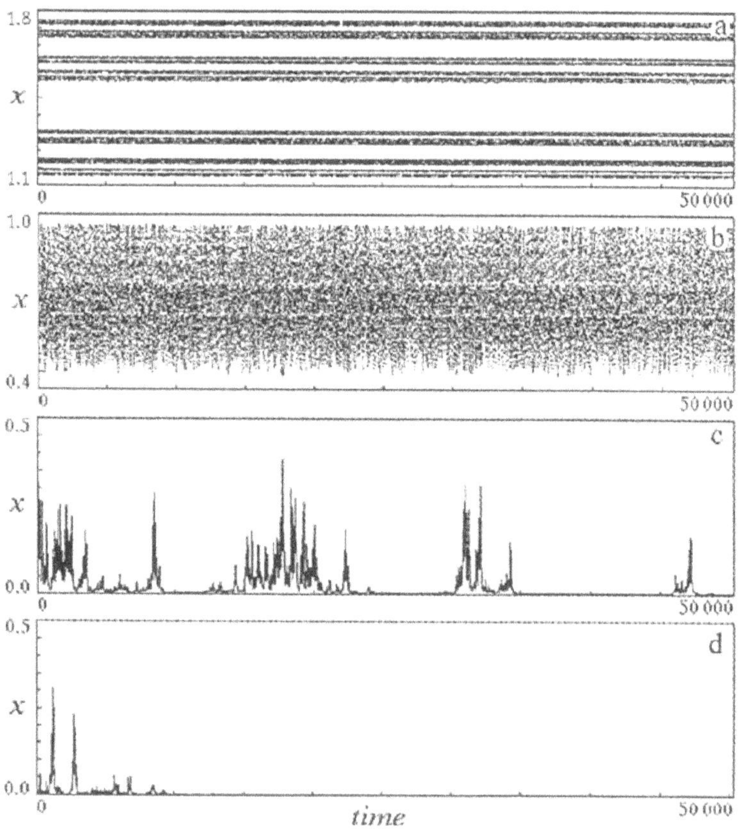

Figure 1: The time series of dynamical variable x of the system (1) in regimes of a) Cantor's attractor, b) IFK–resonances, c) on-off intermittency, d) and zero point attractor.

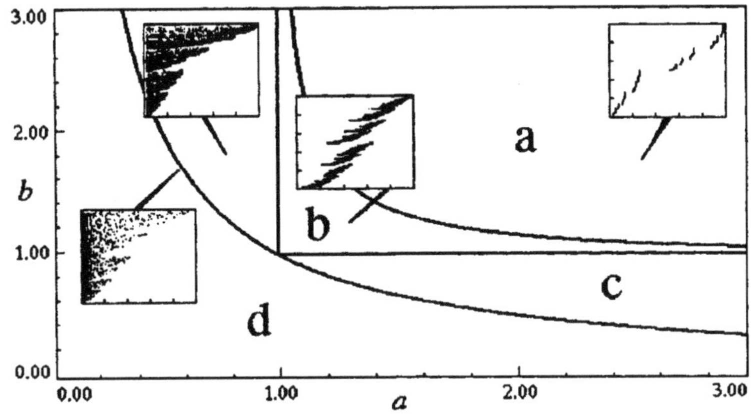

Figure 2: The parameter plane $(a,\ b)$ of the system (1). The domains of four regimes are shown: a) Cantor's attractor, b) IFK–resonances, c) on-off intermittency, d) zero point attractor. In the insertions two dimensional phase portraits on the dynamical variable plane $(x,\ y)$ are presented (see (1) and (5)).

a) Domain with Cantor's attractor:

$$(1/a)^2 + (1/b)^2 < 1. \tag{6}$$

In this domain the attractor of linear system (4) is located inside the interval $[c_b,\ c_a] = [1/(b-1),\ 1/(a-1)]$ (see (2), (3)). The attractor forms in a such way. Let the variable C_0 belongs to the interval $[c_b,\ c_a]$. After the first iteration of (4) it may be found inside the one of two intervals: $C_1 \in [c_b,\ (1/b^2)c_a]$ or $C_1 \in [(1/a^2)c_b,\ c_a]$. With every iteration an amount of intervals is duplicated and its lengthes decrees by factors $1/a^2$ or $1/b^2$. The intervals at every step do not overlap each other and its summary length is less then $(c_a - c_b)$ because of inequality (6).

Consequently, in the domain a) attractor of the system (4) is a Cantor's set with two scales [5]. The attractor of (1) has the same topological structure. It is an image of the Cantor's set under transformation (3).

b) Domain with IFK-resonances:

$$a > 1,\ \ b > 1,\ \ (1/a)^2 + (1/b)^2 > 1. \tag{7}$$

Here the attractor of the system (4) forms in the similar way as above. But now the intervals overlap partially at every time step. As it shown in Refs. [3,4], the overlapping induces the selfsimilar series of peaks and troughs in the experimental probability density $\rho(C)$ (IFK–resonances). (We can not talk about the invariant density in strict mathematical sense because the problem of its existence is not completely solved for today [6].)

Similar peculiarities of probability density one can observe for the system (1). Its empirical distribution $\rho(x)$ in domain under consideration exhibited in Fig. 3a. The specific peaks are seen which correspond to IFK–resonances. In Fig. 3b we ploted the respectiv integral distribution $m(x) = \int_{X_a}^{x} \rho(\tau) d\tau$.

c) On-off intermittency domain:

$$b > 1, \quad 1/b < a < 1 \quad \text{or} \quad a > 1, \quad 1/a < b < 1. \tag{8}$$

The system (1) in this regime switches abruptly from extended periods of stasis to bursts of large variation. Still periods are "laminar" phases and bursts are "turbulent" or better to say "nonlinear" phases (Fig. 1c). On the boundary of this domain, $a = 1/b$, the system demonstrates a critical behavior. Power-lows realize for laminar phases extension distribution (with exponent -3/2) and for growth of mean extension of laminar phase vs. supercritisity (with exponent -1).

Let us suppose that $a < 1$, $b > 1$ and make next substitutions in (1): $x_n = \exp s_n$, $\xi_n = \exp q_n$. Then we obtain:

$$s_{n+1} = s_n + q_n - \ln \sqrt{1 + \exp 2s_n}. \tag{9}$$

The attractor's right point of the system (1) comes to the point $s_b = \ln X_b$ under the substitution and the left one comes to minus infinity (It is seen from (2) that now $X_a = 0$). When variable s_n has large negative value, nonlinear part of (9) is negligible. Hence, system (1) is in laminar phase and map (9) describes a classical one dimensional random walking [7]. For the domain under consideration is valid the inequality $|\ln b| > |\ln a|$ and therefore the diffusion in positive direction dominates in this phase: with random fluctuations the amplitude of x grows. As a result the system enters into a nonlinear phase, $s \sim \ln X_b$. Here the inverted direction of motion begins to dominate because of nonlinearity and it restricts the phase — system returns to a laminar station. This scenario repeats again and again. This is a mechanism of on-off intermittency.

d) Zero point attractor:

$$ab < 1 \tag{10}$$

Let us suppose that $a < b$ and consider the map (9). Now the inequality $|\ln b| < |\ln a|$ is valid and the diffusion in negative direction of s–axis dominates every time. Hence, the amplitude of x tends to zero.

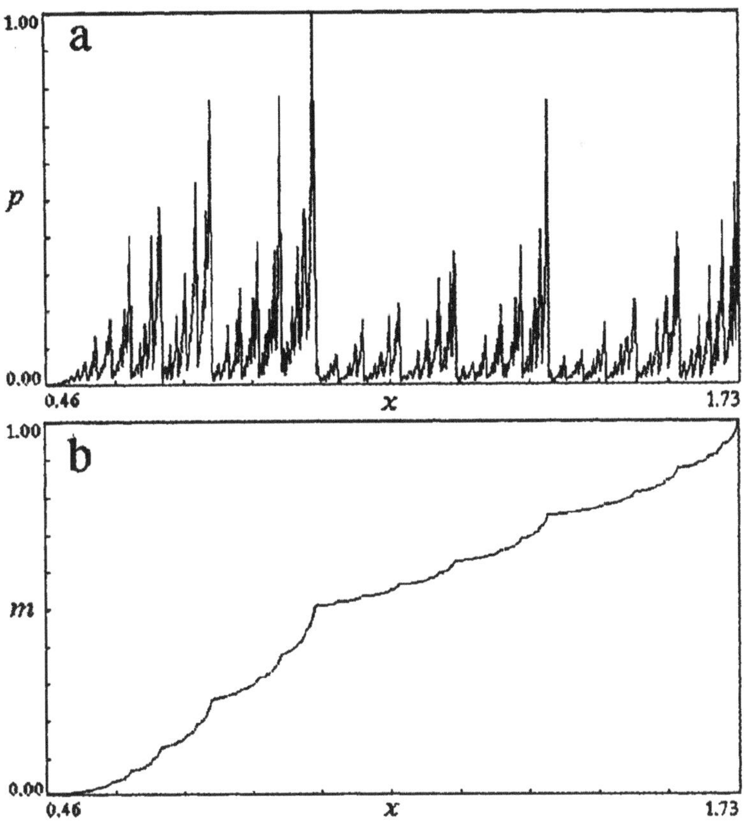

Figure 3: Empirical probability density obtained numerically for the system (1) in regime of IFK–resonances (a) and corresponding integral distribution (b).

Thus for considered system one can observes in changing of parameters the transition from zero point attractor through on-off intermittency and resonances of Irwin–Fraser–Kapral to the fractal attractor. Therefore all of this regimes are the parts of the one picture. A behavior in laminar phases of on-off intermittency is governed by universal lows of random walking. Consequently, our system may serves as a canonical model for this phenomenon.

This work is supported by Russian Foundation of Fundamental Investigations (grants 95-02-05818, 96-02-00717, 96-02-27298).

References

1. N. Platt, E.A. Spiegel, C. Tresser, *Phys. Rev. Lett.* **70**, 279 (1993).
2. J.F. Heagy, N. Platt, S.M. Hammel, *Phys. Rev.* E **49**, 1140 (1994).
3. A.J. Irwin, S.J. Fraser, R. Kapral, *Phys. Rev. Lett.* **64**, 2343 (1990).
4. S.J. Fraser, R. Kapral, *Phys. Rev.* A **45**, 3412 (1992).
5. H. Peitgen, H. Jurgens, D. Saupe, *Chaos and Fractals — New Frontiers of Science* (Springer–Verlag, New York, 1992).
6. J. Stark, *Phys. Rev. Lett.* **65**, 3357 (1990); A.J. Irwin, S.J. Fraser, R. Kapral, *ibid.* **65**, 3358 (1990).
7. W. Feller, *An introduction to probability theory and its applications*, vol. 1 (Wiley, New York, 1970).

PROBLEMS OF THE APPLICATION OF MELNIKOV METHOD FOR CHAOS FORECAST IN DISSIPATIVE DYNAMICAL SYSTEMS

YU.A. TSARIN, V.B. RYABOV, D.M. VAVRIV
Institute of Radio Astronomy, 4 Krasnoznamennaya St. 310002 Kharkov, Ukraine

The applicability of Melnikov criterion in dynamical systems is studied. The problems of two general types are considered: intrinsic applicability of the technique as an analytic tool for the study of chaos threshold conditions and technical difficulties in the solution of particular problems, especially of higher dimensional ones. The importance of balancing the dissipation value which must be small for using the Melnikov method and, at the same time, large enough for making the homoclinic structure attracting is discussed.

1 Introduction

Dynamical chaos is one of typical manifestations of noisy behavior in many physical systems. Although there is a profound difference between chaos, a complicated phenomenon with few degrees of freedom, and infinite-dimensional noise, which is impossible to describe by a set of several equations, such important characteristics of random processes as the power spectrum and autocorrelation function are often indistinguishable for the both phenomena. For the last two decades, a huge amount of experimental and theoretical evidence has been accumulated, evidencing the ubiquity of chaotic phenomena and necessity to take it into account in many theoretical problems and for the development of various physical devices in applications.

One of the principal problems which exist in the theory of deterministic chaos is to predict its appearance. In other words, given a set of differential equations or a map governing the dynamics of a nonlinear system, how to conclude what combination of control parameters would lead to a chaotic behavior? In strongly nonlinear systems the Melnikov's method is most widely used for finding the chaos arising threshold in the control parameter space [1-4]. However, its application for the case of weak nonlinearity requires some additional tricks [3, 4]. At the same time, weakly nonlinear systems are very important both in applications with small degree of nonlinearity and for the case of large values of the nonlinearity parameter, where the interaction of modes plays an important role. It should be noted, that the necessary condition of the appearance of chaos is the availability of more than one oscillatory mode, and the problem of the analysis of chaotic dynamics is intimately related to the phenomenon of modes interaction in nonlinear systems. Such a study is usually conducted beginning from low levels of nonlinearity, with special techniques being used in quasillinear limit. We would like also to stress here that, as it has been shown both theoretically and experimentally, chaotic behavior appears at any, however small, value of nonlinearity parameter [5]. This means that, since there is always some degree of nonlinearity in any system, the deterministic chaoticity is an intrinsic property of almost any physical system with more than one degree of freedom.

In the present work we discuss the problems which are typical for the standard procedure of applying the Melnikov's criterion to strongly nonlinear systems and pay a special attention to some peculiarities characteristic for the weakly nonlinear limit.

2 The Outline of the Melnikov Method

It is well known that the phenomenon of dynamical chaos consists in local instability of phase trajectories and mixing. From this point, special trajectories, homoclinic or heteroclinic, are of particular importance, which are considered as the main source of complexity in dynamical systems. These trajectories are associated with periodic saddle orbits and lie on their invariant manifolds. The availability of such orbits guarantees the local stretching and folding of the phase flow in its vicinity and is necessary for chaotic instability to appear. It is, however, not sufficient, for there is dissipation in the system which prevents the trajectories from the global mixing. The Melnikov's criterion gives the threshold value of the perturbation which is sufficient for balancing the dissipation with mixing and arising of intermingled trajectories.

Technically, the application of the criterion is as follows. Consider a nonautonomous dynamical system

$$\frac{d\mathbf{x}}{dt} = \mathbf{f}(x) + \varepsilon \mathbf{g}(\mathbf{x}, t) \tag{1}$$

which is integrable at $\varepsilon = 0$ and possesses a saddle orbit $\mathbf{x} = \mathbf{X}_0$. In the integrable limit the system also has a homo- or heteroclinic trajectory $\mathbf{x}_0(t)$ which converges to the saddle as time goes to infinity $\mathbf{x}_0(\pm\infty) = \mathbf{X}_0$. Then, at $\varepsilon \neq 0$ the condition of intersection of the manifolds of the saddle is calculated

$$\Delta(t_0) \equiv \int_{-\infty}^{\infty} \mathbf{f}(\mathbf{x}(t-t_0)) \times \mathbf{g}(\mathbf{x}(t-t_0),t)dt = 0. \tag{2}$$

The existence of solutions of this equation evidences the presence of a homo- or heteroclinic structure in the phase space of the perturbed system, and, consequently, the appearance of a non trivial dynamics.

3 Problems of the Method

Although the method seems to be a universal tool for finding the chaos arising conditions, its practical implementation encounters problems, both of fundamental nature and the ones related to technical difficulties. We begin with the description of fundamental problems and then will briefly sketch the technical ones.

3.1 Intrinsic Limitations of the Melnikov's Criterion

The principal drawback of the method is that it gives neither necessary nor sufficient condition for the appearance of a chaotic dynamics. On the one hand, chaotic trajectories may originate from other saddle-type orbits, coexisting in the phase space with the one treated by the Melnikov's method. On the other hand, the homoclinic structure predicted by the method may be non attractive and do not provide the chaotization of motion. We illustrate these effects by an example of quasiperiodically forced Duffing oscillator [3]. Its dynamics in the vicinity of the primary resonance is described by the following set of equations

$$\begin{aligned} \frac{dU}{d\tau} &= \left[\Delta + \beta\left(U^2 + V^2\right)\right]V + \mu\left[P_2 \sin(\Omega\tau) - \delta U\right] \\ \frac{dV}{d\tau} &= \left[\Delta + \beta\left(U^2 + V^2\right)\right]U + P_1 + \mu\left[P_2 \cos(\Omega\tau) - \delta V\right] \end{aligned} \tag{3}$$

where U and V are generalized coordinates, μ is the small parameter. At $\mu = 0$ the system has a closed separatrix loop shown in Fig.1, and, as μ becomes nonvanishing, the Melnikov's method can be used for finding the analytical conditions of chaos onset [3].

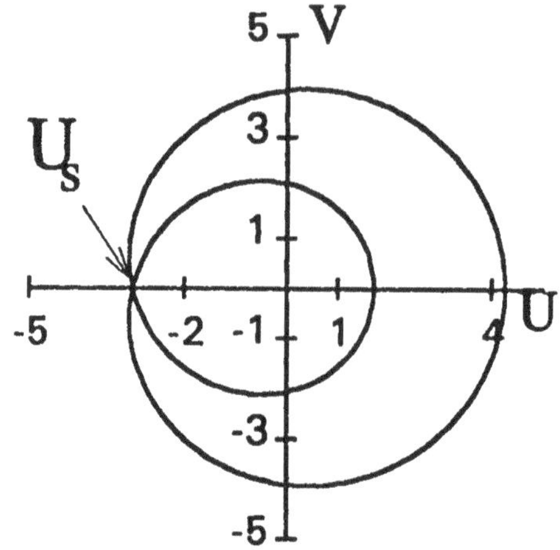

Figure 1: Separatrix loop for the system (3).

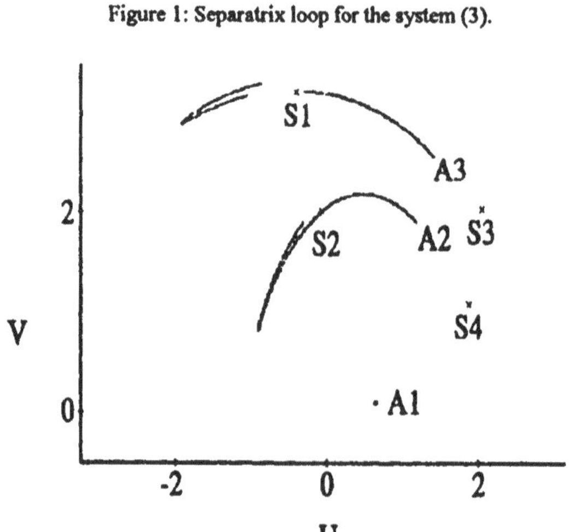

Figure 2: Poincare cross-section for system (3) at $\delta=1$, $\beta=-1.54$, $\Delta=15.05$, $P_2=1.97$, $P_1=3.75$, $\Omega=8.4$. A1-A3 - coexisting attractors, S1-S4 - period-one saddle points.

At a definite combination of control parameters the situation depicted in Fig.2 occurs, where both shortcomings of the Melnikov approach become clearly seen. In Fig.2 two strange attractors A2 and A3 coexist with a single stable periodic orbit A1 and four unstable saddle ones S1-S4. One can see that neither of chaotic attractors coincides with the unstable manifold of the saddle S3 corresponding to the point U_s of the unperturbed system. That means that Melnikov method can not be applied for deriving the conditions of chaotization of motion on the attractors A2 and A3, because each of them is related to another saddle orbit, S1 or S2, which are induced by the external force and are absent in the phase space of the unperturbed system. There is a homoclinic structure in the phase space which is shown in Fig.3. and is associated with the orbit S3 , but it is evidently not attractive.

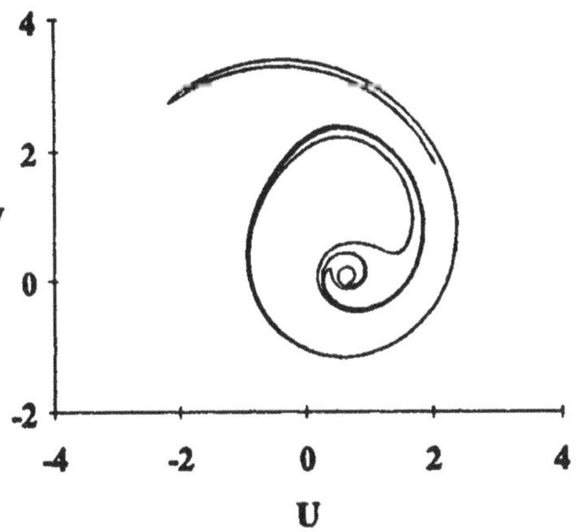

Figure 3: Unstable manifold of the unstable saddle-type orbit S3 from the Fig.2.

An additional problem which also should be mentioned is the role of dissipation. It immediately follows from the Melnikov criterion that the lower is the level of dissipation in the system, the smaller is the threshold value of an external force resulting in the chaotization of the motion. However, the direct numerical experiments indicate that the influence of dissipation is much more complicate. Let us demonstrate this by considering the system (3). At zero perturbation, we have an integrable system which can not demonstrate chaos. If to introduce an external excitation only, the stochastic layers appear in the phase space near unperturbed separatrixes, and their measure grows with the increase in the excitation amplitude P_2. These layers correspond to the motion near the homoclinic structures which appear in accordance with the prediction of the Melnikov method. The introduction of arbitrary small dissipation does not destroy the homoclinic structure, but may completely deteriorate the stochastic dynamics which needs a stronger dissipation to exist. In other words, there exists a threshold in dissipation parameter δ below which chaotic attractors become unstable. The situation is illustrated by the Fig.4,

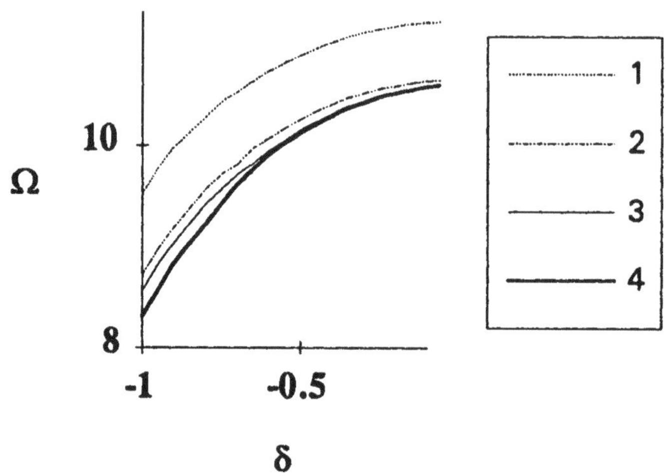

Figure 4: Bifurcation diagram of the system (3). Curve 1-first period doubling, 2-second period doubling, 3-strange attractor appearance, 4-chaos collapse.

where the bifurcation diagram of the system (3) illustrates the collapse of one of existing strange attractors under the decrease of dissipation.

3.2 Technical Difficulties

The problems of this type usually appear when one tries to apply the Melnikov method to the study of higher dimensional systems with the number of degrees of freedom larger than 1.5. In such cases the dimensionality of the separatrixes becomes larger than one and the complexity of the solutions in the unperturbed systems makes the problem substantially more difficult for the analytic solution. Apart from the problem of integration, there also exists the difficulty in defining the unperturbed part itself. It should be noted that for a particular problem there is typically no clear evidence of what terms in the equations must be included into the integrable part and which ones should be considered as perturbation. It is necessary always keep in mind that simple discarding of the "non essential" terms at the initial stage of applying the Melnikov method may result in either the disappearance of the saddle at comparatively small values of the perturbation or in practical impossibility of the analytic integration of the solution on the separatrix with subsequent calculation of the integral (2).

Let us consider, for example the following system, which describes the dynamics of three interacting modes with unidirectional action of one of the modes on the remaining two.

$$\begin{aligned}
\frac{da}{d\tau} &= -\delta_1 a - k_1 b\sin(\psi-\varphi) - P\sin\psi \\
\frac{d\psi}{d\tau} &= -\Delta + \beta a^2 - k_1\frac{b}{a}\cos(\psi-\varphi) - \frac{P}{a}\cos\psi \\
\frac{db}{d\tau} &= -\delta_2 b + k_2 a\sin(\psi-\varphi) \\
\frac{d\varphi}{d\tau} &= -\Delta + \mu - k_2\frac{a}{b}\cos(\psi-\varphi)
\end{aligned} \tag{4}$$

where a and b are amplitudes of two driven modes, and P is the amplitude of external excitation. The phase space of this system is four-dimensional, and if to consider the dissipation δ_i and external force P as the perturbation, we obtain the integrable Hamiltonian system at the initial stage of application of the Melnikov method. The corresponding solution on the separatrix has the form (5), if to introduce the new variables $c = a^2; I = a^2 + b^2; \theta = \psi - \varphi$ and parameters p ,r,and q

$$p = 2k\beta\sqrt{c_0(I_0-c_0)} - \frac{k^2 I_0^2}{c_0(I_0-c_0)}\ ;r = \beta k\left(\sqrt{\frac{I_0-c_0}{c_0}} - \sqrt{\frac{c_0}{I_0-c_0}}\right) = \frac{\beta k(I_0-2c_0)}{\sqrt{c_0(I_0-c_0)}}$$

$$q = \pm\beta\left(2k\left(\beta\sqrt{c_0(I_0-c_0)} - 2k\right)\right)^{\frac{1}{2}}$$

$$
\begin{aligned}
c^h(\tau) &= c_0 + \frac{2p}{q\cosh\left(\sqrt{p}(\tau-\tau_0)\right)-r} \\
\varphi^h(\tau) &= \arctan\left(\frac{-qp^{\frac{3}{2}}\sinh\left(\sqrt{p}(\tau-\tau_0)\right)}{k\sqrt{c_0(I_0-c_0)}\left(q\cosh\left(\sqrt{p}(\tau-\tau_0)\right)-r\right)^2+\frac{qp}{\beta}\left(r\cosh\left(\sqrt{p}(\tau-\tau_0)\right)-q\right)}\right) \\
I^h(\tau) &= I_0 \\
\theta^h(\tau) &= -\varphi^h(\tau)+\theta_0+\left(-\Delta_1+\beta c_0+k\sqrt{\frac{(I_0-c_0)}{c_0}}\right)(\tau-\tau_0)+ \qquad (5) \\
&+3\arctan\left(\frac{qe^{\sqrt{p}(\tau-\tau_0)}}{\beta\sqrt{p}}\right)+\arctan\left(\frac{qc_0e^{\sqrt{p}(\tau-\tau_0)}+2p-rc_0}{\sqrt{p}\left(\beta c_0-\frac{2kI_0}{\sqrt{c_0(I_0-c_0)}}\right)}\right)
\end{aligned}
$$

The calculation of Melnikov integral for this system is a cumbersome problem and we do not present here the final formula which is too large.

4 Peculiarities of Quasilinear Systems

The main difference between quasilinear and all other systems consists in the increased complexity of the integration of trajectories in the phase space due to the necessity to take into account several interacting modes in the analysis of chaotic regimes. This means that in such systems chaos can arise only as a result of interaction of several resonances which may coexist in arbitrarily large numbers at the same value of control parameters. There appears the problem of choice of the particular resonance, responsible for the formation of a given strange attractor, which is in essence the problem of separating the integrable part of the system [6]. In the simplest case of an oscillator with two frequency excitation (Eq. 3) the conditions for the formation of homoclinic structures obtained by the Melnikov technique essentially depend upon which of the spectral components in the external force is considered as perturbation, i.e. upon the initial choice of the governing resonance. This problem, being well defined in the regions of control parameter space where only one resonant condition is met, becomes non trivial in the areas where resonances overlap [5].

5 Discussion of Possible Solution

Melnikov's approach proposes one of the ways for finding the analytic criteria of chaos arising. The present work is devoted to the discussion of some aspects resulting from its nonuniversality and the necessity to use additional tools for the solution of the pointed out problem. We consider two types of difficulties: intrinsic problems of the method itself and technical ones resulting from the complexity of underlying sets of differential equations.

The solutions of the technical difficulties seems can be overcome in the majority of situations by direct numerical calculation of Melnikov integrals. This, however, may lead to the results which are not substantially different from direct numerical integration of the initial differential equations.

The intrinsic limitations of the method itself constitute much more profound problem, and, up to our knowledge, the solution does not exist so far. The only way which seems natural for conducting the Melnikov type analysis is the detailed investigation of all the resonances present in the system, combined with other known techniques, such as the methods of averaging or harmonic balance. This study needs an original approach to be developed for every system investigated and there is no evident way for finding a universal technique.

References

1. V.K.Melnikov, Trans. Moscow Math., 12, p.3 (1963).
2. S. Wiggins, Global bifurcations and chaos: Analytical methods. - New York, Springer-Verlag, 1987
3. V.B.Ryabov, D.M.Vavriv, Phys.Lett.A **153**, 431(1991).
4. K.Yagasaki, Physica D **44**, 445(1990).
5. D.M.Vavriv, V.B.Ryabov, S.A. Sharapov and H.M. Ito, Phys. Rev. E **53**, 103 (1996).
6. A.Yu. Levandovsky, Yu. A. Tsarin and D.M. Vavriv, High order parametric resonances and chaos in dissipative dynamical systems. Chaos, to appear.

NOISE IN BIOLOGICAL SYSTEMS

THE STATUS OF 1/f NOISE RESEARCH IN BIOLOGICAL SYSTEMS: EMPIRICAL PICTURE AND THEORIES

SERGEY M. BEZRUKOV

National Institutes of Health, DCRT, Laboratory of Structural Biology, Bethesda, MD 20892, USA
St.Petersburg Nuclear Physics Institute of the Russian Academy of Sciences, Gatchina, 188350 Russia

The scope of this short review is mostly restricted to fluctuation spectroscopy of ion currents through biological membranes. Examining membrane conductance fluctuations that represent the basic level of 1/f noise generation in biology, we show that this noise is not some inherent property of ion transport; rather, it is generated by independent discrete 'fluctuators'. Ion channels, nanometer-scale protein structures embedded in the membrane lipid matrix, fluctuate between open and closed states producing 1/f noise, while currents through their open steady states are remarkably free of this type of noise. Existing theories do not offer a satisfactory explanation of the phenomenological picture.

1 Introduction

Living organisms are nonequilibrium, nonstationary systems with highly developed hierarchical structures whose multiple subsystems are interconnected through complex mechanical, chemical, and electrical links. Electrical signals produced by these subsystems greatly vary both in amplitudes and in characteristic time scales. In this way, the biological complexity allows the hierarchy of characteristic times necessary for the generation of noise with 1/f type spectra. The dynamic activity of even a relatively simple biological object is a result of many kinetic processes at different 'layers of motion' related to each other by scaling factors[1]. As recently reviewed by Musha and Yamamoto[2], there is a substantial body of data on 1/f-like fluctuations in a wide range of biological systems spanning from cellular to behavioral levels.

The generating mechanisms of these fluctuations can be different in each case, depending on a particular system organization and its place in the organism hierarchy. For example, computer simulations[3,4] show that artificial neuron networks are already complex enough to generate 1/f noise by themselves. On the other hand, the temporal deviations in the heartbeat rate of a healthy human are mostly related to feedback interactions between the natural cardiac pacemaker (the sinoatrial node) and the central nervous system -- recent studies indicate that the heartbeat rate in recipients of heart transplants is much more stable[5].

In general, 1/f (flicker) noise seems to be present at every level where it has been looked for, from electroencephalograms of whole human brain[6] to currents through myelinated axons[7] isolated from individual nerve cells. The next logical question is: *what is the most basic level of 1/f noise generation in biology*, or, in other words, *what is the simplest biological object still capable of 1/f noise generation?*

2 Biological membranes as a source of 1/f noise

Biological membranes, that define boundaries of cells and control voltage gradients and ion fluxes between the cells, seem to represent this basic level. Many of them generate reproducible 1/f noise even under steady-state conditions. The first detailed measurements of flicker noise from a single node of Ranvier of an isolated neuronal axon were reported by Derksen and Verveen thirty years ago[8]. The structure of axons is relatively simple in comparison to the whole neuron (axons do not contain nuclei and other cell machinery); in addition, a three-terminal electrode arrangement with a correlation analysis of the resulting signals was used in these measurements that excluded possible contributions from electrodes or internode axoplasm. The authors were able to conclude that the 1/f noise was mostly related to the flux of potassium ions through the axon membrane.

Since then techniques of noise measurements on biological preparations have been further improved. For example, a difference procedure, which permits compensation of deterministic "drifts" in currents from biological preparations, was developed. As a result, the magnitude of 1/f-type spectral estimates reported for electrical noise of nerve membranes fell by orders of magnitude and Lorentzian noise components were readily measured (see reviews[9-15]). Nevertheless, the results of many recent experiments show that *the phenomenon of membrane 1/f noise does exist and cannot be considered as some artifact from nonoptimal data processing or poor sample preparation.*

3 Ion channels as elementary fluctuators

Ion channels are proteins of about 10 nm size that have a special polar pathway for ion conduction[16]. In biological membranes or when inserted in model lipid films (5 nm thick planar structures), ion channels form conductive sites (Fig. 1) whose properties are defined not only by the proteins themselves, but also by their interactions with surrounding lipids. They switch randomly between different conducting states with dynamics that depend on the cell functional parameters.

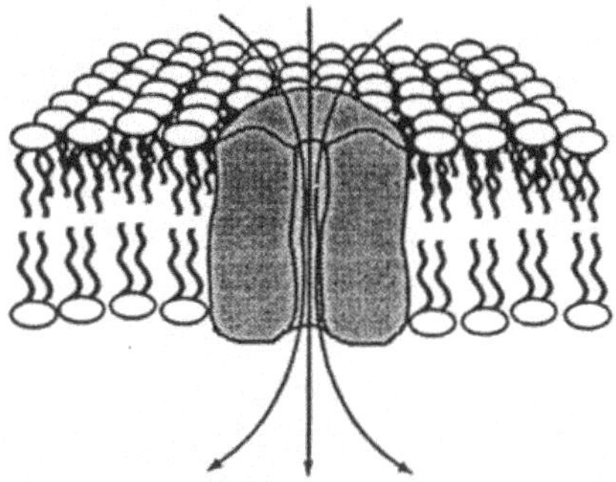

Figure 1: A schematic diagram of a protein ion channel in a membrane lipid matrix.

It was shown that 1/f noise in membrane current or voltage can be generated by artificial systems of planar lipid bilayer membranes with incorporated channel-forming compounds[17-21]. In a pioneering study[17] Sauve and Bamberg used a chemically dimerized form of the popular polypeptide channel-former, gramicidin A. Regular, monomeric gramicidin A incorporated into a lipid bilayer generates noise with a Lorentzian type power spectrum reflecting the association-dissociation reaction of ion channel formation[22]. In contrast to this, the dimerized analog, where association was stabilized by a covalent link, exhibited clear 1/f noise behavior over several decades of frequency. Comparison with the results obtained for large aqueous pores and porous synthetic membranes[23-25] led authors to the conclusion that 1/f noise in ion channel currents was a general phenomenon inherent to ion transport[17].

Several chemically dimerized gramicidin A analogs with differing linking chain lengths were used to study the effects of membrane lipid composition on 1/f noise[18-20]. It was found that 1/f noise intensity was very sensitive to the lipids used for membrane formation. For example, with other conditions similar, the power spectral density of the noise from the dimerized analogs in glycerolmonooleate/cholesterol membranes was 30 times higher than in dioleoyllecithin/cholesterol membranes. In all cases the spectral density was proportional to the number of channels over at least a hundred-fold increase in their density in the membrane. This indicates that *ion channels act as independent conductance fluctuators.*

Lipid bilayers in experiments reported in papers[17-21] were formed with the so-called paint-brush technique[26] which yields membranes with high residual content of nonpolar solvents[27]. In Fig.2 we present results of noise measurements for the "dry" membranes, formed using the monolayer opposition technique[27]. Monolayers were prepared from L-α-diphytanoyl lecithin solution in n-pentane. A dimerized gramicidin analog[a], succinyl-bis-desformyl gramicidin A, was added to the membrane-bathing aqueous solution (1M NaCl) to produce ion channels. The measurements were taken about 1 hour after the membrane formation, to allow for the equilibration between polypeptide in aqueous solution and in the membrane. It is seen that as the transmembrane voltage is increased, 1/f noise emerges from the equilibrium noise background. The noise intensity (normalized to membrane conductance and voltage) is several times smaller than the lowest one reported in references[17-20], thus once again demonstrating that *the parameters of the lipid matrix, that hosts ion channels, are crucially important.*

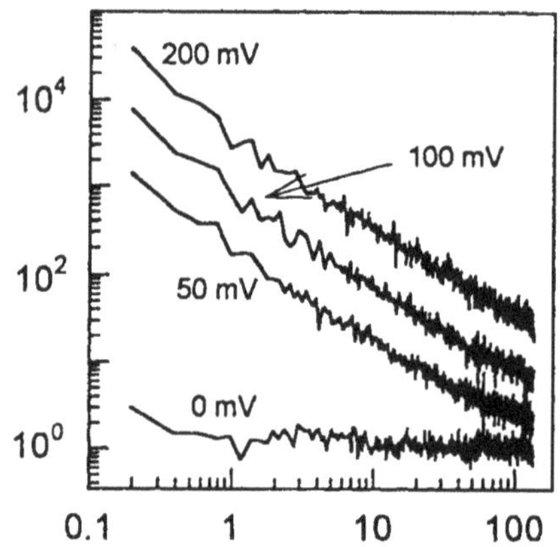

Figure 2: The spectral density of bilayer current noise ($10^{-28}A^2/Hz$) vs. frequency (Hz).

4 Noise of the current through an open channel

Studies performed at the level of a single ion channel give unique information about the mechanisms of noise generation. The possibility that ion transport through a permanently open ion channel is a source of 1/f noise was tested experimentally[18-19] in the following manner. If ion channels act as independent fluctuators, then, given the intensity of 1/f noise in a multichannel membrane, it is easy to calculate the root-mean-square current fluctuation expected from a single channel and compare it to the actual current record. The

[a] Succinyl-bis-gramicidin A is a generous gift by Larissa A. Fonina of the Shemyakin Institute of Bioorganic Chemistry, Moscow, Russia.

number of channels in a particular experiment can be determined as a ratio of the mean current through the multichannel membrane, $\langle I \rangle$, and the mean current through a single channel, $\langle i \rangle$. The mean-square current fluctuation expected from a single channel is:

$$\left\langle (\delta i)^2 \right\rangle = \frac{\langle i \rangle}{\langle I \rangle} \int_{f_l}^{f_h} S_I(f)\, df, \tag{1}$$

where $S_I(f)$ is power spectral density of 1/f noise measured from the multichannel membrane, f_h is the high-frequency cut-off used in the single channel recording to be analyzed, and f_l is the low-frequency limit that can be estimated from the overall time of single channel observation, t_o, as $f_l = 1/2\pi t_o$. The $\langle i \rangle / \langle I \rangle$ term represents the inverse number of simultaneously open channels.

The comparison of this calculation and a single-channel recording obtained from glycerol-monooleate/cholesterol membranes in 1 M KCl at 100 mV transmembrane potential is shown in Fig. 3 (adapted from references[18,19]). It is seen that the noise level calculated assuming the 1/f noise is due only to the current fluctuations of an open channel, is significantly higher than the observed level. It means that 1/f noise in the ion flow through a permanently open channel (if it exists at all) can not account for 1/f noise in the multichannel membrane.

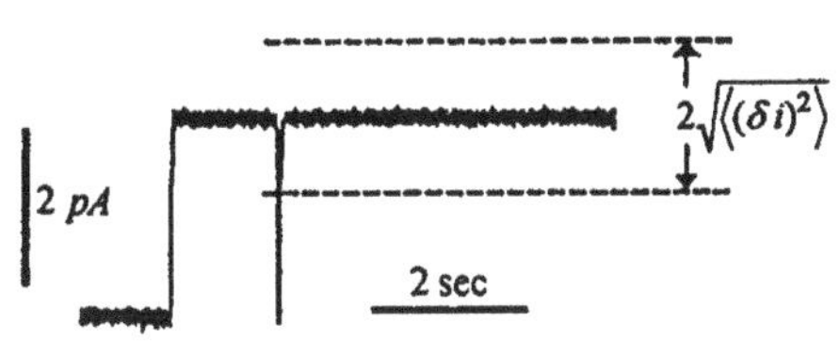

Figure 3: Noise of an open ion channel in comparison with the prediction of Eq. 1.

In fact, currents through permanently open ion channels are free from 1/f noise to a high degree of accuracy[28,29]. The level of flicker noise is so low that fluctuation spectroscopy of ion currents through open channels turns out to be a helpful structural and kinetic tool in ion channel biophysics[30-32]. The conclusion is that *the 1/f noise measured in biological membranes is not an inherent property of ion transport.* Rather, this type of noise is produced by fluctuator dynamics, that is, by random switching of the channels between their different conducting states.

5 Theoretical approaches

Theories developed so far to approach 1/f noise generation in ion conducting membranes can be divided into three main groups:

5.1 Electrochemical relaxation/diffusion models[33,34] (see also review[35] for more references). These models are based on the effects of the type of diffusion polarization or spherical influx into a sink that give rise to $1/t^{1/2}$ kinetics for ion transport. Appealing as they are, they do not account for the main experimental observations. In particular, high-resolution single channel recordings do not show relaxation tails at the opening/closing pulses; currents through permanently open channels are remarkably free from 1/f noise; membrane lipid composition plays a profound role in noise generation (see above).

5.2 Models of channel-lipid interactions[36-38]. Here, the interaction between ion channels and membrane lipids is postulated in such a way that the orientation of hydrocarbon chains of the neighboring lipids influences channel transport properties. The local membrane conductance is proposed to be linearly related to fluctuations in the 'hydrocarbon chain director' without specifying a particular mechanism[36] or in the sense of a time-dependent average by assuming that ion channel remains open for only a short time, smaller than 10^{-4} seconds[37], and that channel formation is coupled to the hydrocarbon chain orientation. Though these models address the important issue of the role of the lipid matrix in 1/f noise generation, unfortunately, in addition to some mathematical difficulties[38], they are in disagreement with recent experimental data. The requirement for conductance of ion channels to be linearly related to a fluctuating continuous parameter - directly or through fast dynamics -- is not consistent with single-channel data. Indeed, single-channel current measurements are usually done with a time resolution that is poorer than 10^{-4} s. It means that, in principle, the channel record of the kind shown in Fig.3 can be comprised of a succession of "unresolved" single-channel events; but, if the assumption of channel formation coupled to some fluctuating continuous parameter were true, the channel record would exhibit substantial fluctuations. Again, experiments[28,29] show that 1/f noise of a single open channel (or what is perceived to be a single channel) is virtually nonexistent.

5.3 Fractal models of channel-forming proteins[39-42]. Protein structures are fractal at least in two ways - in terms of protein backbone sequence and in terms of surface dimension[42]. As an alternative to simple Markov models, these approaches explore the idea that proteins exist in many conformational substates that are kinetically related to each other in a complicated way. Such complexity gives broad distributions of activation energy barriers, or time-dependent energy barriers, for transitions between open and closed states of a channel. Nonexponential kinetics for long-term conductance relaxation and fluctuations then naturally follow. These structural and dynamic interpretations are based on many different observations made on globular proteins. Fractal models are also in agreement with at least some of high-resolution single-channel recordings[39], though the generality of this issue has been questioned[43]. While the existence of complex dynamics in complex protein molecules seems quite plausible, the predictive power of this approach remains unclear. Experiments show that the complex dynamics, i.e., 1/f noise, can be obtained from the simplest channel-inducing polypeptides[17-20] and polyenes[21]. What is more, these dynamics are very sensitive to the composition[18-20] and the method of formation of lipid bilayers (this paper).

6 Random switching: 'weighted diffusion' model

As we have argued in the preceding section, there is no satisfactory explanation of membrane 1/f noise at present. Some of the theoretical approaches discussed above seem to be potentially useful, but before they can be tested even on a qualitative level, they should be brought in agreement with the modern experimental findings.

Interpretations of 1/f noise as a result of summation of several simple relaxation processes with Lorentzian spectra require the assumption of uniform or, at least, broad distributions of activation energies for ion channel open/closed transitions. For ion channels, the situation is probably even more challenging than that in solid state physics. For example, in the case of Lorentzians with exponentially distributed corner frequencies[44], the mechanism of activation energy broadening should be so 'delicate' as not to distort the open/closed channel equilibrium that defines the Lorentzian amplitudes.

As one of the outcomes, the search for alternatives to uniform activation energy distributions has led to the 'weighted diffusion' approach[45-49]. Free, unbiased diffusion of a particle with a spatial weighting function assigned to its instantaneous (and varying in time) position constitutes the essence of these models. In this way they are different from the recently described[50] mechanism of noise generation from the linear diffusion equation, where the system is driven externally by white noise boundary conditions. Analytical[45-47, 49] and computer simulation[48] methods showed a number of different possibilities for obtaining flicker[45,47-49], 'universal' diffusional[46], and simple Lorentzian[49] noise from the weighted diffusion approach.

In what follows we consider yet another unexplored option. To visualize the model, imagine a single ion channel undergoing free two-dimensional diffusion in the plane of a membrane. The ion channel is always fully open unless it diffuses into some 'special' areas of the membrane where it is closed (local membrane properties of 'special' areas trigger its transition to the closed conformation). Our computer simulations[51] show that a variety of random profiles for the 'special' areas generate time-invariant noises with $1/f^{\alpha}$ power spectra.

Surprisingly, a very simple and perhaps even physically plausible profile (the projection of a random polymer coil on the membrane surface) gives $\alpha = 1.02 \pm 0.02$. The profile, a particular example of which is presented in Fig. 4, is

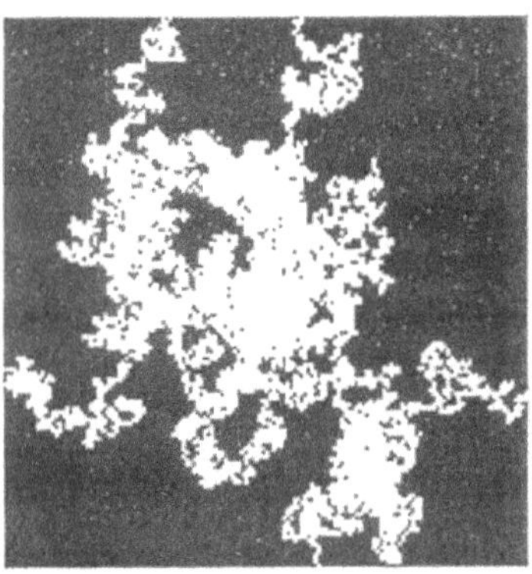

Figure 4: The profile generated by Brownian walkers for weighted diffusion simulations.

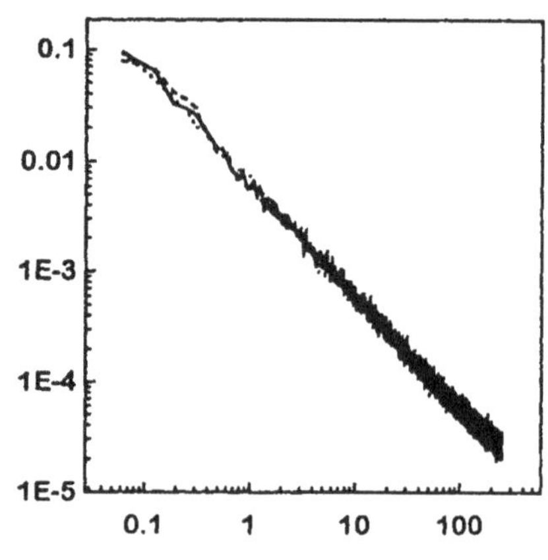

Figure 5: Spectral density of 'channel noise' obtained from weighted diffusion simulations.

constructed as an overlapping white traces of five random Brownian walkers (drunken ants) starting their trips in the center of 150 by 150 lattice. A channel does not conduct in the white areas. The spectra of the 'ion channel current' diffusing on such a lattice are shown in Fig. 5. Periodic or reflecting boundary conditions were used in simulations, no systematic difference was detected. Each spectrum represents not only new 'channel' runs but also a new set of Brownian walkers' traces. 'Single-channel' records intended to illustrate self-simularity at different time scales are shown in Fig. 6. A small-amplitude white noise is added to the records to enhance their likeness to real data obtained in single-channel experiments.

Computer simulations were checked and calibrated using a number of spatial weighting functions for which analytical solutions are readily available. For example, black and white stripes generated a 'universal' diffusional $1/f^{3/2}$ behavior at high frequencies[46]; a sine-wave profile gave a pure Lorentzian spectra with corner frequencies that were in excellent agreement with the straightforward calculations[49].

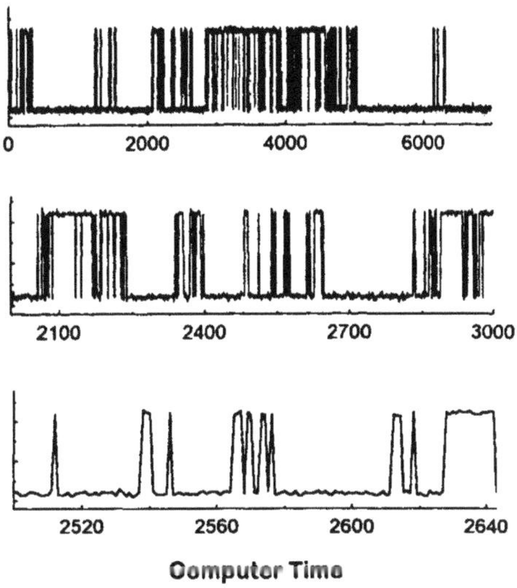

Figure 6: Simulated single-channel recordings at three time scales differing by a factor of 7.

7 Conclusions

We regard biological membranes as the 'basic' biological level of 1/f noise generation. Together with ion-transporting protein structures embedded in their lipid matrix, they represent the simplest biological objects still producing this type of noise[b]. The main points of this short review may be summarized as follows:

[b] In some sense a power-law scaling recently found for DNA base sequences[52-53] may be considered to be the most fundamental level of biological 1/f fluctuations. It should be kept in mind, however, that DNA reflects the enormous complexity of the whole organism, so it should be regarded as a very complex system falling in the category considered by West and Shlesinger[1].

- 1/f noise in biological membranes is not a fundamental (inherent) feature of ion transport.
- 1/f noise is generated by discrete fluctuators -- ion-transporting protein structures (channels) randomly switching between different conductance states.
- Complex dynamics of these structures are not yet completely understood. However, we demonstrate that the dynamics are highly sensitive to the membrane lipid matrix (its composition and the membrane formation technique); we also emphasize the potential capacity of 'weighted' diffusion in 1/f noise description.

Acknowledgments

We thank V.A. Parsegian, I. Vodyanoy, D.C. Rau, J.K. Chik, and J.J. Kasianowicz for fruitful discussions and comments on the manuscript. This work was supported by a grant from the ONR.

References

1. B.J. West and M. Shlesinger. The noise in natural phenomena. *American Scientist* **78**, 40 (1990).
2. T. Musha and M. Yamamoto. 1/f-like fluctuations of biological rhythm. In: *Noise in Physical Systems and 1/f Fluctuations*, ed. V. Bareikis and R. Katilius, 22 (World Scientific, Singapore, 1995).
3. M. Nakao, T. Takahashi, Y. Mizutani, and M. Yamamoto. Simulation study on dynamics transition in neuronal activity during sleep cycle by using asynchronous and symmetry neural network model. *Biol. Cybernetics* **63**, 243 (1990).
4. M. Usher, M. Stemmler, and Z. Olami. Dynamic pattern formation leads to 1/f noise in neural populations. *Phys.Rev.Lett.* **74**, 326 (1995).
5. P.A. Shapiro, R.P. Sloan, E. Bagiella, J.T. Bigger, and J.M. Gorman. Heart rate reactivity and heart period variability throughout the first year after heart transplamtation. *Psychophysiology* **33**, 54 (1996).
6. T. Yoshida, S. Ohmoto, and S. Kanamura. 1/f frequency fluctuation of human EEG and emotional changes. In: *Proc. Int. Conf. on Noise in Physical Systems and 1/f Fluctuations*, ed. T. Musha, S. Sato, and M. Yamamoto, 719 (Ohmsha, Ltd., Tokyo, 1991).
7. A.A. Verveen and H.E. Derksen. Fluctuations in membrane potential of axons and the problem of coding. *Kybernetik* **2**, 152 (1965).
8. H.E. Derksen and A.A. Verveen. Fluctuations of resting neural membrane potential. *Science* **151**, 1388 (1966).
9. C.F. Stevens. Inferences about membrane properties from electrical noise measurements. *Biophys.J.* **12**, 1028 (1972).

10. F. Conti and E. Wanke. Channel noise in nerve membranes and lipid bilayers. *Quart.Rev.Biophys.* **8**, 451 (1975).
11. E. Neher and C.F. Stevens. Conductance fluctuations and ionic pores in membranes. *Ann.Rev.Biophys.Bioeng.* **6**, 345 (1977).
12. L.J. DeFelice. *Introduction to Membrane Noise* (Plenum, New York, 1981).
13. B. Neumcke. Fluctuation of Na and K currents in excitable membranes. *Int.Rev.Neurobiol.* **23**, 35 (1982).
14. H.M. Fishman. Relaxations, fluctuations and ion transfer across membranes. *Prog.Biophys.Molec.Biol.* **46**, 127 (1985).
15. H.M. Fishman and H.R. Leuchtag. Electrical noise in physics and biology. *Curr. Topics Membrane Transp.* **37**, 3 (1990).
16. B. Hille. *Ionic Channels of Excitable Membranes* (Sinauer Associates, Sunderland, 1992).
17. R. Sauve and E. Bamberg. 1/f noise in black lipid membranes induced by ionic channels formed by chemically dimerized gramicidin A. *J.Membrane Biol.* **43**, 317 (1978).
18. S.M. Bezrukov, G.M. Drabkin, L.A. Fonina, A.I. Irkhin, E.I. Melnik, and A.I. Sibilev. Experiments on 1/f noise in lipid bilayer membranes. *Preprint 598 of the Leningrad Nuclear Physics Institute*, Leningrad, 1980.
19. S.M. Bezrukov, G.M. Drabkin, L.A. Fonina, A.I. Irkhin, E.I. Melnik, and A.I. Sibilev. Fluctuation spectroscopy of ionic currents through model membranes: 1/f noise problem. In: *Bilayer Lipid Membranes*, 64 (Far East Science Publishers, Vladivostok, 1983).
20. A.I. Irkhin, S.M. Bezrukov, and E.I. Melnik. Properties of ionic channels formed by gramicidin A dimeric analogs in lipid bilayers. *Biologicheskie Membrany* **1**, 739 (1984).
21. S.M. Bezrukov and R.A. Brutyan. Fluctuations of the lipid bilayer electroconductance at one-sided application of polyene antibiotics. *Biophysics* (Engl. Transl.) **32**, 573 (1987); *Biofizika*, **32**, 526 (1987).
22. R.E. Koeppe II and O.S. Andersen. Engineering the gramicidin channel. *Annu.Rev.Biophys.Struct.* **25**, 231 (1996).
23. D.L. Dorset and H.M.Fishman. Excess electrical noise during current flow through porous membranes separating ionic solutions. *J.Membrane Biol.* **21**, 291-309 (1975).
24. L.J. DeFelice and J.P.L.M. Michalides. Electrical noise from synthetic membranes. *J.Membrane Biol.* **9**, 261 (1972).
25. M.E. Green and M. Yafuso. A study of the noise generated during ion transport across membranes. *J.Phys.Chem.* **72**, 4072 (1968); corr: *J.Phys.Chem.* **73**, 1626 (1969).
26. P. Mueller, D.O. Rudin, H.Ti. Tien, and W.C. Wescott. Methods for the formation of single bimolecular lipid membranes in aqueous solution. *J.Phys.Chem.* **67**, 534 (1963).
27. M. Montal and P. Mueller. Formation of bimolecular membranes from lipid monolayers and a study of their electrical properties. *Proc.Natl.Acad.Sci.USA* **69**, 3561 (1972).
28. S.M. Bezrukov, A.I. Irkhin, and A.I. Sibilev. An upper estimate for 1/f noise in ionic conductors from experiments with a molecular microcontact. *Physics Lett.* **123**, 477 (1987).
29. S.M. Bezrukov and I. Vodyanoy. Electrical noise of the open alamethicin channel. In: *Proc. Int. Conf. on Noise in Physical Systems and 1/f Fluctuations*, ed. T. Musha, S. Sato, and M. Yamamoto, 641 (Ohmsha, Ltd., Tokyo, 1991).
30. S.H. Heinemann and F.J. Sigworth. Open channel noise. V. Fluctuating barriers to ion entry in gramicidin A channels. *Biophys.J.* **57**, 499 (1990).

31. S.M. Bezrukov and J.J. Kasianowicz. Current noise reveals protonation kinetics and number of ionizable sites in an open protein ion channel. *Phys.Rev.Lett.* **70**, 2352 (1993).
32. S.M. Bezrukov, I. Vodyanoy, and V.A. Parsegian. Counting polymers moving through a single ion channel. *Nature (London)* **370**, 279 (1994).
33. E. Frehland. Diffusion as a source of 1/f noise. *J.Membrane Biol.* **32**, 195-196 (1977).
34. M.E. Green. A model for channel noise, including the effect of diffusion. *Acta Biotheoretica* **27**, 61 (1978).
35. B. Neumcke. 1/f noise in membranes. *Biophys.Struct.Mechanism.* **4**, 179-199 (1978).
36. I. Lundstrom and D. McQueen. A proposed 1/f noise mechanism in nerve cell membranes. *J.Theor.Biol.* **45**, 405 (1974).
37. J.R. Clay and M.F. Shlesinger. Unified theory of 1/f noise and conductance noise in nerve membrane. *J.Theor.Biol.* **66**, 763 (1977).
38. M.B. Weissman. Models for 1/f noise in nerve membranes. *Biophys.J.* **16**, 1105 (1976).
39. L.S. Liebovich and T.I. Toth. Fractal activity in cell membrane ion channels. *Ann.N.Y.Acad.Sci.* **591**, 375 (1990).
40. T.G. Dewey and J.G. Bann. Protein dynamics and 1/f noise. *Biophys.J.* **63**, 594 (1992).
41. L.S. Liebovitch. Interpretation of protein structure and dynamics from the statistics of the open and closed times mesured in a single ion-channel protein. *J.Stat.Phys.* **70**, 329 (1993).
42. T.G. Dewey. Fractal analysis of proton exchange kinetics in lysozyme. *Proc.Natl.Acad.Sci.USA* **91**,12101 (1994).
43. M.S.P. Sansom, F.G. Ball, C.J. Kerry, R. McGee, R.L. Ramsey, and P.N.R. Usherwood. Markov, fractal, diffusion, and related models of ion channel gating. *Biophys.J.* **56**, 1229 (1989).
44. R. Sauve and G. Szabo. Interpretation of 1/f fluctuations in ion conducting membranes. *J.Theor.Biol.* **113**, 501 (1985).
45. J.M. Richardson. The linear theory of fluctuations arising from diffusional mechanism -- an attempt at a theory of contact noise. *Bell Syst.Tech.J.* **29**, 117 (1950).
46. K.M. van Vliet and J.R. Fasset. Fluctuations due to electronic transitions and transport in solids. In: *Fluctuation Phenomena in Solids*, ed. R.E. Burges, 267 (Academic Press, New York, 1965).
47. M.B. Weissman. A model for 1/f noise from singularities in the fluctuation weighting function. *J.Appl.Phys.* **48**, 1705 (1977).
48. Z. Gingl, L.B. Kiss, and R. Vajtai. $1/f^k$ noise generated by scaled brownian motion. *Solid State Comm.* **71**, 765 (1989).
49. S.M. Bezrukov and I. Vodyanoy. Noise in biological membranes and relevant ionic systems. In: *Advances in Chemistry Series. 235. Biomembrane Electrochemistry*, 375 (American Chem.Soc., Washington, DC, 1994).
50. H.J. Jensen. 1/f noise from the linear diffusion equation. *Physica Scripta* **43**, 593 (1991).
51. S.M. Bezrukov and M.A. Pustovoit. Manuscript in preparation.
52. R.F. Voss. Evolution of long-range fractal correlations and 1/f noise in DNA base sequences. *Phys.Rev.Lett.* **68**, 3805-3808 (1992).
53. C.-K. Peng, S.V. Buldyrev, A.L. Goldberger, S. Havlin, F. Sciortino, M. Simons, and H.E. Stanley. Long-range correlations in nucleotide sequences. *Nature (London)* **356**, 168 (1992).

STORAGE CAPACITY OF ASSOCIATIVE MEMORY WITH NON MONOTONE NEURONS

IOAN OPRIŞ

Department of Anatomy and Neurobiology, College of Medicine, University of Tennessee at Memphis, 855 Monroe Ave., Memphis, TN 38163

The mechanisms of associative memory with enhanced storage capacity occurring in neural networks with non monotone neurons generate interesting questions for artificial intelligence and neuroscience. The questions addressed in this paper focus on the Hopfield energy landscape and stability of non monotone neuron networks, the bifurcation diagram approach of fixed point attractor multiplication, the biophysical relevance of these types of neurons and the noise induced effects.

1 Introduction

In the last two decades a lot of attention has been given to the emergence of "collective phenomena" in a variety of relatively simple systems [1-5] mimicking the brain's neuronal networks. There are several reasons which motivate the investigation of memory storage properties of such networks. These include the impressive increase in the computer power and speed on one hand, and the developments in the theoretical understanding of neural phenomena which attempt to narrow the gap between neural computing and neurophysiology or neuropsychology on the other hand. In this regard the understanding of storage capacity mechanisms involved in the neural computing seem to be of wide interest. The associative memories with enhanced storage capacity occurring in neural networks with non monotone neurons will be discussed in this paper.

One of the most essential part of the neural network is the neural cell, commonly called "neuron". This is a complex nonlinear system [6,7] which do the processing of neural activity. It can be represented by simplified elements (binary, analog, nonlinear or integrate and fire neurons). In this paper we deal with the networks of interconnected elements modeled by analog neurons [5]. The behavior of the analog neuron is given by an input-output function called transfer function (fig.1.). The synapse function is modeled by a modifiable weight(coupling) associated with each connection. Each neuron converts the pattern of incoming activities into a single outgoing activity to other units. It performs this conversion in two stages: first, it multiplies each incoming activity by the weight of the connection and adds together all these weighted inputs to get a quantity called "total input", then, the neuron uses an input-output function that transforms the "total input" into the outgoing activity.

The weights are modified according to a learning rule. In our case, the Hebbian rule has the well known interpretation: if two connected neurons are activated at a given moment, the connection between them is reinforced. In all other cases the connection is not modified. The neural network models of the associative memory are dynamical systems with associated attractors to the cognitive events. An associative memory is a content addressable memory in which different input patterns become associated to one another [4]. The memories are defined as dynamically stable attractors and the retrieval (recollection) process as a down hill motion in the energy landscape [4,5], the energy function playing the role of a Lyapunov function.

In 1990 Morita et al. [8] have introduced an interesting model in which the elements of the network obey a non monotone input-output relation rather than a conventional sigmoid one. The associative memory of these systems have some remarkable storage properties [8-16], such as an enhanced storage capacity, an improved memory attractor convergence which reduce the spurious attractors and a rich diversity in dynamical behavior ranging from fixed point attractors to chaotic dynamics. Another associative model with enhanced storage capacity and improved convergence was proposed by Opris [17]. The high storage capacity in this case is enabled by the self-consistent noise of the competing patterns.

2 Solved Aspects

Let us consider some solved aspects of associative memories exhibiting an enhanced storage capacity. The first one deals with the non monotone neurodynamics, conjectured by Morita [8-10]. He imagined his partial reversed method as an algorithm to improve the recollection dynamics based on the fact that "thermal noise" avoids the local minima of the energy and get the network state into stable attractors. The algorithm of the partial reversed method comprises two steps at each stage: in the first step the network evolves according to a conventional dynamics and in the second step the outputs of some neurons are reversed. Numerical experiments have shown that:

a) the critical overlap is smaller than that of conventional dynamics, which means that the recollection ability of the network is raised;

b) the number of time steps required to complete the recollection process is decreased;

c) the storage capacity is enhanced compared with that of the conventional case;

d) the noise variance is reduced after the second step.

The non monotone gain function introduced by Morita et al.[9,10] (fig.1), based on their partial reversed algorithm can be rewritten as a product between the ordinary sigmoid function and a function resembling the 'Mexican hat' shape [17]. This way of expressing the input-output function is showing us that the second function plays an important role in the regulatory mechanism of the global activity of the network[28].

The absolute storage capacity of the associative memory with non monotone neurons was analytically derived by Kobayashi [14] following the techniques of Gardner [18]. The maximum value of this parameter is around 10 for an input threshold of h=1.22. Comparing these results with those of Gardner, a five times enhancement in the storage capacity, given by the non monotone conjecture, for well suited values of input, is quite exciting. After the matching between Gardner's model [18] and those for analog networks [9,11,15,16], as it was expected, the storage capacity can take values around 0.7 for a continuous time dynamics. These results have already been obtained by Shiino and Fukai[16], and by Yoshizawa et al.[10]. The discrete time dynamics gave lower performances also to Morita [8,9] for his partial reversed method. By taking the equilibrium limit of the neurodynamics recursion relations we find an enhanced memory capacity of 0.22 [11-13], compared with the conventional monotonic one [2,3] of 0.138.

Two methods of investigation the retrieval process in the general case are mentioned here. The first one is a generalization of the Amari-Maginu neurodynamics[20]. It employs the signal-to-noise ratio analysis [11-13,20], and applies to any analog neural network. This theory was extended further to the non monotone neuron network. The phase diagram of storage capacity plotted versus input threshold, regarded as non monotonicity parameter evidences various retrieval regimes. The results have been compared with the spin-glass variant of Little-Hopfield model [21] and checked by Monte Carlo simulations[13]. The number of neurons used in the simulations were 5000, 7000 and 9000. The second investigation considers a self-consistent signal to noise ratio method [17,18], and takes into account the fatigue effect of the neuron [18]. A discrete time master equation framework[17] has been used to carry out the analysis. We note that in the low temperature limit and fatigue absence, for a sigmoid transfer function, the storage capacity approaches 1. Also, we have plotted the phase diagrams showing first and second type phase transitions. All these schemes have been imagined for any general odd input-output function.

3 Unsolved Aspects

3.1 The mechanism of storage capacity

The corresponding Hopfield energy landscape. The corresponding Hopfield energy landscape [4] of non monotone neural network is very difficult to be defined; this happens because the non monotone input-output function does not have an inverse. If this difficulty can be avoided, then the energy landscape should exhibit more global minima for memory attractors, and less meta stable states for spurious memories, as it has been pointed out by Morita[8,9]. What would happen if the non monotone behavior of neurons will be implemented in an Ising-like neural network?. It is difficult to say, because the analysis can be done only numerically. There will probably be also some interpretation obstacles, but in this way we probably can get the corresponding global minima of the energy landscape, etc., because the theoretical framework is well known.

The bifurcation diagram approach. If we choose 'h' from fig.2, as control parameter one can characterize the graded change from monotone to nonmonotone behavior. By plotting the number of global minima versus control parameter 'h' we expect to get a bifurcation diagram showing thus the increase of fixed point attractors(memories). The multiplication of fixed point attractors occurs when one pass from the sigmoid type of input-output function of the neuron to the nonmonotone one. This being a nonlinear phenomenon it is expected that the multiplication takes place through a Feigenbaum-like bifurcation scenario [6]. The multiplication number being around 5, may confirm this hypothesis. If the superpositions of many attractors together with the static noise effects and chaos regims are taken into account, this may be a valid idea. In order to perform it we need to find a global minimum and see if this splits into two or more minima. If this happens we are very near the answer to our question. Anyway, an overlap bifurcation diagram [28] will certainly give some good insights as Bolle et all. [28,29] already reported.

3.2 The stability of neural networks with non monotone neurons

A Lyapunov function or a Hopfield type Lyapunov function requires the inverse of input-output function [5]. As far as I know, this is very difficult to be defined and it was already mentioned (because the non monotone input-output function does not have an inverse). But if we can reconsider this point in the framework of nonlinear theory as it was suggested by Aoyagy[22], which have built a Lyapunov function for a nonlinear model of Van der Pol coupled oscillators associative memory, then it may be a way. The paradox which arises here is that the storage capacity of coupled nonlinear oscillator neural network is much less than that of Hopfield model. The stability of neural networks with non monotone neurons is a not yet answered question, mainly because the dynamics is so rich.

3.3 Biophysical relevance for non monotone neurons

The answer to this question is partially given by recent experiments from cat auditory cortex in which non monotone firing rate of neurons in response to level stimulus has been recorded[30]. These experiments also points to the correlation between non monotone firing rate and a lateral inhibition mechanism. The electro-neurophysiology of neuronal membrane gives us additional insight by evidencing various ionic currents which exhibit nonlinear dependence on biophysical variables; among these effects can be mentioned those that modulates the firing activity, yielding thus a non monotone firing rate. In this respect Horikawa [23] has shown that two coupled FitzHugh-Nagumo neurons exhibit non monotone firing rate. Abbott [7] has already shown that the physiological neuron model of Hodgkin and Huxley preserves some of the nonlinear features of FitzHugh-Nagumo neuron [24]. Extending this model, Hindmarsh and Rose[25] got a bursting neuron model which mimics the firing of real neuron from thalamus [26]. A stochastic nonlinear neuron model [27] may also exhibit some non monotone firing rate. Finally, I would like to emphasize that the non monotone like behavior of the neurons may be considered as an "effect of regulatory mechanism" of the global activity in the network[28].

3.4 Stochastic neural network with non monotone neurons

If we add a noise term in the dynamics equation we get a stochastic neural network model with non monotone neurons. Usually the noise contributes to the decreasing of storage properties but there may arise some noise induced effects. A clear cut answer cannot be given for the same reason -the missing of Lyapunov function- required as energy function by the Fokker-Planck equation of the network.

4 Conclusion

It is not completely clear what the mechanism behind this memory enhancement is, however, I have only highlighted few possible keys to solve this question. As I already outlined, some of them yield to paradoxes. The nonlinear approach based on bifurcation diagrams seem to catch some of this rich dynamical behavior at both levels: neuron and network.The non monotone neuron approach evidences also a relationship between local and global properties of neural ensemble. Finally, it has to be mentioned the noise induced effects which may play an important role in such complex systems. The answer to the questions raised by non monotone input-output function seems to be useful not only from the point of view of artificial intelligence but also may give a clue for neurophysiological investigations based on simultaneous multi-unit recording.

References

1. J.W. Clark, Phys. Rep. **158, 2**, 91 (1988).
2. D.J. Amit, *Modeling brain functions*, (Cambridge University Press, 1989).
3. J. Hertz A. *et al*, *Introduction to the Theory of Neural Computation*, (Addison Wesley, 1991).
4. J.J. Hopfield, Proc. Natl. Acad. Sci. U.S.A. **79**, 2554 (1982).
5. J.J. Hopfield, Proc. Natl. Acad. Sci. U.S.A. **81**, 3088 (1984).
6. J. Guckenheimer and P. Holmes, *Nonlinear oscillations, dynamical systems and bifurcation of vector fields*, (Springer, 1986).
7. L. F. Abbott, *Talk Presented at XI Sitges Conference on Neural Networks, Sitges, Spain*, (1990).
8. M. Morita *et al*, Trans.IEICE **J73-D-II(2)**, 242 (1990).
9. M. Morita, Neural Networks **6**, 115 (1993).
10. S. Yoshizawa *et al*, Neural Networks **6**, 167 (1993).
11. H. Nishimori and I. Opris, Neural Networks **6**, 1061 (1993).
12. H. Nishimori and I. Opris, in *Proc. IEEE ICNN, San Francisco*, 351 (1993).
13. H. T. Nishimori *et al Computer Aided Inovation of New Materials II*, (North-Holland, 383-388, 1993).
14. K. Kobayashi , Network **2**, 237 (1991).
15. N. Brunel and R. Zecchina, Phys. Rev. E **49, 3**, 1823 (1994).
16. M. Shiino and T. Fukai , *Private communication at Tokyo Institute of Technology*, (1993).
17. I. Opris, Phys. Rev. E **3**, 2169 (1995).
18. D. Horn and M. Usher, Phys. Rev. A **40, 2**, 1036 (1989).
19. E. Gardner, J. Phys. A **21**, 271 (1988).
20. S. Amari and K. Maginu, Neural Networks **1**, 63 (1988).
21. J. F. Fontanari and R. Koberle, Phys. Rev.A **36**, 5 (2475) 1987.
22. T. Aoyagi, Phys. Rev. Lett. in press , (1995).
23. Horikawa Y., in *Proc. IEEE ICNN, San Francisco*, 471, (1993).
24. R. FitzHugh, Biophys. J. **1**, 445 (1961).

25. J.L. Hindmarsh and R.M. Rose, Proc.Roy.Soc.Lond. **B221**, 87 (1984).
26. R.M. Rose and J.L. Hindmarsh, Proc.Roy.Soc.Lond. **B225**, 161 (1985).
27. T. Ohira and J. D. Cowan, Neural Computation **7**, 518 (1995).
28. D. Bolle and B. Vinck, Physica A **223**, 293 (1996).
29. D. Bolle and R. Erichsen Jr, J. Phys.A: Math. Gen. **29**, 2299 (1996).
30. M. B. Calford and M. N. Semple, J. Neurophysiol. **73**, 1876 (1995).

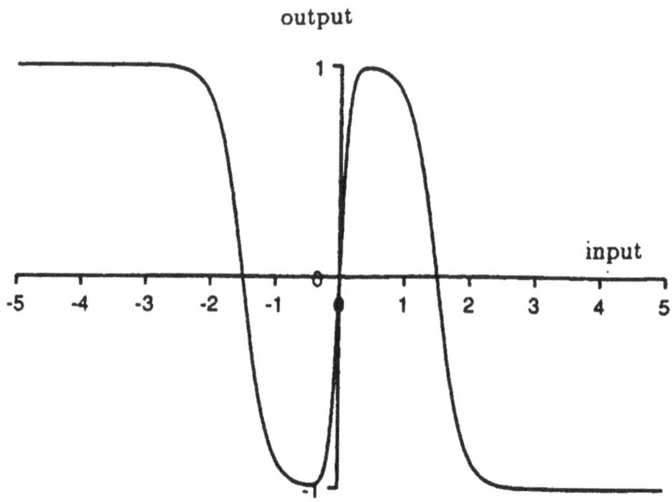

Fig. 1 The non monotone transfer function.

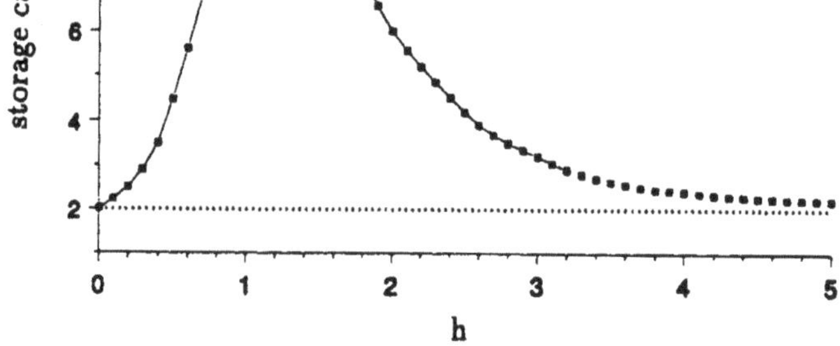

Fig. 2 The storage capacity versus non monotonicity.

NOISE IN HIGH-T_c SUPERCONDUCTORS

RANDOM TELEGRAPH VOLTAGES IN HIGH–T_c SUPERCONDUCTORS

GRZEGORZ JUNG[a] and YAKOV YUZHELEVSKI
Department of Physics, Ben Gurion University, 84105 Beer Sheva, Israel

BONAVENTURA SAVO and CORRADO COCCORESE
Dipartimento di Fisica, Universita' di Salerno, 84081 Baronissi (SA) Italy

VILYAM ASKHENAZI and BORIS YA. SHAPIRO
Department of Physics, Bar-Ilan University, 52100 Ramat Gan, Israel

Random telegraph voltages in high-T_c superconducting films have been investigated experimentally. The mechanism laying behind the two-level fluctuations in granular films is well understood. Plausible models have been proposed to explain the experimental observations of telegraph voltages in oriented and epitaxial high–T_c films in the current induced dissipative state only. The noise seen in the presence of applied magnetic field still awaits an explication.

1 Introduction

High temperatures of operation, combined with strong anisotropy and relatively low pinning energies of high critical temperature superconducting (HTSC) materials result in pronounced magnetic flux noise. Voltage noise is generally traced back to an indirect processes involving the fluctuator mechanism governing the dynamics of random flux processes and the detector mechanism coupling these fluctuations to the observable voltages.

The fluctuating component of a voltage drop across dc current biased HTSC samples frequently takes form of bias and magnetic field dependent random telegraph voltage signals (RTS)[1]. In this paper we discuss the experimental aspects of random telegraph noise manifestations in HTSC thin films. We present the events that we well understand[2,3,4,5,6], followed by the description of the events that we think we understand, and concluded with the RTS aspects that constitute for us the Unsolved Problems of Noise in HTSC.

2 What we see experimentally.

In the experiments we are measuring the fluctuations of a voltage developing across dc current biased HTSC samples. We have detected voltage telegraph fluctuations with switching rates ranging from mHz up to MHz frequencies,

[a]also with Instytut Fizyki PAN, Warszawa, Poland

in various HTSC systems possessing different geometrical arrangements and structural properties. The RTS events discussed in this paper disappear at temperatures exceeding the sample critical temperature. This fact enables one to univocally associate RTS generation with an exclusively superconducting mechanisms. The observed events are thermally activated in the entire temperature range of our experiments, from T_c down to 4.2 K.

The fingerprints of random telegraph voltage noise in HTSC are strong and characteristic dependencies of the RTS waveform on bias conditions[3,6]. The most puzzling feature of RTS manifestations in HTSC is however, their appearance in large superconducting samples. HTSC telegraph noise is clearly associated with spatially extended macroscopic two-level fluctuator possessing dimensions comparable with a sample size [6].

3 What we understand.

Paradoxically, we understand best the RTS signals appearing in the system that is most complicated, i.e., in a disordered granular HTSC sample. In the following we shall discuss only in brief the RTS events in granular systems. For the detailed description we guide readers to our previously published papers.

We ascribe RTS voltages switching at UHF rates in granular films[2] to the interrupted random walk of vortices[5]. Generation of low frequency RTS in granular films involves thermally activated jumps of flux lines that convert into observable voltages within a current biased intrinsic dc quantum interferometer[4]. The switching rate of is controlled by a stress imposed on the pinning sites by a screening current circulating in an itragranular loop containing a Josephson junction biased by the current flow [3].

4 What we think we understand.

Despite numerous experimental observations of macroscopic RTS voltages in oriented and epitaxial HTSC films we are still unable to suggest a reliable physical mechanism that would explain the amplitude and switching rates of these events. We observe a markedly different noise behavior in the absence and in the presence of weak perpendicular magnetic field. Shown in Fig.1 is the dependence of RTS amplitude on current flow, measured at 77 K at zero applied field. The linear increase of RTS amplitudes with increasing current, symmetric with respect to the direction of the current flow, points out to a mechanism consisting in resistance fluctuations converted into RTS voltage by a dc current flow. Observe that in the simplest case of flux-flow dissipation the change ΔN in the number of flowing vortices changes the flux-flow resistance

$$R_f = \frac{B\Phi_o}{w\eta} = N\frac{\Phi_o^2}{w\eta}, \tag{1}$$

(where η is the viscosity coefficient, Φ_o the flux quantum, N the total number of flowing vortices, w the width, and l the length of the strip), by $\Delta R \propto \Delta N$, and produces a voltage signal with an amplitude $\Delta V \propto \Delta N$. The correspond-

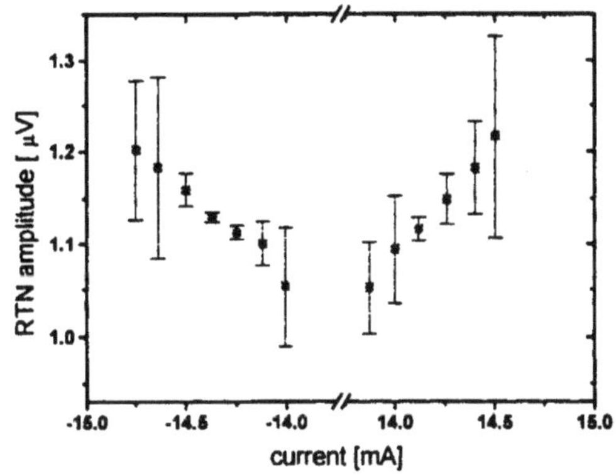

Figure 1: The dependence of telegraph noise amplitude on current flow.

ing experimental dependencies of average RTS lifetimes in up and down RTS voltage states on current flow are shown in Fig. 2. For thermally activated RTS switching the average lifetime, in the first order approximation, is

$$\tau = \tau_0 exp\left(\frac{U(I)}{kT}\right) = \tau_0 exp\left(\frac{U_0}{kT}\right) exp\left(\frac{I}{kT}\frac{\partial U(I)}{\partial I}\right) \tag{2}$$

Thus a plot in Fig. 2, shows directly the dependence of the activation energy on current flow. One easily notices a clear difference between the linear behavior of τ_{dn}, and nonmonotonic dependence of τ_{up} in Fig. 2. Clearly the lifetimes of two RTS states are governed by physically different mechanisms. We tentatively propose a scenario in which the *up* RTS voltage state corresponds to flux in motion, while the *down* voltage develops during the times when flux is immobile. Such mechanism is consistent with the linear amplitude characteristics, see Fig. 1, and with an exponential decrease of τ_{dn} with increasing current

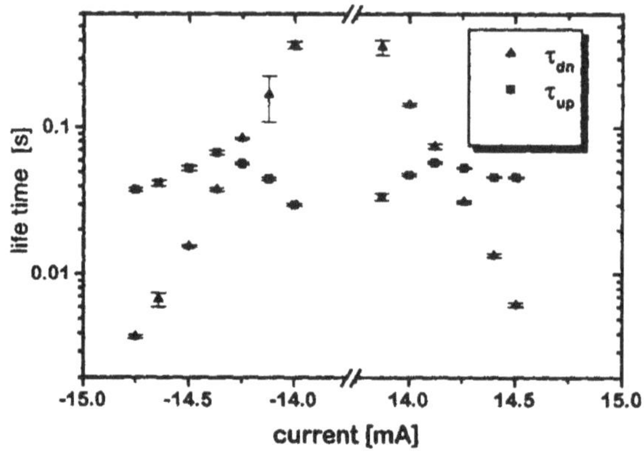

Figure 2: The dependence of RTS average lifetimes on current flow at 77 K and B=0.

in Fig. 2. The linear decrease of the activation energy associated with τ_{dn} can be attributed to the increasing strength of the Lorentz force with increasing current flow. The nonmonotonic behavior of τ_{up} may be seen a superposition of two opposing effects; a linear decrease of the life time due to increased velocity of flight and the increase in the lifetime due to decrease of the pinning probability at high velocities.

The Lorentz force action moves the bundle across the stressed energy well at a distance x where the bundle becomes free to move. The moving bundle produces the excess voltage corresponding to the RTS amplitude $\Delta V = V_{up} - V_{dn}$. The free energy change associated with the transition of a bundle into the state of flow is $U(I) = BJxS_b d$, where $B = N\Phi_0/S - b$, is the magnetic induction, N the number of vortices in a bundle of a surface S_b, and $J = I/wd$ the current density in a strip of a thickness d. Consequently, the slope of the linear characteristic in Fig. 2 is

$$\frac{\partial U}{\partial I} = \frac{\partial}{\partial I}\left(\frac{N\Phi_0}{S_b}\frac{I}{wd}xS_b d\right) = N\Phi_0\frac{x}{w} \tag{3}$$

The experimentally determined slope is $\partial U/\partial I = 2.9 \pm 0.110^{-18} J/A$, i.e.,$xN =$ 75nm. Assuming that the size of a pinning site is of the order of the coherence length ξ we find that the bundle contains N 1000 flux quanta.

At this point we should verify the correctness of the approach using Eq. 1. The experimental data render the viscosity coefficient h of the expected value

$\eta \propto 10^{-8}$ SI units. However, the evaluated time of flight of vortices across the strip is order of magnitude shorter then τ_{up}. This apparent discrepancy can be eliminated by adopting the model in which the vortices move in channels across the sample width. The channel opens and closes in a random way and thus τ_{up} corresponds to the average time of opening of the channel while during τ_{dn} the channel remains closed. In real samples the two edges of the strip that are supposed to control the dynamics of channel opening are not equivalent. Thus the noise characteristics should be asymmetric with respect to the direction of the current flow. Clearly, in our case the picture is symmetric, see Figs 1 and 2, and the detailed mechanism causing the randomness in channel opening and closing still has to be enlighten.

5 What we do not understand.

When the sample is brought into the dissipative state in a presence of magnetic field the situation dramatically changes. The amplitude characteristics becomes nonlinear and, moreover nonmonotonous, see Fig. 3. The picture resembles very much the one seen in the case of a Josephson based detector [4]. However, we have found that the current position of the maximum RTS amplitude is, within an experimental error, insensitive to the applied magnetic field. This fact strongly contradicts the Josephson mechanism. The situation

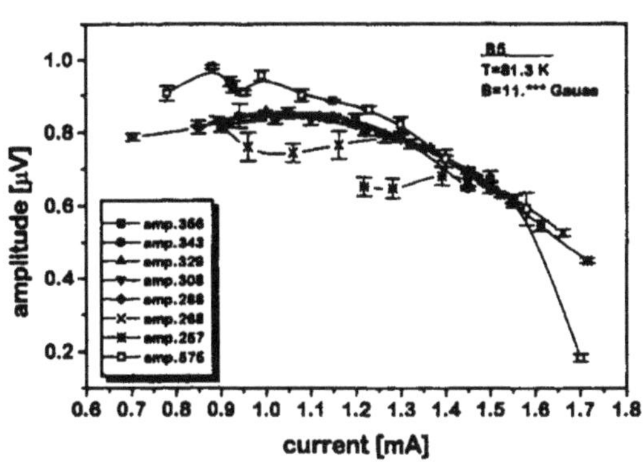

Figure 3: The dependence of RTS amplitudes on current flow at various magnetic field

concerning the lifetimes is even more complicated. The characteristics depend on mutually interacting magnetic field and transport current in a nontrivial way. As an example we illustrate the current dependence of the switching rates in Fig. 4. To make the experimenter life even more difficult the RTS waveforms in the presence of magnetic field are modulated in frequency by yet another telegraph signal. That is the reason why we have report two distinct rates for the down state in Fig. 4.

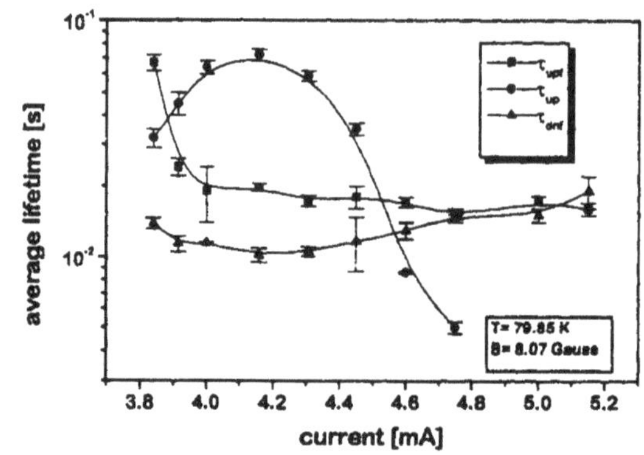

Figure 4: The dependence of RTS lifetimes on current flow at a fixed magnetic field

We conclude that we are still unable to propose any plausible explanation for neither for the peculiarities of RTS manifestations in the presence of applied magnetic field nor for the origin of frequency modulation observed.

References

1. L.B. Kiss and P. Svendlingh, *IEEE Trans.* **ED 41**, 2112 (1994), and references therein.
2. G. Jung, S. Vitale, J. Konopka, and M. Bonaldi, *J.Appl.Phys.* **70**, 5440, (1991)
3. G. Jung and B. Savo, *Phys. Rev.* B **53**, 90 (1996)
4. M. Bonaldi *et al.*, *Physica B* **194–196**, 2037, (1994)
5. V.D. Ashkenazy it et al., *Sol. State Comm.* **98**, 517 (1996) *Phys. Rev.* B **45**, 7495 (1992)
6. G. Jung and B. Savo, *J. Appl. Phys.*, in print (1996)

FLICKER NOISE AND FRACTAL STRUCTURE NEAR THE PERCOLATION THRESHOLD FOR $YBa_2Cu_3O_7$ EPITAXIAL FILMS

A. BOBYL, D. SHANTSEV
St.Petersburg State Tech.University, St.Petersburg 195251, Russia

M. BAZILJEVICH, H. BRATSBERG, Yu. GALPERIN, T.H. JOHANSEN
Dep. Phys., University of Oslo, P.O.B. 1048 Blindern, N-0316 Oslo 3, Norway

M. GAEVSKI, R. SURIS
A.F. Ioffe Physical-Tech.Institute, St.Petersburg 194021, Russia,

V. GASUMYANTS, R. DELTOUR
Universite Libre de Bruxelles, B-1050 Brusseles, Belgium ,

I. KHREBTOV, V. LEONOV
S.I. Vavilov State Optical Institute,St.Petersburg 199034, Russia.

Typical size of structural microdefects in epitaxial $YBa_2Cu_3O_7$ films (~ 0.1-0.3 μm) is usually much less than the spatial resolution of experimental methods used for their characterization (~ 1-3 μm). In this case one can expect that the model based on resistor networks is valid for a description of both resistance and noise. However, in the vicinity of superconducting transition in an strongly inhomogeneous superconductor (near the percolation threshold) this model fails, and *a specific description of electric noise* should be used. Such a decription (which is, as far as we know, absent at present time) is definetly an unsolved problem of noise in HTSC.

Another unsolved problem is a reliable test procedure to descriminate between different theoretical models to describe noise properties of thin films.We present a semi-empirical computer model which fits this task. It is a network model based on the effective medium approximation.

1 Introduction

Due to the substrate/film lattice mismatch and large difference in thermal expansion coefficients of $YBa_2Cu_3O_7$ films and substrata, the films are known have a block structure. This can be confirmed by electron microscopy, X-ray diffraction, as well as follows from experimental data on flicker noise in the normal state[1] . The latter allows one to investigate the energy distribution structural defects with internal degrees of freedom located at the blocks' boundaries. At high enough temperatures, close to T_c, such defects can switch between two (or more) meta-stable states causing fluctuations in local parameters of a superconductor (SC). This is why the dynamical defects are often called the elementary fluctuators. Typical size of the blocks is of the order of film thickness (~0,1-0,3 μm). In this connection, three important questions concerning flicker noise can be formulated:

1. What is the characteristic spatial scale of inhomogeneity which is responsible for T-dependencies of resistance R and voltage noise S_v of a HTSC microstrip?
2. Can a group of correlated blocks form a bottleneck which controls the properties of the macroscopic sample? What is the probability for such an event?
3. How one can test the models describing dependencies $R(T)$ and $S_v(T)$?

Looking for answers to these questions we have met several unsolved problems of noise (UPoN). To analyze those the following methods were used. We have measured the T-dependencies of R (from 10^2 to $10^{-5}\Omega$) and S_v in $YBa_2Cu_3O_7$ microstrips at different bias currents I and in magnetic fields H up to10 kG. These dependencies are interpreted on the basis of distributions of local resistivity $\rho(x,y)$ and current density $j(x,y)$. The latters were measured at fixed I_{expr} and $H=0$, as well as calculated for arbitrary I and H. The most difficult point is a fractal structure of the SC cluster in the temperature region close to the percolation threshold. In this region, the experimental dependencies deviate significantly from the predictions of the effective medium approximation (EMA). On the other hand, to obtain satisfactory results with the help of computer models, one needs to decrease the network scale that, in its turn, leads to an exponential increase in the computer time.

2 Spatial scales

The lowest scale is the typical size of the above mentioned blocks. In high quality films their boundaries are transparent for carriers and contain ~ 10^6 fluctuators for every block with energies 0.1-0.3 eV.

The next scale is the spatial resolution of characterization methods. We used two approaches having similar resolution (≥ 1-2 μm): (i) calculation of spatial distributions of the critical temperature (T_c-map), $\rho(x,y)$ and $j(x,y)$ from T-dependencies of the electron-beam-induced voltage (EBIV)[2]; (ii) imaging of magnetic flux distribution based on the magneto-optic Faraday effect that allowed to determine the spatial distribution of the critical currents far below T_c[3]. Thus, in both methods, there are about ~10 blocks between two adjacent experimental points. Consequently, it is clear that any further improvement of the spatial resolution will require an analysis of a new (and unsolved) fundamental problem - _electrodynamics of noise in a SC chaotic strongly inhomogeneous medium_[4].

The largest scale in question is the size of a microstrip. We have studied the samples of 50x500 μm which were described by a matrix of up to 25x250 values of local parameters.

Thus, the answer to the question 1 depends both on the film structure and on the experimental resolution. In general, UPoN are due to various mesoscopic phenomena, such as noise in 2D channel[5], noise depending on the parity of the number of electrons in single-electron SC devices[6], etc. In the case of HTSC film, another situation is probable - occurrence of a bottleneck when characteristics of the whole macroscopic sample are controlled by a microscopic region concentrating all current passing through the film[7]. Such situations require vigilance since they can appear in an uncontrollable way.

3 Bottleneck

The presence of large structural defects blocking the current path can be established by SEM or by EBIV images. Besides, they can lead to an unusually high sensitivity of noise to applied magnetic field[7]. It is rather difficult to reveal the situations where the large-scale correlation is the most important one. Fig.1 shows the correlation functions obtained from T_c-maps by the expression

$$G(r) = \frac{\overline{T_C(R+r)\,T_C(R)} - \overline{T_C}^2}{\overline{T_C^2} - \overline{T_C}^2} \quad . \qquad (1)$$

Here the average is calculated over all the points R of the sample. We observe an exponential decay of $G(r)$ with the distance, the charateristic length being dependent on the growth conditions. We believe that the

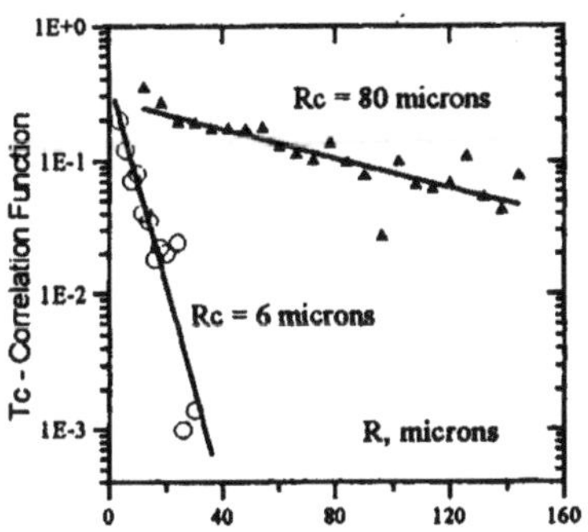

Fig.1 Correlation functions of T_c-spatial distribution for YBaCuO films grown on MgO (Δ) and $NdGaO_3$ (o) substrata.

correlation in this case is due to the clusters of dislocations of ~ 80 μm size in the MgO substrate observed by X-ray studies. To reduce the possibility of bottleneck formation, the microstrip's size should be larger than this scale.

4 A test for theoretical models

Theoretical models describing transport and noise properties can be used to obtain an information about the structure. Therefore it is reasonable to check model applicability based on the relations for classical inhomogeneous media[8]:

$$R = \sum_{ij} \langle r_{ij} \rangle \left(\frac{i_{ij}}{I}\right)^2, \qquad S = \langle dR^2 \rangle = \sum_{ij} \langle dr_{ij}^2 \rangle \left(\frac{i_{ij}}{I}\right)^4 \quad . \qquad (2)$$

However, the following problems appear: (1) determination of local currents i_{ij} is difficult because of above mentioned UPoN; (2) insufficient accuracy of the methods for determination of local parameters; (3) to test a model in a wide range of I and H, one must introduce some assumptions about I,H-dependencies of local SC . Thus, *the absence of a reliable test procedure for theoretical models* is UPoN too. To illustrate this, we present attempts to explain $R(T)$ and $S_v(T)$ characteristics using the model described in the following subsections.

4.1 A model for voltage noise in an inhomogeneous superconductor

To work out a model for the voltage noise in an inhomogeneous SC film we have to specify three following points: 1) origin of the noise in a local parameter (according to our assumption, in the critical current of a weak link); 2) influence of the fluctuations in a local parameter on the local resitance element, r_{ij} (or, in the non-linear case, on the shape of the local I-V curve); 3) effect of local fluctuations in r_{ij} on the integral noise S_v. Note that all three stages are T-dependent. Therefore, for proper description of $S_v(T)$, the contributions to T-dependence from all the stages should be taken into account. However, at all three stages there exist some problems:

Stage 1: There are at least two candidates for the microscopic origin of the noise: (i) structural defects (fluctuators) located at the block boundaries, and (ii) hopping of flux vortices between the centers of pinning. What we need, is the excitation energy distribution of such degrees of freedom, which determines both the T-dependence of noise and the deviations of its spectrum from the $1/f$ law. This problem is far from its solution for both cases. Indeed, a first-principle calculation of the energy distribution for the structural defects requires full account of the structure and concentration of point defects at the block boundaries which seems almost impossible[9]. In principle, such a distribution can be obtained from from experimental frequency and temperature dependencies of noise in the normal state[1] or from the experiments on internal friction[10]. However, one should bear in mind that the distribution depends on the treatment in course of the measurement since $YBa_2Cu_3O_7$ film is a thermodynamically unstable and strained system.

Stage 2: So far there is no generally accepted theory even for linear dissipation in HTSC below T_c; this is even more true for the noise properties.

Stage 3: Since, firstly, I-V characteristics of SC below T_c are strongly nonlinear and, secondly, HTSC films are very inhomogeneous, the determination of the integral S_v requires the solution of a fundamental and unsolved problem - *theory of noise in a macroscopically inhomogeneous nonlinear medium*. Indeed, the well-known effective medium approximation (EMA), successfully used for description of transport properties in inhomogeneous linear media, cannot be generalized to the case of nonlinear medium[11] and fails to provide a self-consistent description of the noise[12]. Therefore, the only reliable way out here is a computer calculation of noise in a network of nonlinear resistors.

Despite all these complications, we propose the following version of EMA. Taking into account spatial distribution of T_c in the film we assume that its fragments with local $T_c(\mathbf{r})<T$ are normal and described by a linear *I-V* curve: $V=IR(H,T)$. In zero external magnetic field, the finite resistance of a fragment below T_c is presumed to be related to weak links (generally, regions with suppressed superconducting order parameter[13]) between microblocks. To describe *I-V* curve in this case, we employ the model of resistor-shunted junction (RSJ)[14], which yields $V=R_n(I^2-I_c^2)^{1/2}$. Here R_n is the normal resistance of the weak link, while $I_c = I_c(T,H)$ is its critical current. Fluctuations in local I_c at $I_c<I$ lead to the voltage fluctuations

$$S_V = \frac{R_n^2 I_c^2}{I^2 - I_c^2 + \tilde{I}^2} S_{I_c} \tag{3}$$

which are very large at $I \approx I_c$ (small item $\tilde{I}^2$ is introduced into the denominator to take account of "rounding" of the *I-V* curve resulting from thermally-activated phase slippage[15]). The term associated with fluctuations in R_n is omitted since it turned out to be not necessary to describe experimental results.

At a given T, the SC film is considered as a two-component system consisting of normal regions with linear *I-V* curve ($V=R_0I$) and SC regions with nonlinear *I-V* curve in the RSJ form with different values of I_c depending on the local T_c-value, $I_c=I_{c0}(1-T/T_c)$. Here I_{c0} is the critical current at $T=0$. R_n of the weak links are supposed to be the same for all the fragments. Though, strictly speaking, EMA can be only applicable to weakly nonlinear media[11], we use standard EMA expression which for our case takes the form

$$\int_{T_c>T} \frac{\rho_e - \rho_s(I,I_c)}{\rho_e + \rho_s(I,I_c)} f(T_c)\, dT_c + \frac{\rho_e - \rho_0}{\rho_e + \rho_0} \int_{T_c<T} f(T_c)\, dT_c = 0 \tag{4}$$

where ρ_e is the effective resistivity of medium, while ρ_0, ρ_s are the resisitvities of the normal and SC fragments (with given I_c), respectively. $f(T_c)$ is the T_c-distribution function over all the film. The current I_s through a SC fragment having the critical current I_c is given (within the EMA approach) as

$$I = I_e \frac{2\rho_e}{\rho_e + \rho_s(I,I_c)} \tag{5}$$

where I_e is this current for the homogeneous case. Together with (4), this formula provides two equations to find ρ_e. Solution of these equations can be simplified if one replaces ρ_s and I_s in the formulas by their values averaged over SC fragments. To calculate S_v of the whole medium we use the expression similar to (2) where the values of local currents are taken from formula (5) and from the analogous one for normal fragments. Finally, we obtain

$$S_V \propto \int_{T_c>T} \frac{R_n^2 I_c^2 I_s^2}{I^2 - I_c^2 + \tilde{I}^2} S_{I_c} f(T_c)\, dT_c + \left(I_e \frac{2\rho_e}{\rho_e + \rho_0} \right)^4 S_{R_0} \int_{T_c<T} f(T_c)\, dT_c \quad (6)$$

So we outline the model which, despite its very rough approximations, can be used when direct calculations of networks are too complicated.

4.2 Fractal structure of superconducting cluster and current distribution near T_c

As a first step for calculation of S_v in a network model, one should determine the spatial distributions of SC regions and of the current density, $j(x,y)$. In this subsection we demonstrate the fractal structure of SC cluster near the percolation threshold and the evolution of j-distribution throughout the SC transition.

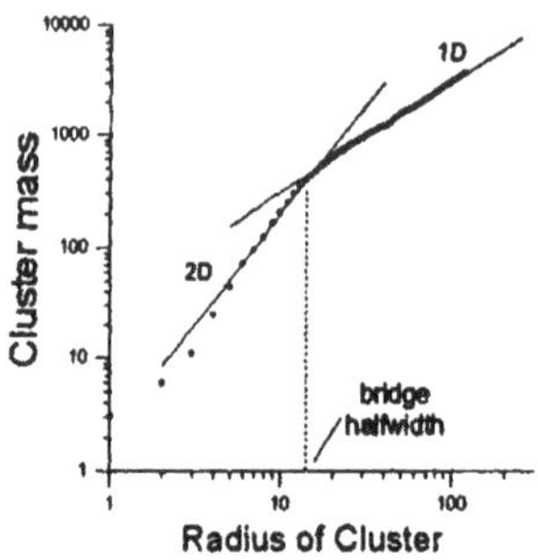

Fig.2 SC cluster mass *vs* its radius near the percolation threshold calculated from T_c map for sample on $NdGaO_3$.

SC cluster. When the fraction of SC fragments p=0.59 one can analyze a fractal structure of 2D SC cluster at the percolation threshold, as well as in the range of spatial scales between the experimental resolution 2μm and the sample's size (see fig.2). It is known that for an infinite random 2D system at the percolation threshold the SC cluster has fractal dimensionality D_f=1.89 [16]. D_f can be defined as the critical exponent connecting the mass s of the cluster confined in a circle of the radius L and the value of L : $s \propto L^{D_f}$. If the system is finite and 2L is larger than its minimal size, the system becomes effectively one-dimensional, and D_f is equal to 1. A crossover from 2D to1D behaviour takes place when the radius L is or the order of the bridge's halfwidth. Fitting the data with straight lines provides D_f=1.95 for the 2D region and D_f=1.02 for the 1D region. Deviation from the classic result for 2D systems (D_f= 1.89) is related (as we believe) to the presence of a spatial correlation in T_c at small distances (see fig.1). As a result, the percolative cluster is slightly more dense than for a pure random system.

Evolution of Current Density Maps. Using images of magnetic field distribution and applying critical state model, the spatial distribution of I_c in the film has been calculated. Together with T_c-mapping data, this provides the full information necessary to describe *I-V* curves of the fragments for all the temperatures (R_n are assumed to be the same and *T*-independent for all the fragments). The current density distributions are then calculated by solution of a set of nonlinear equations and three of them at different *T* are shown in Fig.3. As *T* is lowered the path of the current through the film becomes more

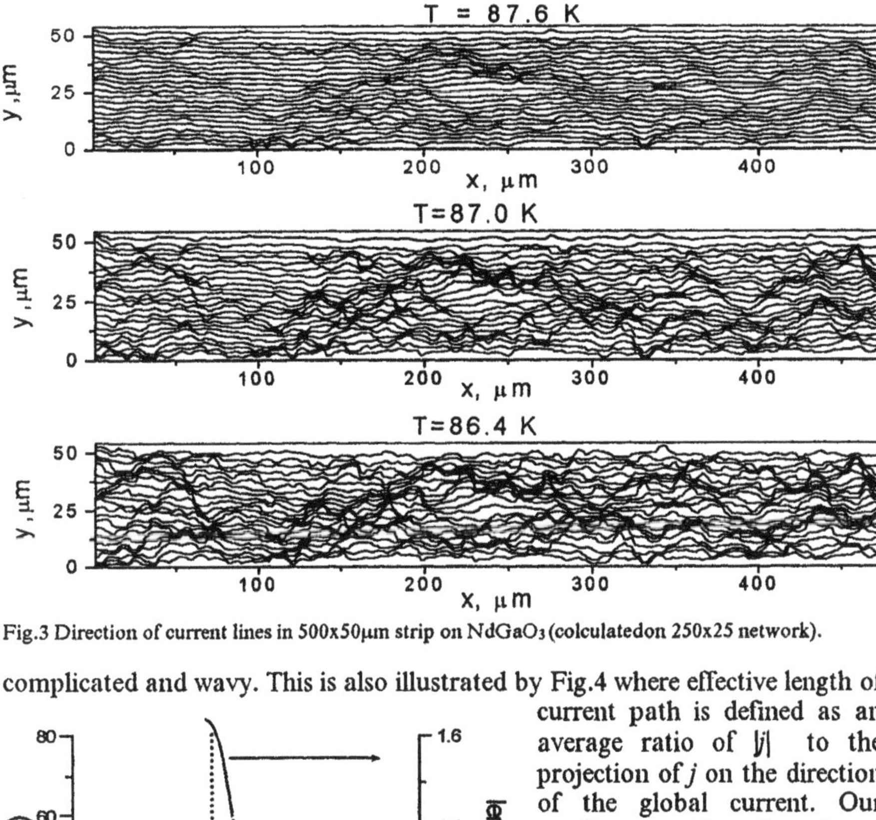

Fig.3 Direction of current lines in 500x50μm strip on $NdGaO_3$ (colculatedon 250x25 network).

complicated and wavy. This is also illustrated by Fig.4 where effective length of current path is defined as an average ratio of $|j|$ to the projection of j on the direction of the global current. Our results provide the lower estimate since they are limited by 2μm resolution .

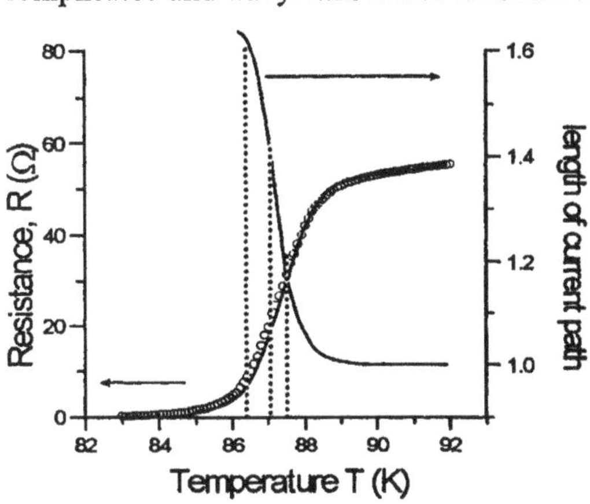

Fig.4 *T*-dependence of the average length of the current path normalized to that for homogeneous current flow and transition curves: experimental (circles) and calculated (line). Dashed lines mark the points on the curve for which current density distribution of Fig.3 are shown.

4.3 Analysis of noise experiments and comparison with models.

Voltage noise of the network has been calculated as follows. Random variations in the critical currents were introduced as: $I_c \rightarrow I_c(1+\delta)$, $\delta << 1$, and a variation of the integral voltage δV_{Ic} was then calculated. In the same manner, we calculate a change of voltage δV_R induced by variations in the resistances of normal resistors. Then experimental $S_v(T)$ was fitted by a linear combination of $(\delta V_{Ic})^2(T)$ and $(\delta V_R)^2(T)$ with arbitrary coefficients. These two coefficients (which are in fact the normalized spectral densities of I_c and R fluctuations) are the only two fitting parameters, since R_0 and R_n have been determined independently by fitting of the $R(T)$ curve. The final version of the computer program is under development now. Preliminary results for YBaCuO film are presented in Fig.5. They can be

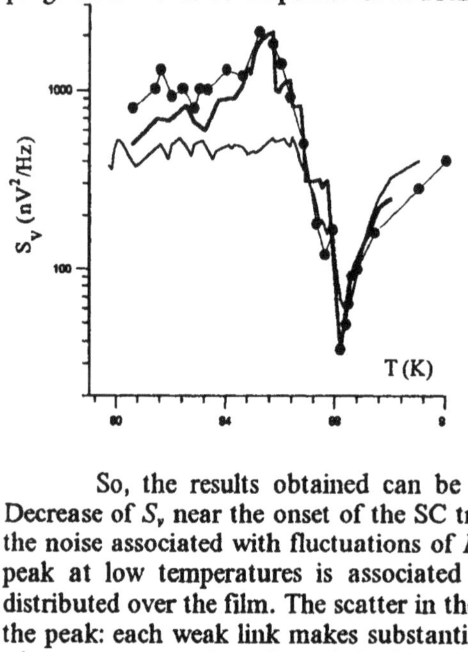

Fig.5. Spectral power density network of nonlinear of voltage noise: experiment (circles), calculations using resistors (solid line). Calculation within EMA-based model is also showen.

compared with the experimental T-dependence of S_v at frequency 10 Hz measured in a YBaCuO/NdGaO sample in zero magnetic field. Here the function $f(T_c)$ needed for the calculations has been taken from the data on T_c-mapping. It can be also derived from transition curve[17].

So, the results obtained can be qualitatively interpreted as follows. Decrease of S_v near the onset of the SC transition is related to the decrease of the noise associated with fluctuations of R in normal regions (see also [18]). A peak at low temperatures is associated with fluctuations of weak links I_c distributed over the film. The scatter in the values of I_c leads to large width of the peak: each weak link makes substantial contribution to the noise at the T when current passing through it is equal with its critical current, $I_c(T)=I$. Individual sharp peaks observed in some samples seem to arise from individual weak links located in the bottleneck of current path.

It should be noted that our model fails to explain the maximum of $S_v(T)$ corresponding to $R=0.02\ R(T>T_c)$. We believe that it is due to the fact that the model cannot take into account actual distribution of currents in the film. The distribution, however, has special features in the temperature region where there exists percolation through regions of SC fragments $(T<T_c)$, but there is no percolation through regions of "SC" weak links ($I_c>I$). In this case most of weak links tend to carry current which is a little bit less than their critical current. Thus, they tend to reside in the most nonlinear point of their I-V curve. This tendency, existing only in nonlinear networks, leads to a dynamical instability and, hence, to a very high noise. This speculation is supported by one more fact. It has been found that in samples where the spatial correlation in T_c is weak, the peak of noise is wide and pronounced, while in the samples with strong correlation there are usually many individual sharp peaks. This is clearly illustrated by comparison of Fig.6 and Fig.1 for two samples grown on different substrata. In the last case of highly correlated disorder, a redistribution of current in the film is more complicated (which is somewhat similar to the situation in 1D systems). Since the current I through a given weak link is almost independent of T, this weak link gives rise to a sharp peak of noise at the T where $I_c(T)=I$. For the whole sample one would have a sequence of such peaks at different T. On the contrast, in the absence of spatial correlation in SC parameters, current redistribution is possible. Then the system of weak links becomes self-adjusting and the effect of dynamical instability comes into play. In this case one would expect one wide peak for the whole sample.

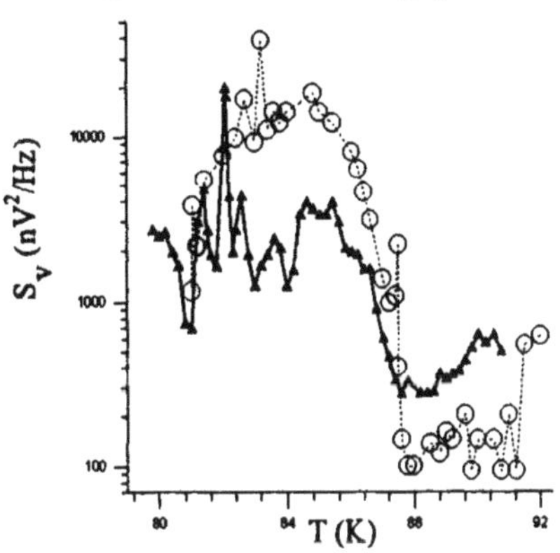

Fig.6. Spectral power density of voltage noise for the same samples as in fig.1: grown on MgO (Δ) and $NdGaO_3$(o) substrata

Acknowledgments

The work is supported by Russian National Program on SC project #94048 and Research Council of Norway.

References

[1] A.V. Bobyl et.al., J.Appl.Phys. (1996)
[2] A.V. Bobyl *et.al.*, *Scanning microscopy* **10** (1996)
[3] M. Baziljevich *et.,al.*, *Journal de Physique IV* **6**, C3-259(1996).
[4] I.M. Lifshits et.al., Introduction to the Theory of Unordered System (Nauka, Moscow, 1982).
[5] V.I. Kozub and A.A. Krokhin, *J.Phys.: Condens. Matter* **5**, 9135(1993).
[6] Ulrik Hanke, Yu.M.Galperin, and K.A.Chao,*Appl.Phys.Lett.* **65** 1847(1994)
[7] A.V. Bobyl *et.al.*, *Physica* C, **266**, 33 (1996).
[8] D.J. Bergman and D. Stroud, Solid State Phys. **46**, 147(1992).
[9] A.M.Stoneham, *Theory of defects in solids* (Clarendon press, Oxford, 1975).
[10] J.C. Phillips, *Physica* C **221**, 327 (1994).
[11] O.Levy, D.J.Bergman, *J. Phys.: Condens. Matter* **5**, 7095 (1993).
[12] D.J.Bergman, *Phys. Rev. B* **39**, 4598 (1989).

[13] J.Halbritter, *Phys. Rev. B* **48**, 9735 (1993).
[14] K.K.Likharev, *Dynamics of Josephson Junctions and Circuits* (NY, Gordon and Breach, 1986).
[15] V.Ambegaokar and B.I.Halperin, *Phys. Rev. Lett.* **22**, 1364 (1969).
[16] J. Feder, *Fractals* (Plenum Press, New York, London 1988)
[17] A.V. Bobyl *et.al.*, *Physica* C **247**, 7(1995).
[18] L.B.Kiss, P.Svedlindh, *IEEE Trans. Electr. Dev.* **41**, 2112 (1994)

THE ROLE OF INTERGRAIN CONTACTS IN THE RESISTANCE NOISE OF HTSC

L.K.J. VANDAMME
Department of Electrical Engineering
Eindhoven University of Technology
5600 MB Eindhoven, The Netherlands

The relative noise, C, in the transition temperature region of modern high temperature superconducting compounds (HTSC) increases over nine decades with decreasing temperature as $C \propto R^{-\gamma}$ with $1.4 < \gamma < 2.8$. Percolation and temperature fluctuation models have been proposed to explain the experimental results. Here some other possibilities are proposed to explain the observed trend. Some models are based on the temperature dependence of the 1/f noise parameter α. Others are based on the coalescence of multi spot intergrain contacts within the HTSC.

Introduction

The leading candidate for the explanation of the low critical current density problem in high temperature superconductivity is the weak link phenomenon[1]. The layers are composed of grains of very anisotropic compounds with touching boundaries of different sizes. The grain boundary is considered as a non superconducting region. The super current tunnels through this barriers. The result is a superposition of many local I-V characteristics and fluctuation behaviour which smears out. Critical currents reduce with barriers thickness at a given temperature. Good grain alignment is necessary for an improved critical current density. Above the critical current density the material loses its interest for electric wires. Noise has been used as an indicator of improvements in the HTSC technology[2].

The resistance noise in the normal state is well described by $S_R = R^2C/f$ in the Ohmic region. In samples of the same volume and at the same temperature a reduction factor of about 10^{-10} in the relative 1/f noise value C has been observed since 1989 and an increase of 10^4 A/cm^2 up to 10^6 in current density.[2] Another striking result, the relative noise in the transition temperature region of modern samples increases over nine orders of magnitude with decreasing temperature as $C \propto R^{-\gamma}$ with $1.5 < \gamma < 2.8$ (see fig. 1).

Existing models based on percolation, temperature fluctuations or flux fluctuations explain partially the experimental results[2,3,4]. Temperature fluctuations can lead to a 1/f spectrum only in a very limited frequency range and under very limited conditions[5]. In the p-noise percolation model[2] the noise is considered as an intrinsic property of a network between conducting grains where the resistance noise is ascribed to the random switching of certain junctions. The p-noise percolation model predicts $\gamma \approx 2.7$ in 3D and $\gamma \approx 1.5$ in 2 dimensions[2]. All switching events together can result in a 1/f like spectrum assuming Lorentzian spectra with the correct weight function[4]. In the classical percolation model only the number of weak links becoming short-cuts increases between superconducting grains and the remained normal conductor intergrain junctions provoke the 1/f noise. At every different temperature we have a different network but no switching in time. This model predicts $\gamma = 1$ for 1D (chain). In spite of the success in explaining partially the normalized noise versus resistance with temperature as a parameter, in the above mentioned models there still remain unsolved problems. For example: (i) the percolation models are linear, however, HTSC layers show a non ohmic current-voltage characteristic, (ii) Is there a changing intergrain contact behaviour with temperature ? Does the intergrain contact remain interface or constriction dominated ? Does the number and area of the conducting paths between grains change with temperature ?

Models

Here I hope to generate new unsolved problems by presenting other possible explanations based on experimental facts which might also explain the peculiar characteristics in HTSCs:

(i) The first hypothesis is based on experimental facts where the 1/f noise parameter α in different metals seems to be inversely proportional to the resistivity squared[6,7]. For a small single crystal without grain boundary complications we expect[5] $R \propto \rho$ and $C \propto \alpha/N$. With an α increasing with decreasing temperature like $\alpha \propto \rho^{-2}$ we find $C \propto R^{-\gamma}$ with $\gamma = 2$. Hence we explain $C \propto R^{-2}$, if the number of electrons does not change with temperature and the average mobility increases with decreasing temperature due to the onset of superconductivity over a coherence length smaller than the length of the sample. The problem is the fact that an α independent of temperature has been observed in the c direction and an α strongly reducing with increasing temperature in the a-b plane of a $YBa_2Cu_3O_{7-\delta}$ single crystal.[8]

(ii) For samples consisting of touching grains we might explain the resistance and noise of the complete sample by taking the average behaviour at a boundary and multiply the result with a fixed geometry parameter[9]. In contrast to percolation models, if the geometry of the micro contacts between grains does not change with temperature we have a fixed geometry coefficient between the behaviour of two-touching grains and the behaviour of the whole system. Without assuming classical or p-noise percolation we can expect for the resistance and relative noise of a film dominated contact at the weak link between two touching grains[10,11]

$$R = \rho_f t/A \text{ and } C = \alpha/n_f At \text{ or } C = \alpha q\mu_f R/t^2 \qquad (1).$$

If the resistance reduces with temperature mainly due to a change in the resistivity ρ_f of the weak link, while the cross section A, thickness t and mobility in the interface layer remain constant we find if $\alpha \propto \rho_f^{-2}$ holds

$$C \propto R^{-1} \quad \text{or} \quad C \propto R^{-2} \tag{2}$$

depending on whether or not μ_f is constant or $R\mu_f$ is constant.

(iii) If the resistance and noise stem from the interface contribution at the grain boundaries where $R \propto T^{\beta}$ and $C \propto \alpha/N_i \propto T^{-x} \propto R^{-\gamma}$ with $\gamma = x/\beta$. This means $1.5 < x/\beta < 2.8$. Arguments for interface dominated contacts are in[10,11].

(iv) For samples where hypothesis (i) and (ii) are not applicable there still is a possibility to explain the experimental results with $1.4 < \gamma < 2.8$ like in fig. 1. By assuming multispot nano-constriction contacts between the touching grains with a temperature and current dependence of the number of conducting channels in parallel and their average diameter, we also obtain $1 < \gamma < 3$. This can be considered as a new type of percolation phenomena in which coalescence between conducting paths is possible into a reduced number of paths with larger cross sections at increasing temperatures increasing the resistance and reducing the relative noise. The electric conduction area is smaller than the apparent conducting area between two grains. Multispot contacts with a relative small number (k) spots in parallel with radius a (a-spots) lying relatively far apart (several times the average diameter 2a) behave like k parallel individual spots. The resistance and relative noise of a simple multi spot contact between two grains are given by[12,13].

$$R = \rho/\pi k a \quad \text{and} \quad C = \alpha/20\pi n k a^3 \tag{3a), (3b}$$

where ρ is the resistivity in the bulk of the grain and n the free charge carrier density. If we consider α as a constant the general relation for a simple multispot contact C versus R shows a dependence as[13].

$$C \propto R^{-\gamma} \text{ for } k \propto a^{x} \text{ with } x = -(\gamma+3)/(\gamma+1) \text{ or } \gamma = -(x+3)/(x+1) \tag{4}$$

For $1.5 < \gamma < 3$ holds $-1.8 < x < -1.5$ and with $\gamma = 3$ we have $x = -1.5$. The value of the exponent γ is very sensitive to x. Arguments for a change in k and a can be found in ref. 11 and 14.

(v) Arguments against model (iv) are that changes in resistance and noise with temperature become a pure geometrical phenomenon, the so-called coalescence effect with increasing temperature, ignoring the temperature dependence of ρ and α. These assumptions are against experimentally observed trends[1,8]. Therefore we propose a blend of the models (i) and (iv) in which the total cross section $\pi k a^2$ of the multi nanoconstriction contact between touching grains remains constant with temperature and also the concentration of free carriers in the regions where superconduction is not yet established. We assume: $k \propto a^{-2}$ or $k \propto T^{\delta}$ with $\delta < 0$ and hence $a \propto T^{-\delta/2}$; $\rho \propto T^{m}$; $\alpha \propto T^{-\ell}$ with $m \geq 1$and $\ell > 0$. This results in the following dependencies for R and C on temperature and for the C versus R relation:

$$R \propto T^{m-\delta/2} \text{ and } C \propto T^{-\ell+\delta/2} \text{ and } C \propto R^{-\frac{\ell-\delta/2}{m-\delta/2}} \qquad (5).$$

The exponent $\gamma = (\ell-\delta/2)/(m-\delta/2)$. Independent of the value of the exponent δ, for $m = \ell$ holds $\gamma = 1$. This includes also model (iv) with $m = \ell = 0$. For $\gamma > 1$ and $\delta < 0$ it is necessary to have $\ell > \gamma m$. In the temperature range $100\ K < T < 150\ K$ the noise parameter in single crystals changes by a factor 14 which[8] points to $\ell \approx 6$.

Kiss, Svedlindh et al.[2] observed above a critical temperature and resistance value $C \propto R^{-\gamma}$ with $0 < -\gamma < 2$. If the concentration and α remains constant with temperature we expect $-\gamma = 1$ or $\gamma = 0$ whether or not $\alpha\mu$ is constant or α and $R\mu$ are constant with temperature.

Conclusion:

The large amount of models to explain the experimental results are indications that the dependence $C \propto R^{\gamma}$ is ill understood. In addition to the existing models: classical percolation, p-percolation and temperature fluctuation induced 1/f noise in HTSC, we propose a few new models explaining the experimentally observed trends. Two types of models have been proposed. The first type (i,ii) is based on the temperature dependence of the 1/f noise parameter α which is possible for homogeneous crystals and for compounds even where the contact resistance and noise between grains are dominated by a grain boundary layer. In the second type (iv) of model, α and the concentration of free carriers are kept constant but the resistance and noise are dominated by constriction dominated multi spot contacts between touching grains. Model (v) is a blend of (i) and (iv), an essential point in this model is that resistance increases with temperature partially by the coalescence of several spots into less spots with slightly larger diameters and partially by an increase in the resistivity.

Acknowledgement

I would like to thank Prof. L.B. Kiss from Szeged University in Hungary for valuable discussion and for providing fig. 1. Part of this work has been supported by the Human capital and mobility ELEN programme.

References

1. T.P. Sheahen, Introduction to high-temperature superconductivity, Plenum Press - New York (1994).
2. L.B. Kiss and P. Svedlindh, *IEEE Trans.on Electron Devices*, **41**, pp. 2112-2122 (1994).
3. S. Jiang, P. Hallemeier, C. Surya and J.M. Phillips, *IEEE Trans.on Electron Devices,* **41**, pp. 2123-2132 (1994).
4. G. Jung, B. Savo and A. Vecchione, *Europhys. Lett.*, **21**, 947 (1993).
5. F.N. Hooge, T.G.M. Kleinpenning and L.K.J. Vandamme, *Rep. Prog. Phys.***44**, pp. 479-532 (1981).
6. D.M. Fleetwood and N. Giordano, *Phys. Rev.* **B27**, 667-671 (1983).
7. L.K.J. Vandamme, Proc. Int. Conf. Noise Montpellier, 1983. Eds. J.P. Nougier and G. Lecoy. Elsevier, Amsterdam, pp. 183-192.
8. F. Ying, D.L. Zhang and J.W. Xiong, *Phys. Rev.* **B51**, 1334 (1995).
9. L.K.J. Vandamme, *Electrocomponent Science and Technology*, **4**, 171 (1977).
10. C.L. Lin, C.C. Chi, C.C. Chen and M.K. Wu, *Physica* **C235-240**, 1789 (1994).
11. A. Marx, et al. *Inst. Phys. Conf. Ser.* **148**, 1295, 1995, IOP Publishing Ltd.
12. L.K.J. Vandamme and R.P. Tijburg, *J. Appl. Physics*, **47**, 2056-2058 (1976).
13. L.K.J. Vandamme, PhD thesis Eindhoven Univ. of Technology 1976, p. 78-84.
14. M. Hatle, K. Kojima and K. Hamasaki, *IEICE Trans. Electron.* **E-77**, 1169 (1994).

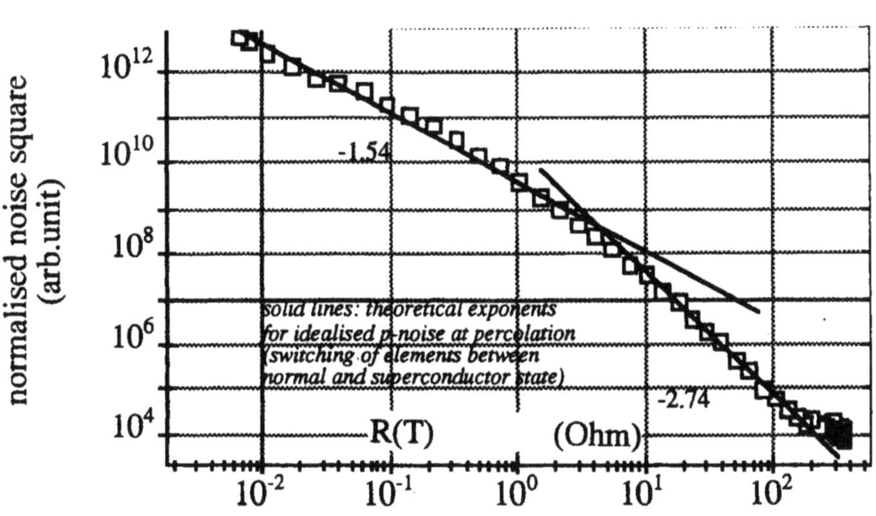

Fig. 1. Scaling of the normalized noise $C = (fS_v/V^2)N$ versus the resistance.

CURRENT CONTROLLED PERCOLATION EXPONENTS IN THE NOISE OF HIGH-TEMPERATURE SUPERCONDUCTOR THIN FILMS

LASZLO B. KISS and PETER SVEDLINDH
Uppsala University, Institute of Technology, Box 534, Uppsala, S-751 21 Sweden

LODE K.J. VANDAMME
University of Eindhoven, Department of Electrical Engineering, POB 513, Eindhoven, MB5600 The Netherlands

CHRIS M. MUIRHEAD
University of Birmingham, School of Physics and Astronomy, Edgbaston, Birmingham, United Kingdom

ZDRAVKO IVANOV and TORD CLAESON
Chalmers University, Department of Physics, Gothenburg, Sweden

Current-dependent investigations of the conductance noise in Yttrium based and Tallium based high-Tc-superconductor films provide important information about the validity and limitations of present models based on percolation. The well known scaling effects between the resistance and the noise exist only under the spacial condition when the control parameter is the temperature. Moreover, in Ta based films, at low current densities the classical percolation noise scaling exponents, while at high current densities, the p-noise scaling exponents have been observed. This effect indicates that the low-frequency contribution of the elementary fluctuators of p-noise are strongly suppressed due to high current densities. Several new unsolved problems can be add to the old questions.

1 Introduction

High temperature superconductor (HTSC) materials (see [1-4] and references therein) are strategically important object of technological research, because future low-temperature electronics will probably be based on these materials. Since 1989, we have had the opportunity to follow the improvement of HTSC film technology via noise and conductivity measurements [2]. The strong interrelation between the quality of technology and the strength of the noise in the normal conducting state (described by the Hooge parameter based on atomic numbers) is very obvious. The improvement of the Tc by a few Kelvins and the increase of the critical current density from 10^4 to 10^6 A/cm^2 has been accompanied by a radical reduction of the noise: a reduction by over 10^{10}, see Figure 1. However, noise investigations and the understanding the mechanism of noise generation in these materials have also other important roles in these materials:

i) Proper conductance noise models of these materials can help to find a proper noise model of HTSC active devices;

ii) Proper understanding of the noise generation can help to find various alternatives for noise reduction.

The most important temperature region of conductivity noise is the low-temperature part of the conductor-superconductor transition regime. In this regime the normalised voltage noise at fixed dc measuring current is rapidly increasing when the temperature is lowered.

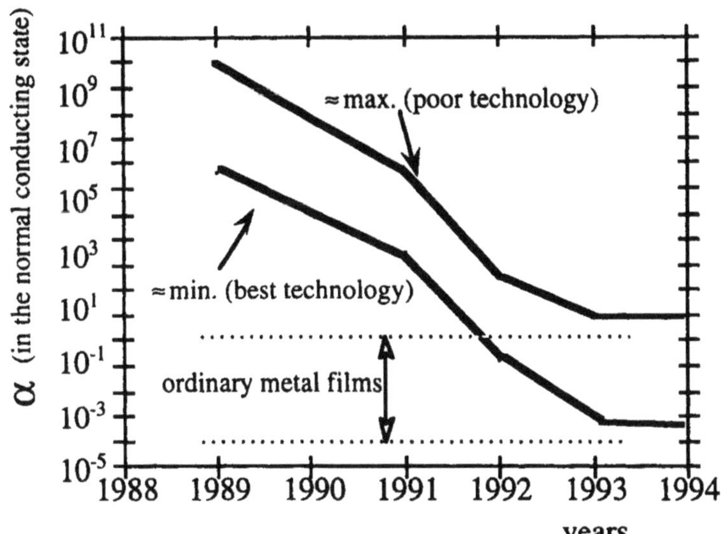

Figure 1: Reduction of the Hooge parameter (in the normal conducting state) versus the improvement of HTSC technology [2].

Thorough studies [1,2] of the conductance noise of Yttrium based HTSC superconductor films (YBCO films on various substrates with various kinds of deposition technology) have indicated that, in the low-temperature region of the conductor-superconductor transition region, the temperature dependence of the noise is controlled by percolation effects. This conclusion is based on the fact that all the reliable experimental noise data yield universal noise exponents which are determined by the geometrical dimension of the current transport only. Namely, the normalised resistance noise spectrum at a fixed current density, a fixed frequency f_0 and varied temperature is scaling as a power function of the resistance [1,2]:

$$\frac{S_R(f_0;T)}{R^2(T)} = \text{constant} * R^X(T) \quad , \tag{1}$$

where the exponent X has universal negative values in the temperature range of percolation. This behaviour indicates that the sample is a random mixture of conductor and superconductor elements and that the increase of the conductivity by decreasing temperatures is due to the increasing number fraction p of the superconducting elements.

Noise exponents around -1 have been identified as *classical percolation noise* phenomena [1,2]: the noise is originating from the independent resistance fluctuation of normal conductor elements of the film (see Fig.2.a). On the other hand, noise exponents around -1.5 and -2.7 have been interpreted as a newly discovered effects [1,2], namely, *p-noise at percolation:* a given number of elements is switching between normal state and superconducting state (see Fig.2.b). According to thorough experimental investigations, high quality film samples with a low Hooge-parameter in the normal state usually show *p-noise* while poor quality samples usually show classical percolation noise [1,2].

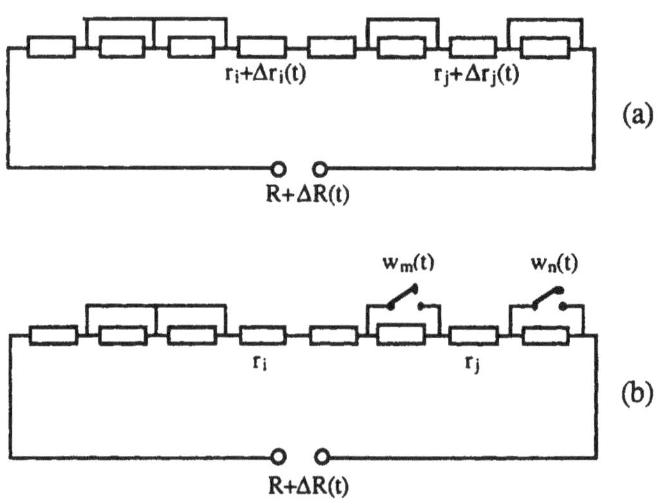

Figure 2: One-dimensional illustration of the origin of classical percolation noise (a) and the origin of *p-noise* which also implies universal scaling at the percolation limit.

2 Unsolved problems, motives and new investigations

The fact that, in the transition temperature regime, the HTSC materials have a strongly non-Ohmic characteristics implies serious questions about the applicability and limitations of the linear percolation models and the relevant universal exponents [2,3]. A relevant experimental result can be seen on Fig.3. Interestingly, in the same region where the normalised resistance noise is scaling with the resistance, the differential resistance R_d is also scaling with the resistance. This effect and the general problem of noise in the percolating non-Ohmic HTSC materials doe not have any theoretical explanation.

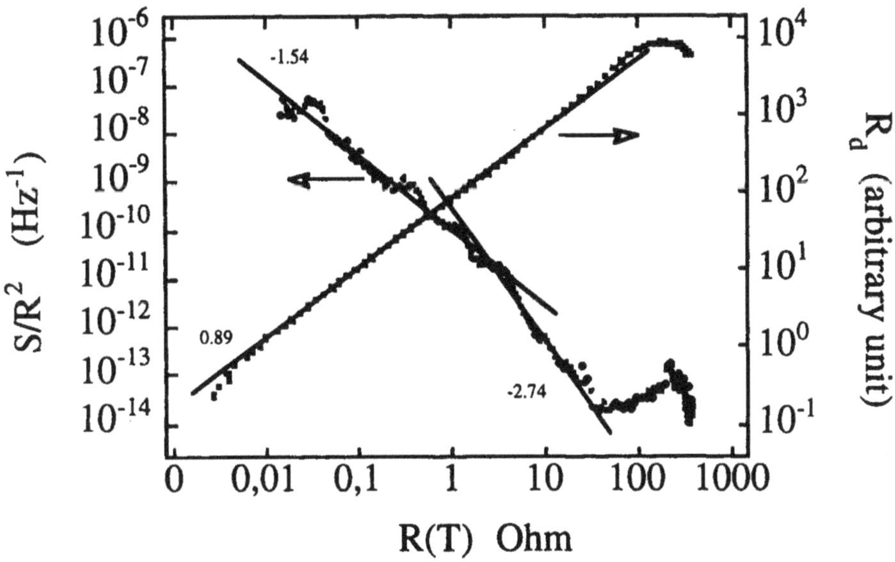

Figure 3: Three-dimensional and two-dimensional scaling (exponents ≈ -2.7 and ≈ -1.5) of the normalised noise as a result of *p-noise* at percolation in an YBCO film. The control parameter is the temperature. In the same temperature range, the differential resistance R_d is also scaling with the resistance.

Recently, in order to investigate the problem of noise at nonlinear percolation more thoroughly, we have designed a high stability low noise current noise source which can provide a dc current with an ultra low noise in the range of $2*10^{-6}$A - $64*10^{-3}$ A. the first samples on which we have carried out investigations, were very high quality, laser ablated Yttrium-based films (Univ. of Birmingham) and Tallium-based films (Chalmers Univ.) have been measured and compared at various current densities. In YBCO, we have found basically p-noise phenomenon at all current densities (Fig.4).

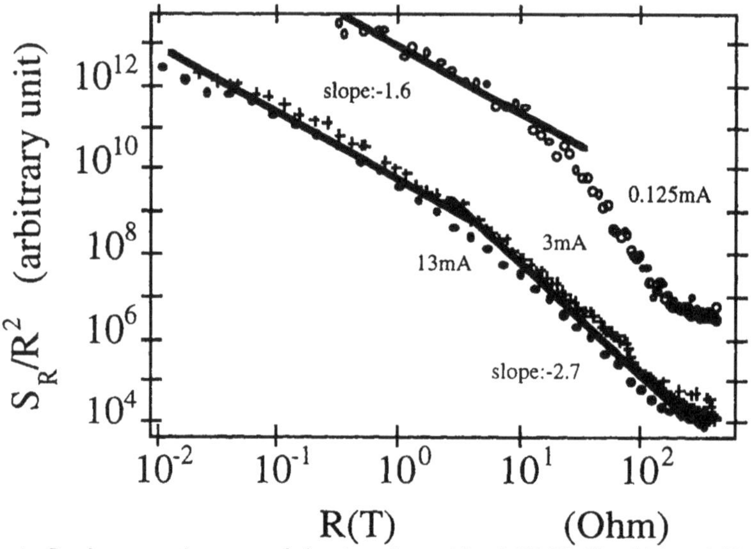

Figure 4: Results on *p-noise* at percolation in a laser ablated YBCO film (Univ. of Birmingham). Though, at small currents, the 3-dimensional scaling does not exist, the scaling exponents are basically not affected by the current density, especially in the 2-dimensional case.

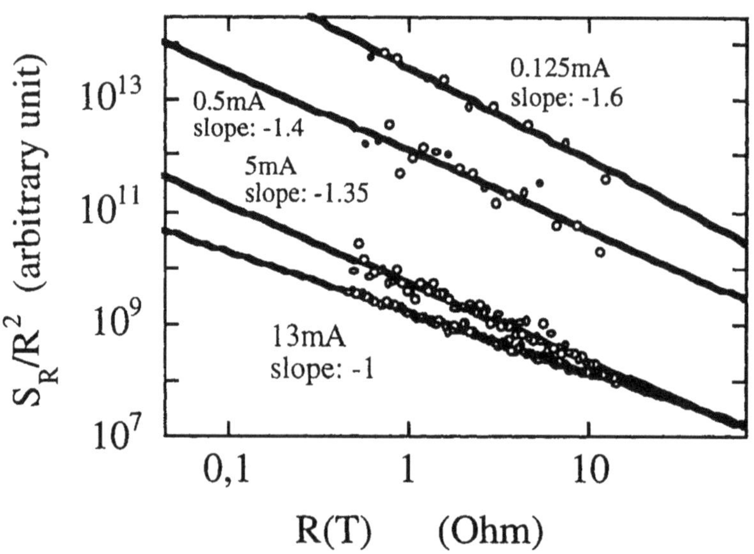

Figure 5: Results on noise-resistance scaling at percolation in a laser ablated Tallium-based film (Chalmers Univ., Sweden). The 2-dimensional p-noise exponent (≈ -1.5) found at low current densities cannot be found at high currents. The exponent gradually shifts toward the classical exponents and at high current densities, the noise is purely a classical percolation noise.

On the other hand, in the Tallium-based films, at small current densities, *p-noise* has been the dominant effect while at high current densities we have observed a smooth transition of the exponents into *classical percolation noise* exponents, see Fig. 5. It is obvious from the plotted data that the low-frequency noise contribution of the *p-noise* fluctuators is strongly diminishing with increasing current density and this is the reason why the *classical noise* (generated by the conductance fluctuations of the normal conducting elements) becomes dominant at high currents.

We have to emphasise that there is no theoretical model or picture which is able to explain the different behaviour of the two HTSC materials. These facts are emphasising the following unsolved problems [4]:

1. What is the microscopic generation mechanism of p-noise?

2. The present models of percolation [1] are linear models. Why do they work on the contrary of the fact that the HTSC samples are strongly non-Ohmic and how to introduce a nonlinear model?

3. Why do have p-noise fluctuators a different current density dependence in the two different materials?

Acknowledgements

This work has been supported by EAST-ELEN (EU); TFR (Sweden), Royal Society (UK) and OTKA (Hungary).

References

[1] L.B. Kiss and P. Svedlindh, *Phys.Rev.Lett.* **71**, 2817 (1993)
[2] L.B. Kiss and P. Svedlindh, *IEEE Trans. on Electron Devices* **41**, 2112 (1994)
[3] L.K.J. Vandamme, in this proceedings
[4] G. Jung, in this proceedings

1/f NOISE AND LATTICE DYNAMICS

DYNAMICS OF SPATIAL - TEMPORAL 1/f NOISE MECHANISM IN CHARGE DENSITY WAVES

I. BLOOM
E. E. Dept., Technion - Israel Institute of Technology,
32000 Haifa, Israel.
Email: bloom@ee.technion.ac.il

Charge Density Waves is one of a few interesting study cases in the solid state where one finds competition between randomness and interaction in a system with many degrees of freedom. This wave-like macroscopic entity has many configurational rearrangements in the presence of randomly placed defects, thus when sliding, it exhibits rich and complicated dynamics with huge low frequency noise, memory and hysteretic effects, glassy behavior, chaos and other features - a celebration to any physicist. In this paper we present some of the main disputes on the poorly understood low frequency noise mechanism and related phenomena of Charge Density Waves.

1 Introduction

In this paper we do not attempt to present or site all experimental and theoretical findings, rather we mainly review the controversial issues and the unsolved questions related to the 1/f noise mechanism of the CDW.

Since the discovery of Charge Density Waves (CDW) some thirty five years ago, there have been a vast number of experimental findings as well as models suggested to describe the rich complicated phenomena observed.[1] One of the robust features of the CDW is the low frequency (1/f - like) noise present when the CDW is sliding. Since this phenomenon is only one out of many, all related to the same entity - namely the CDW, one cannot isolate and investigate it without addressing the other phenomena and the relation between them.

In the ideal case of CDW in semi-one dimensional metallic crystals below the transition temperature, an incommensurate periodic elastic deformation of the lattice occurs, simultaneously, a gap is opened at Fermi level and many electrons condense to a lower energy state below the gap, created by the periodic potential. It can be described as perfect parallel planes of wave fronts with transletional invariance, free to move either direction. In the absence of defects, even an arbitrary small electric field induces CDW current - namely the condensed electrons with the periodic lattice deformation slide together. This is a unique case in nature where one finds a coherent conduction of many electrons as one entity with macroscopic dimensions.

In the presence of randomly placed defects the CDW is pinned. Each defect tries to dictate the local CDW phase and amplitude. Since the CDW has a finite deformability, it cannot simultaneously minimize the energy at every pinning site. As with other systems combining such "frustration" with randomness, the CDW is expected to have many metastable configurations (MCs).

When applying an external electric field, high enough to overcome the defects pinning force, the CDW starts sliding, introducing a non-linear excess conductivity and accompanied by low frequency noise fluctuations with a spectrum S(f) ~ 1/f, called Broad-Band Noise (BBN). In addition, a periodic signal, called Narrow-Band Noise (NBN), is detected in the MHz range due to wavefronts passing the sample's contacts. The NBN frequency is proportional to the excess CDW current. When applying a dc + ac bias a mod-lock occurs, causing a dramatic increase in the NBN peaks' sharpness as well as a reduction in the BBN magnitude.[2]

Pulse overshoot experiments[3-5] as well as wavelength direct view using x-ray[6] indicate that a dc bias causes a CDW wavelength shortening upstream along the electric field direction (global deformation), accompanied by a large polarization. Reversing the bias causes counter polarization and a transient excess current. Returning to equilibrium does not erase the last polarization. These long memory and hysteretic effects remind the behavior of other glassy systems. It is reasonable to believe that each rearrangement of the CDW (MC) has a different charge distribution with different polarization, thus jumping between MCs must cause measurable fluctuations that are manifested as low frequency noise.

Low-frequency voltage and current noises as well as other macroscopic measurable quantities are found sensitive in probing the dynamics of this system: ac dielectric response at the pinned state,[7] ac response of Young's modulus and internal friction,[8,9] NBN frequency,[2,10] and probably sample's length.[11] Since CDW phenomenon is essentially a strong coupling between electronic and elastic entities, i.e. electrons charge-waves and phonons elastic-waves, configurational rearrangements and fluctuations in one must directly affect the other. Indeed, it was found that when the CDW starts sliding, Young's modulus is reduced and the internal mechanical friction is increased.[12,13] Also, an hysteretic bias dependence of sample's length was found.[11]

Many attempts have been made to model and simulate the dynamics of CDW,[14-20] most are classical phenomenological models containing many loosely interacting sites with nonlinearity, periodic potential and external force. None of them, as far as I know, were able to produce all CDW phenomena in one model just by changing parameters.

2 Controversial issues and the unsolved questions

An important unanswered question arises: is the noise at a bias above threshold an equilibrium fluctuations (where the current plays as a probe for the fluctuations - similar to the 1/f noise in simple metallic ohmic resistors) or a dynamically driven noise mechanism, where the excess current helps creating the fluctuations. When the

CDW is pinned, only normal electrons current is sensed, hence the system is not sensitive enough to pinned CDW fluctuations. Though fluctuations have been seen[21] in small samples below threshold - an indication that the pinned CDW is jumping between configurations. Above the threshold field the siding CDW introduces excess nonlinear current accompanied with a large BBN.[22] In some cases the MCs available to the CDW at the sliding state is also found at the pinned state[7,23,24] so the large BBN is mainly a manifest of the equilibrium fluctuations (maybe with faster dynamics). But mesoscopic samples,[22,24] where one can observe discrete microscopic fluctuating entities, reveal irreversible time traces in the sliding fluctuations[24] - an indication of nonequilibrium dynamics that cannot exist at equilibrium. Another indication for the nonequilibrium dynamics is the memory loss of the slow fluctuating mechanism when a large ac bias is applied.[22,24,25] It is possible that in TaS_3 the BBN is mainly an equilibrium fluctuations while in $NbSe_3$ it is mainly a dynamical phenomenon.

The specific origin of 1/f noise mechanism at the sliding state of the CDW is not known, also it is not clear if there is one or more and if all CDW materials (like $NbSe_3$, TaS_3, $K_{0.3}MoO_3$ and more) have the same origin(s). There have been a few specific suggestions. The simplest suggestion is that the threshold electric field fluctuates due to pinning forces fluctuations at the randomly placed defects.[7,23,26] These quasi-equilibrium fluctuations were found to be true in TaS_3, though part of the noise came from a different source.[26] Another dynamical process that is widely mentioned is the phase slip of the CDW at strong pinners and electrical contacts locations.[15,27-30] This mechanism can exist only at the sliding state and cannot explain low frequency fluctuations measured at the pinned state using dc[21] or ac bias.[22] Another interesting picture is illustrated to explain the three level cyclical time trace measured:[24] since all cases of irreversible time traces in the noise were explained as a displacement of some entity along the current, a model of nonequilibrium CDW defect drag was proposed.[22] Comparison between elastic properties under ac bias, dielectric response at the pinned state, BBN in the sliding state and overshoot current experiments indicate that at least in TaS_3 all show broad distribution of time constants therefore all come from a common relaxational origin.[9] The precise relaxational process was not specified.

Another dispute over the BBN origin concerns velocity shear versus velocity temporal fluctuations: according to the phase-only model[16,18] the sliding state is a unique configuration, so the CDW slides with the same velocity everywhere, without temporal fluctuations, thus there is no BBN. According to ref. 27 experimental results indicate that BBN is associated with different time-average CDW velocities at different locations of the sample causing velocity shear and low frequency noise. The same group demonstrated that the BBN of a sample with a thickness step parallel to the long axis, is much bigger that the sum of BBNs of two samples separated from the previous one, each with a rectangular cross-section - an indication that when there is no velocity shear between various sections of the sample the BBN is much lower. Theoretical arguments[17] agree that the sliding CDW is broken into small segments with different time-averaged CDW velocities. However, it is not clear how the velocity shear can cause 1/f-like fluctuations. Furthermore, according to ref. 10 direct simultaneous measurements of BBN and instantaneous frequency fluctuations of NBN show strong cross-correlation between the two quantities, indicating that the BBN is mainly a manifest of temporal velocity fluctuations of the CDW throughout the whole sample, rather than some velocity shear mechanism.

The common phenomenological explanation among theoreticians, of the 1/f noise mechanism in CDW, is based on more general arguments:[14-20] a random system with spatial interaction due to the stiffness of the CDW and the competing forces between pinning sites, are believed to be characterized by a very rough free energy landscape in phase space.[31,32] The CDW is presented as a point wandering in this energy landscape that contains energy wells ("attractors") of all depths that exist close to any point in this space.[20] Since each location in phase space is a unique CDW rearrangement with a specific polarization, the wandering causes the low frequency noise. Two forces drive the system to wander: thermal random force as well as external bias. Many models and simulations are based on the above description, though there is no agreement on many important assumptions.

One of the main disputes is: how many degrees of freedom needed for the model? Can we describe CDW phenomena, and especially the low frequency fluctuations, by CDW phase degree of freedom,[14,18,19] or do we need CDW amplitude too?[15,17,28,33] While phase-only models produce some of the CDW phenomena, it was not able to exhibit the sliding state low frequency fluctuations,[16,18] as well as other CDW characteristics,[28,34] like switching behavior, hysteresis and chaos. Since phase slip is believed to be involved in the BBN mechanism one cannot describe the dynamics of CDW using phase-only model, because at the phase slip location amplitude variations must occur. There was no attempt so far to simulate the CDW dynamics using phase and amplitude together in one model.

Another related topic is the behavior of the CDW near threshold: is it a critical phenomenon or not? Is there some typical length scale that diverges at the vicinity of threshold field? If the system is critical then all the CDW volume must start sliding together, alternatively - in a non-critical system the onset of threshold is probably characterized by percolation paths where the CDW conduction starts. Phase-only models simulations indicate that indeed it is a critical phenomenon,[14] also by formulating a connection between BBN and a dynamic coherence length and measuring the BBN[35] it was shown that this length diverges when approaching threshold. On the other hand general arguments indicate[15,17] that at the threshold the critical behavior is destroyed since the sample is broken into small segments with different time-averaged CDW velocities, thus there is no divergence in any typical length. Also - if critical behavior is assumed using a phase-only model then infinite CDW strains will occur when increasing sample's volume.[15,17] We believe that a critical onset of sliding must be characterized by a large sharp peak in the BBN at the threshold - which was never experimentally observed as far as I know. It is not clear what is the relevant length scale for the phenomenological model and particularly for the critical behavior: phase-phase, amplitude-amplitude or velocity-velocity correlation length. It is not clear what is the relations between these lengths and what is the relations to a discrete fluctuator typical size.[22]

An extension to the equilibrium vs. driven mechanism problem presented above is: can one predict CDW fluctuations using deterministic dynamic equations or does one need to add a random term[36] (Langevin) to detect BBN fluctuations in simulations. Most of the models presented so far used deterministic equations without adding a Langevin term to the force. Since it is believed that there are energy wells of all depths in phase space, one cannot neglect thermal fluctuations. It was shown experimentally[36] that when a mod-lock occurs, although the NBN peaks sharpen and the BBN is reduced, there still exist NBN amplitude fluctuations that cannot be explained using deterministic dynamics. We can speculate that BBN of CDW can be invoked by some collaboration between thermal noise that is amplified by the nonlinear dynamics, and resulting in 1/f - like rather than white spectrum.

Can we formulate one detailed model that produces all CDW phenomena just by changing parameters? How much will the simulated spectrum of the fluctuations be affected by the details of the model? These questions are connected to the broader ever lasting challenge: can we formulate a simple phenomenological model for 1/f noise that loosely include all physical cases in nature with similar dynamics, i.e.: CDW, type II superconductors in the mixed state, spin density waves, various magnetic systems and other glassy systems where one finds long range interaction, randomness and external driving force.

3 Speculations and future research

Fig. 1 shows an illustration of a specific rearrangement of the sliding CDW in a non-ideal crystal containing various defects. We do not claim that this is the only precise description of CDW but we speculate that most of it is reasonably accurate. The figure shows a deformable CDW sliding from one contact to the other, with phase tear, velocity shear etc. The proportions are exaggerated a little for clarity. Strong pinners cause CDW pinning in some surrounding volumes, while weak pinners cause long range deformation in their vicinity. Electrons near the contacts must change their identity since there are no condensed CDW electrons outside the sample. At these locations a phase slip occurs, wavefronts appear at the bottom location and disappear at the top location. Due to a specific distribution of defects, it is energetically favorable, in this specific example, for two segments to slide at different velocities, causing velocity shear at the boundary. Any nonideality of the CDW (deformation, tear, shear, slip etc.) costs energy. Since external force (i.e. electric field) and thermal

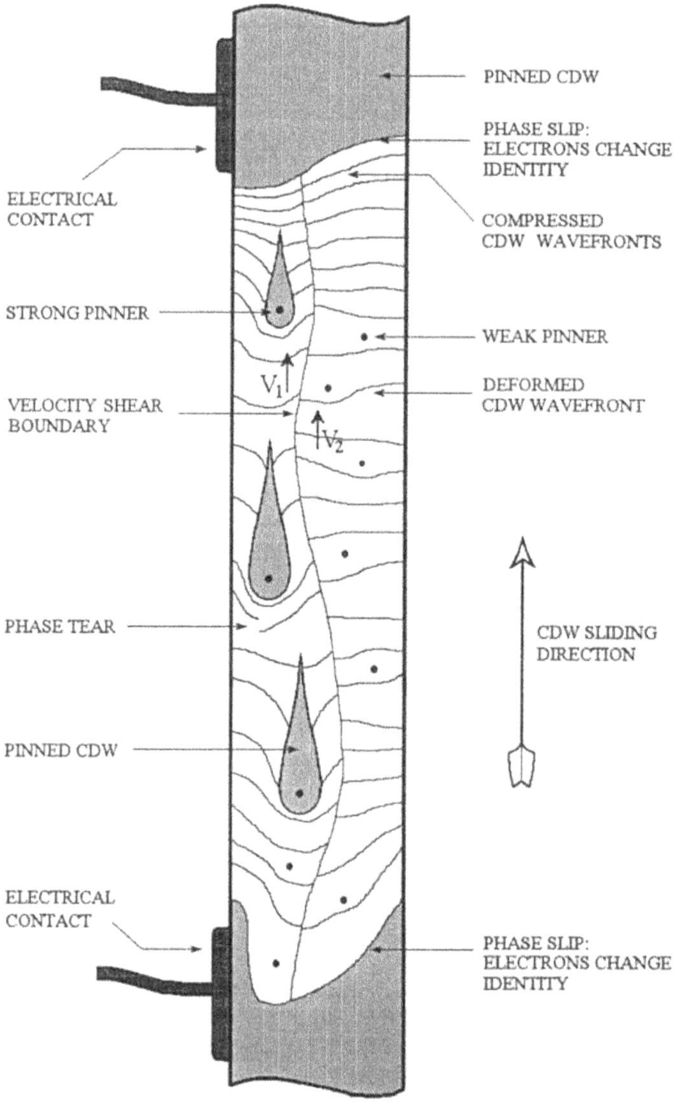

Fig.1: An illustration of a configurational rearrangement of the sliding CDW in the presence of randomly placed defects.

force act on the system, the CDW can fluctuate by jumping to another rearrangement - that might be different from the one illustrated in this figure.

Much theoretical as well as experimental works are needed to be done in order to deepen our understanding on the 1/f noise mechanism of CDW. Theoretical work is needed to improve detailed CDW models and include phase as well as amplitude degrees of freedom. On the other hand, general dynamical models of fluctuating mechanisms that develop 1/f noise spectrum, are to be pursued. These models should include randomness, long range interactions, nonlinearities, external force and random thermal noise. We believe that these models can give some insight to CDW noise mechanism as well as other noisy physical systems.

Further experimenting with CDW is needed: measuring simultaneous sample's length fluctuations and BBN and correlating the two can give more insight into the relation between elastic and electronic proparties as well as better understanding of the CDW metastability. Experimenting with split edges CDW samples and driving each edge with a different CDW current can give some ideas on velocity shear and phase slip and their part in the BBN mechanism. More experiments on mesoscopic CDW samples are needed: one can find direct connection between a single fluctuator observed in the noise and a specific single intentionally planned defect in the crystal. A more pretentious idea is to directly view CDW rearrangements, a picture that might look like Fig. 1. Such an experiment is not far from reaching since current microscopy techniques used in STM, AFM or SEM are well established for atomic-size viewing.

4 Conclusions

So far, although many pieces of the puzzle are known, one cannot pinpoint one specific origin of the low frequency fluctuations of the sliding state of CDW. Theoretical and experimental results indicate that most CDW phenomena can be described using a classical rather then quantum mechanism. The origin of the noise involves competition between interaction and randomness. Metastability of the configurational rearrangements of the CDW over a randomly placed defects plays an important roll in the mechanism. Random thermal noise as well as external electric field cause the bouncing between these configurations.

Acknowledgments

The author is grateful to M. B. Weissman and A. C. Marley for many stimulating discussions. This work was partially supported by the NSF grant No. DMR8920538 and the Rothschild Foundation. The author is also acknowledges the support from the Physics Dept. of the University of Illinois.

References

1. For a review see: G. Gruner, "The Dynamics of Charge-Density Waves", *Rev. Mod. Phys.*, **60**, 1129 (1988).
2. S. Bhattacharya et. al., *Phys. Rev. Lett.* **59**(16), 1849 (1987).
3. J.C. Gill, *Solid State Comm.*, **39**, 1203 (1981).
4. G. Mihaly and L. Mihaly, *Solid State Comm.*, **48**(5), 449 (1983).
5. J. Levy and M.S. Sherwin, *Phys. Rev.* B. **43**(10), 8391 (1991).
6. D. DiCarlo et. al., *Phys. Rev. Lett.* **70**(6), 845 (1993).
7. S. Bhattacharya et. al., *Phys. Rev.* B. **40**(8), 5826 (1989).
8. L.C. Bourne et. al., *Phys. Rev. Lett.* **56**(18), 1952 (1986).
9. R.L. Jacobsen et. al., *Phys. Rev.* B. **43**(16), 13198 (1991).
10. I. Bloom et. al., *Phys. Rev.* B. **50**(16), 12218 (1994).
11. S. Hoen et. al., *Phys. Rev.* B. **46**(3), 1874 (1992).
12. J.W. Brill, *Solid State Comm.*, **41**(12), 925 (1982).
13. J.W. Brill and W. Roark, *Phys. Rev. Lett.* **53**(8), 846 (1984).
14. D.S. Fisher, *Phys. Rev.* B. **31**(3), 1396 (1985).
15. S.N. Coppersmith, *Phys. Rev. Lett.* **65**(8), 1044 (1990).
16. P.B. Littlewood, *Phys. Rev.* B. **33**(10), 6694 (1986).
17. S.N. Coppersmith and A.J. Millis, *Phys. Rev.* B. **44**(15), 7799 (1991).
18. A.A. Middleton, *Phys. Rev. Lett.* **68**(5), 670 (1992).
19. C.R. Myers and J.P. Sethna, *Phys. Rev.* B. **47**(17), 11171 (1993).
20. P.Sibani and P.B. Littlewood, *Phys. Rev. Lett.* **71**(10), 1482 (1993).
21. S.V. Zaitsev-Zotov and V.Y. Pokrovskii, *JETP Lett.*,**49**(8), 514 (1989).
22. I. Bloom et. al., *Phys. Rev.* B. **50**(8), 5081,(1994).
23. A.C. Marley et. al., *Phys. Rev.* B. **44**(15), 8353 (1991).
24. I. Bloom et. al., *Phys. Rev. Lett.* **71**(26), 4385,(1993).
25. H.T. Hardner et.al., *Phys. Rev.* B. **46**(15), 9833 (1992).
26. A.C. Marley and M.B. Weissman, *Phys. Rev.* B. **52**, 7965 (1995).
27. R.E. Thorne et. al., *Phys. Rev.* B. **35**(12), 6348 (1987).
28. R.P. Hall et. al., *Phys. Rev.* B. **38**(18), 13002 (1988).
29. M.P. Maher et. al., *Phys. Rev. Lett.* **68**(20), 3084 (1992).
30. M.P. Maher et. al., *Phys. Rev.* B. **43**(12), 9968 (1991).
31. A.C. Marley et. al., *Phys. Rev.* B. **49**(23), 16156 (1994).
32. I. Bloom and A.C. Marley, *ICNF95 conference preceedings,* p. 99 (1995).
33. S. Ramakrishna et. al., *Phys. Rev. Lett.* **68**(13), 2066 (1992).
34. J. Levy et. al., *Phys. Rev. Lett.* **70**(17), 2597 (1993).
35. M.O. Robbins et. al., *Phys. Rev. Lett.* **55**(25), 2822 (1985).
36. S. Bhattacharya et. al., *Phys. Rev.* B. **38**(14), 10093 (1988).

SIMULATION OF PHONON-NUMBER FLUCTUATIONS AND $1/f$ NOISE

N. FUCHIKAMI
Department of Physics, Tokyo Metropolitan University, Minami-Ohsawa, Hachioji, Tokyo 192-03, Japan

S. ISHIOKA
Department of Information Science, Kanagawa University at Hiratsuka, Hiratsuka, Kanagawa, 259-12, Japan

We discuss the possibility of $1/f$ fluctuations in phonon numbers. Some experimental studies and computer simulations reported so far are briefly surveyed. Our simulations were performed in collaboration with H. Kawamura and D. Choi of our group. Results obtained for Fermi-Pasta-Ulam Hamiltonian show that the spectrum of phonon number fluctuations of each mode is generally Lorentzian. For low frequency modes, the relaxation time can be very long so that the spectrum looks like $1/f^2$.

1 Introduction

All semiconductor materials exhibit $1/f$ noise in the electrical conductance. On the origin of the noise, there has long been a controversy: mobility noise or carrier number fluctuations. Without arguing this point, if we assume that the conductance noise is due to the mobility noise, then the latter can be interpreted by a further assumption that the phonon numbers of each mode fluctuate with a $1/f$ spectrum[1]. In the present paper, we take up this second assumption and ask 'The $1/f$ phonon-number fluctuations, is that true or false?' No theory has ever succeeded in predicting the $1/f$ phonon-number fluctuations. So this is an open question. The question leads us to numerical experiments on lattice vibrations.

Experimental studies and computer simulations in previous work are briefly reviewed in Section 2 and 3, respectively. Our computer simulation is described in Section 4. The last section is devoted to summary and discussions.

2 Review of Optical Experiments

Some optical experiments with the use of insulators seem to be a direct proof of the $1/f$ phonon-number fluctuations.

2.1 Brillouin Scattering in a Single Crystal

The intensity fluctuations of a laser light scattered by a quartz crystal have a $1/f$ spectrum in the frequency range $10^{-3.8} < f < 10^{-1.8}$ (in Hz)[2]. The results can be explained if the energy fluctuations are $1/f$ -like for the individual phonon mode that is involved in the light scattering.

2.2 Attenuation in an Optical Fibre

The intensity fluctuations of a light at the output of the quartz fibre have a $1/f$ spectrum in the range $1 < f < 10^2$ (in Hz)[3]. That can also be understood by assuming the phonon number fluctuations.

In the spectra of these experiments, no plateau has been observed in the low frequency region, that means the fluctuation process might be non-stationary or the relaxation time is too long to be observed.

3 Numerical Experiments — Review of Previous Work

If the intensity fluctuations observed in the optical experiments are attributed to the phonon-number fluctuations, $1/f$ -like behavior can be an intrinsic character of weakly anharmonic lattices. However, there is no theory that can predict $1/f$ phonon-number fluctuations. Several attempts have been made to simulate the lattice vibrations numerically. Those simulations are performed for hamiltonian systems describing a nonlinear chain of particles.

3.1 3rd Order Potential

Among the model systems for which the simulations have been performed, Fermi-Pasta-Ulam (FPU) lattices are the most popular. One-dimensional FPU hamiltonian with 3rd order nonlinear potential was employed by Koch *et al.*[4] They obtained a $1/f^2$ -like spectrum for the lattice size (*i.e.* the number of particles) $N = 64$, where the initial conditions are such that the lowest one, two or three mode-energies are excited. No plateau is observed in the low frequencies.

3.2 4th Order Potential

FPU hamiltonian with 4th order potential was simulated by Fukamachi[5]. The initial condition is such that the coordinate of each particle is random and its velocity is zero. For $N = 8$ was obtained a $1/f$ spectrum in a wide range of frequency. There is no plateau observed in the low frequency region. Maybe this is the only one successful result obtained for the hamiltonian system of anharmonic lattice.

3.3 3rd plus 4th Order Potential

FPU hamiltonian with 3rd order potential together with 4th order one was employed by Musha *et al.*[6] for $N = 128$. The nonlinear parameters are chosen as fitting the quartz crystal for which they observed $1/f$ fluctuations from optical experiments. The power spectral density (PSD) of phonon energy belonging to the mode with the wave length 4 (in unit of particle distance) is Lorentzian with its low frequency tail being elevated. It is, however, unclear whether the tail is $1/f$ -like or not.

3.4 Lennard-Jones Potential

A one-dimensional chain with Lennard-Jones potential was simulated by Koch *et al.* for several initial conditions[7]. The PSD of the lowest mode energy is $1/f^{\alpha}$ -type with no plateau in the low frequency region. The value of the spectral exponent α seems to be $\lesssim 2$ in all tested cases.

4 Numerical Experiment — Our Study

From the numerical experiments so far reported, it is difficult to deduce any definite conclusion on the general features of the phonon-number fluctuations. Most of the results seem to indicate a $1/f^2$ spectrum but a $1/f$ spectrum is also observed in some cases. A more systematic study is thus necessary.

We made a simulation[8] employing the same type of FPU hamiltonian as used in ref.5 where the $1/f$ fluctuations are reported. We thoroughly investigated the following points.

1) System-size dependence of fluctuations.

2) Temperature/nonlinerity dependence of the fluctuations.

3) Mode dependence of the fluctuations.

We are also concerned with the relaxation time of the system because, as mentioned already, the PSD observed in the optical experiments does not saturate to white noise in the low frequency region. Our initial conditions are different with those in the past simulations. We started from the most probable states in thermal equilibrium, *i.e.*, we adopted initial conditions such that the energy is equally distributed to each vibrational mode. Therefore our PSD is expected to be free from transient effects.

4.1 FPU Hamiltonian and the Computational Methods

One-dimensional FPU hamiltonian with 4th order nonlinear potential is given by

$$H = H_0 + V , \tag{1}$$

$$H_0 = \frac{1}{2M}\sum_{i=1}^{N} p_i^2 + \frac{M\omega_0^2}{2}\sum_{i=1}^{N}(q_{i+1} - q_i)^2 , \tag{2}$$

$$V = \frac{\beta}{4}\sum_{i=1}^{N}(q_{i+1} - q_i)^4 , \tag{3}$$

where

$$q_{i+N} \equiv q_i , \quad p_{i+N} \equiv p_i . \tag{4}$$

The lattice, for a given N, is specified by model parameters (M, ω_0, β). The dynamical state of the system is specified by the value of the total hamiltonian H. We introduce the nondimensional quantities

$$\omega_0 t \to t , \quad \frac{\sqrt{\beta}}{\omega_0\sqrt{M}} q_i \to q_i , \quad \frac{\beta}{\omega_0^4 M^2} H \to H , \tag{5}$$

and obtain a reduced hamiltonian with $M = 1$, $\omega_0 = 1$, $\beta = 1$ in eqs.(1)$\sim$(3). Then the relation

$$\begin{aligned}&(\textit{the energy of the reduced system})\\ &\quad = \frac{\beta}{\omega_0^4 M^2} \times (\textit{the energy of the original system})\end{aligned} \tag{6}$$

holds so that the number of independent parameters is only one: the total energy of the reduced system. Note that the energy of the original system is proportional to the temperature of the system. Therefore the energy of the reduced system is proportional to both the nonlinearity and the temperature of the original system.

We are concerned with fluctuations of the phonon number n_m in each vibrational mode m. The phonon number corresponds to the mode energy E_m in the classical limit. If the lattice is purely harmonic, the state of the system $\{q_j(t), p_j(t)\}$ can be expressed as a superposition of the normal modes with eigenfrequency $\Omega_m/2\pi$ as

$$\left.\begin{aligned} q_j(t) &= \sum_{m=1}^{N} A_m \cos\left(\frac{2\pi m j}{N} - \Omega_m t - \varphi_m\right), \\ p_j(t) &= \sum_{m=1}^{N} A_m \Omega_m \sin\left(\frac{2\pi m j}{N} - \Omega_m t - \varphi_m\right), \\ &\qquad\qquad (j = 1, 2, \cdots, N), \end{aligned}\right\} \tag{7}$$

where

$$\Omega_m \equiv 2\sin\left(\frac{\pi m}{N}\right), \qquad (m = 1, 2, \cdots, N). \tag{8}$$

The amplitude A_m and the phase φ_m are constants determined by the initial conditions. Then the total energy with no anharmonicity is given by the sum of the mode energies:

$$H_0 = \sum_{m=1}^{N} E_m, \tag{9}$$

where

$$E_m = \frac{1}{2} N \Omega_m^2 A_m^2. \tag{10}$$

The mode energy E_m can be rewritten as

$$E_m = \frac{1}{2N} |\eta_m + i\Omega_m \xi_m|^2, \tag{11}$$

in terms of the Fourier transforms of $q_j(t)$ and $p_j(t)$:

$$\xi_m(t) \equiv \sum_{j=1}^{N} e^{i\frac{2\pi m j}{N}} q_j(t), \qquad \eta_m(t) \equiv \sum_{j=1}^{N} e^{i\frac{2\pi m j}{N}} p_j(t). \tag{12}$$

When the hamiltonian includes anharmonic terms, some ambiguity cannot be avoided in defining the 'phonon mode' and its energy. Here we take the simplest definition that is usually employed: The 'mode energy' in weakly nonlinear lattices is given by eq.(11) with eqs.(12) and (8). This means that the state of the anharmonic lattice at any moment $\{q_j(t), p_j(t)\}$ is regarded as a superposition of the normal modes of the harmonic lattice (eq.(7) with A_m and φ_m which are not constant any more). The sum of E_m's is therefore not equal to H but to the value of H_0 at that moment. A drawback involved in

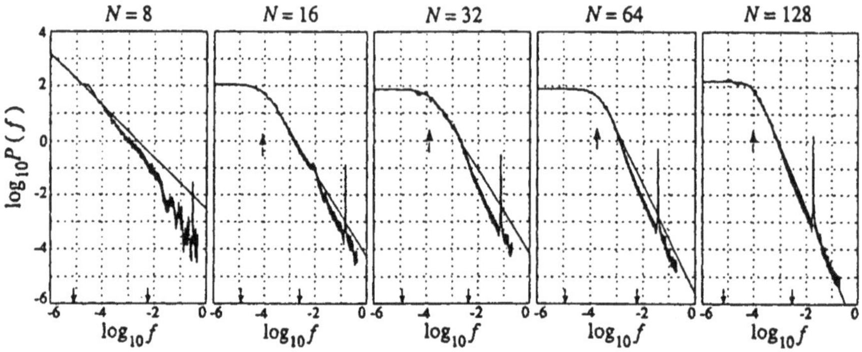

Figure 1: PSD of the lowest mode energy averaged over N_s sample processes with different initial conditions. $N_s = 50$ for the particle number $N = 8, 16, 32, 64$; $N_s = 48$ for $N = 128$. The average mode energy is fixed at $e_0 = 0.2$. The line $F(f) = A/\{1 + (f/f_S)^\alpha\}$ is obtained by fitting the PSD in the frequency range between the two ↓ marks. The 'shoulder' at $f = f_S$ is indicated by a ↑ mark. The observation time is $T = n \times m \times dt$ where n is the number of time-series data used to obtain the PSD in each process, $m \times dt$ the sampling time with $dt = 0.0075$. $n \times m = 2^{17} \times 140$ for the system size $N = 8$; $n \times m = 2^{16} \times 160$ for $N = 16$; $n \times m = 2^{15} \times 320$ for $N = 32, 64$; $n \times m = 2^{16} \times 320$ for $N = 128$.

this definition is the origin of the unphysical sharp peak of PSD in the high frequency region [8].

We chose the initial condition such that each mode energy has the same value e_0 because the energy is equally distributed among the normal modes in thermal equilibrium:

$$E_m\ (t=0) = e_0 \quad \text{for } m = 1, 2, \cdots, N-1; \tag{13}$$

$$E_N\ (t=0) = 0 \quad : \text{no translational motion of the center of mass.} \tag{14}$$

This corresponds to the initial state

$$\left.\begin{aligned} q_i(0) &= \sqrt{\frac{2e_0}{N}} \sum_{m=1}^{N-1} \frac{1}{\Omega_m} \cos\left(\frac{2\pi m i}{N} - \varphi_m\right), \\ p_i(0) &= \sqrt{\frac{2e_0}{N}} \sum_{m=1}^{N-1} \sin\left(\frac{2\pi m i}{N} - \varphi_m\right), \\ &\qquad (i = 1, 2, \cdots, N), \end{aligned}\right\} \tag{15}$$

where φ_m is a uniform random number in the interval $(0,\ 2\pi)$.

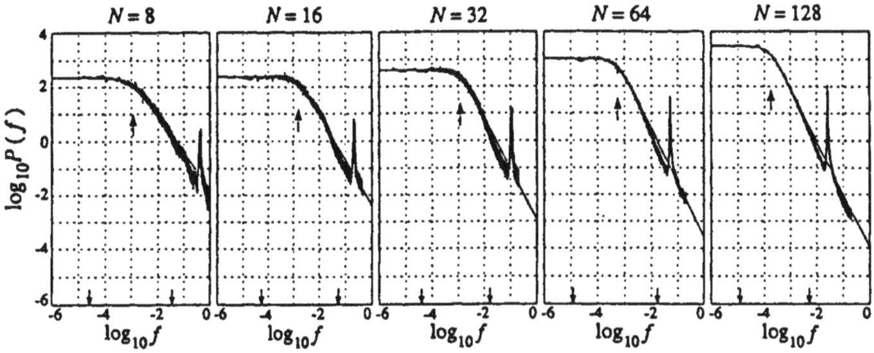

Figure 2: Same as Fig.1 but $e_0 = 1.0$. $N_s = 50$ for $N = 8, 16$; $N_s = 47$ for $N = 32, 128$; $N_s = 49$ for $N = 64$. $n \times m = 2^{17} \times 80$ for $N = 8$; $n \times m = 2^{17} \times 160$ for $N = 16$; $n \times m = 2^{17} \times 320$ for $N = 32$; $n \times m = 2^{15} \times 320$ for $N = 64, 128$.

We performed the simulation varying the system size N and the initial mode energy e_0. For each combination (N, e_0), the PSD is averaged over $47 \sim 50$ runs with different initial conditions (*i.e.*, different random number sets $\{\varphi_m\}$). The averaged PSD $P(f)$ is then fitted with the line $F(f) = A/\{1+(f/f_S)^\alpha\}$ so that $\sum_i\{P(f_i) - F(f_i)\}^2$ is minimized in the fitting range $\mathrm{Max}\{f_{\min}, 10^{-1.5}f_S\} < f < 10^{1.5}f_S$. The simulation was performed using the 4th order Runge-Kutta method with the time step $dt = 0.0075$.

4.2 Numerical Results

We first focus on the lowest mode energy E_1. Figs. 1 and 2 show the PSD of E_1 for $e_0 = 0.2$ and $e_0 = 1.0$, respectively.

The behavior of the PSD for $(N = 8, e_0 = 0.2)$ is similar to that obtained by Fukamachi[5] at least in the low frequency region: the PSD is $1/f$ -like with no saturation to white noise.

Unfortunately that is an exceptional case. General features of the fluctuations are as follows: First, the PSD is flat at low frequencies, which means that the process is stationary. The inverse of the frequency at the 'shoulder', $\tau_c \equiv 1/f_S$, is a relaxation time of this stationary process. Second, the spectral exponent α is not 1 but close to 2.

It is now important to determine how the shoulder frequency f_S and the exponent α depend on the system size N and the (nondimensional) total energy.

In Fig.3, we present log-log plots of f_S against N for different values of e_0. For a fixed value of N, smaller values of e_0 yield smaller values of f_S.

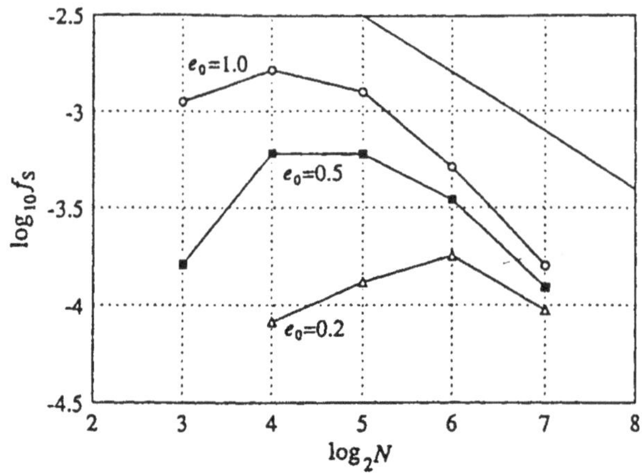

Figure 3: Log-log plot of the shoulder frequency f_S vs N for various initial values of the mode energy e_0. The straight line represents the relation $f_S \propto 1/N$.

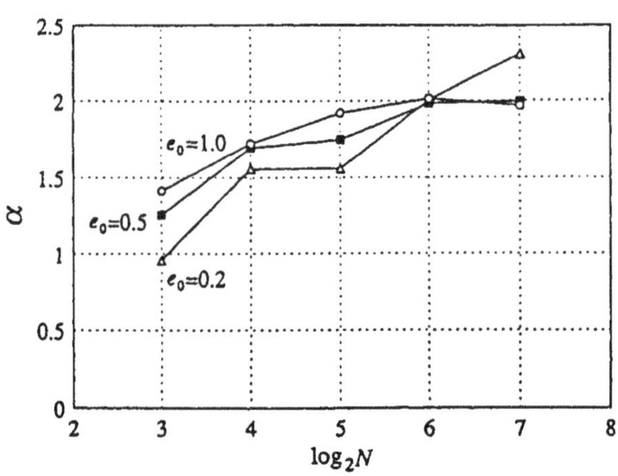

Figure 4: Log plot of the spectral exponent α vs N for various initial values of the mode energy e_0.

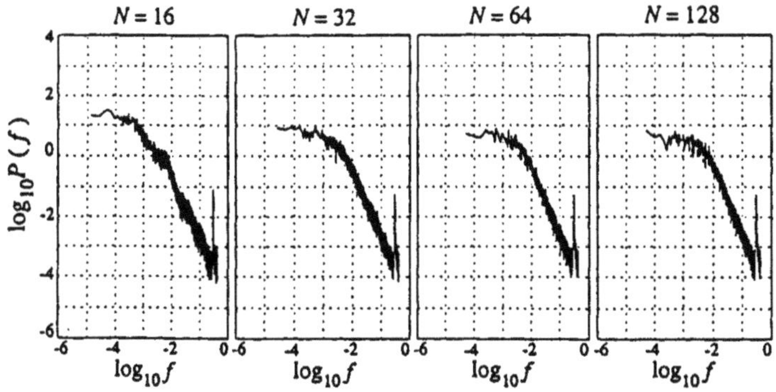

Figure 5: PSD of higher mode energies E_m, where the wave number $k = m/N$ is fixed as $k = 1/8$. $e_0 = 0.2$. Each PSD is obtained by averaging over $N_s = 10$ sample processes with different initial conditions. $T = n \times 160 \times 0.0075$. $n = 2^{16}$ for $N = 16$; 2^{15} for $N = 32$; 2^{14} for $N = 64, 128$. For $N = 8$, see Fig.1.

We can see two aspects of this from eq.(6). When the vibrational energy (or temperature) is fixed, the relaxation time $1/f_S$ becomes longer as the nonlinear parameter β decreases. When β is fixed, on the other hand, smaller energies (or lower temperatures) lead to longer relaxation times. These trends are physically reasonable. It is interesting that f_S depends on the system size nonmonotonically. The nonmonotonic behavior might be explained in terms of two competing factors.

1) An increase in N makes the relaxation time $1/f_S$ longer because $1/f_S$ is the time necessary for the system to lose the memory of the initial state so that the PSD becomes white.

2) An increase in N makes the frequency spacing smaller so that the resonance conditions necessary for mixing mode energies are more easily satisfied.

The convex curve of f_S versus N indicates that the second factor is effective until it is overwhelmed by the first one. f_S must decrease faster than $1/N$ as N increases if we observe the power law in macroscopic lattices. This is because the phonon number fluctuations are meaningless in the frequency region higher than the relevant mode frequency which is equal to $1/N$ for the lowest mode.

If we apply the present simulation to, for example, a quartz crystal in which $1/f$ fluctuations are observed[2], the parameter values are[6] $M = 10^{-25}$ kg, $M\omega_0^2 = 60\,\mathrm{Nm}^{-1}$, and $\beta = 25 \times 10^{21}\,\mathrm{Nm}^{-3}$. Then we should read $t/\omega_0 \sim t \times 4 \times 10^{-14}$ s for t and $(M^2\omega_0^4/\beta)e_0 \sim e_0 \times 1.4 \times 10^{-19}$ J for e_0. Room temperature then corresponds to $e_0 \sim 0.03$. It is difficult to extrapolate the curves in Fig.3 reliably to the macroscopic range $N \sim 10^8$ and $e_0 \sim 0.03$. However, it seems to be reasonable that the shoulder frequency f_S in macroscopic systems can be

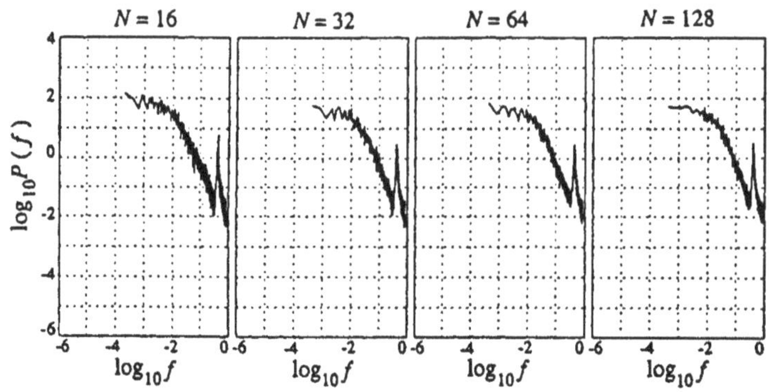

Figure 6: The same as Fig.5 but $e_0 = 1.0$. $T = n \times 80 \times 0.0075$. $n = 2^{13}$ for $N = 16$; 2^{12} for $N = 32, 64, 128$.

very low.

The exponent α is plotted in Fig.4. We see that α increases from $\alpha \sim 1$ to $\alpha \sim 2$ as N increases from 8 to 64 or 128. An exceptional case in Fig.4 is the datum for $e_0 = 0.2$ with $N = 128$ which is even larger than 2. Actually, the estimation of α is, unlike f_S, sensitive to the method of fitting, especially when the frequency range of the flat spectrum is not wide enough. However, we can see that the spectrum is closer to $1/f^2$-like than $1/f$ for large N.

Fluctuations of higher mode energies are also investigated. We fix the wave number $k = m/N$ at $1/8$ and present the PSD of the mode energy $E_m = E_{N/8}$ in Figs. 5 and 6 for $e_0 = 0.2$ and 1.0, respectively. The spectrum for large N appears to be typical Lorentzian, varying from $1/f^2$-like to constant as the frequency decreases.

Finally, we present in Fig.7 the spectrum for $e_0 = 0.03$ (corresponding to quartz crystal at room temperature) with $N = 128$. We see that the spectra of E_1 (Fig.7(a)) and E_{16} (Fig.7(b)) are $1/f^2$-like again and in the former case, the PSD has not yet become a white spectrum during the observation time $T = 2^{16} \times 80 \times 0.0075$ because e_0 is very small.

5 Summary and Discussion

After all, our computer experiment showed that the phonon-number fluctuations obey the $1/f^2$-law in a wide frequency range except when the system size is small. The spectrum is generally Lorentzian, namely saturates to white noise at low frequencies, and so the fluctuation process is stationary. But our simulation suggests the possibility that the relaxation time for the low frequen-

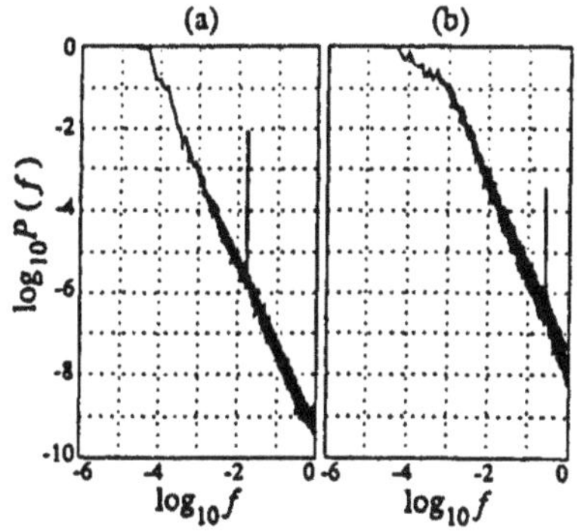

Figure 7: PSD for $N = 128$ and $e_0 = 0.03$. Each PSD is obtained by averaging over $N_s = 10$ sample processes with different initial conditions. $T = 2^{16} \times 80 \times 0.0075$. (a) Mode energy E_1. (b) Mode energy E_{16}.

cy modes can be very long if the system size is extrapolated to a macroscopic one. A simulation on a different type of high-dimensional hamiltonian system (although not corresponding to the lattice vibrations) also yields a Lorentzian spectrum [9]. These results seem to be consistent with Aizawa's stagnant layer theory [10]. He argued the long-term tail of chaotic hamiltonian based on KAM theorem and suggested that the $1/f^2$ fluctuations are universal for hamiltonian systems.

Our study is restricted on the system size, the dimension of the system and the type of anharmonicity. Can these factors play any role to shift the spectrum from $1/f^2$ to $1/f$?

Previously, we proposed a simple stochastic model of a quantal harmonic oscillator contacting with a heat bath [11]. The problem was described as a random walk on a ladder of energy levels $E(n) = h\nu(n + 1/2)$ $(n = 0, 1, 2, \cdots)$. The jumping rate $r_\uparrow(n)$ which is the transition rate from the state n to $n + 1$ was assumed as $\propto (1 + n)^\rho$. It was found that when $\rho = 3 \sim 4$, the spectrum is $1/f$ -like, while for $\rho = 1$ it becomes Lorentzian with a long relaxation time. In weakly-nonlinear hamiltonian systems, one phonon process is dominant, which corresponds to the case $\rho = 1$ in the stochastic model. The results obtained from the stochastic model do not directly depend on the system size, the dimensionality or the type of the potential. From these considerations, we feel that the fluctuations of each mode energy are generally $1/f^2$ like instead of $1/f$ at least in hamiltonian systems.

If the phonon number really exhibits $1/f$ fluctuations, what can change the intrinsic $1/f^2$ -like behavior into $1/f$ -like? We have no answer to this question at present. Our final hope is to expect some dramatic effect of weak interactions of anharmonic lattice with electrons, photons or heat bath. So far we have assumed non-dissipative systems, *i.e.*, hamiltonian systems. Concerning this point, Anton *at al.*'s paper [14] interests us. They solved spatially discretized Burgers equation numerically. When the excitation force is a deterministic sine wave, a $1/f^2$ -like spectrum is obtained, while for white Gaussian noise, the PSD approaches to $1/f$ -like.

In connection with the phonon number fluctuations, only a few studies on non-hamiltonian systems have been reported[11,12,13]. We consider that more extensive research on weakly dissipative systems with many degrees of freedom is needed.

References

1. T. Hooge in *Proc. Int. Conf. Noise in Physical Systems and* $1/f$ *Fluctuations*, eds. V. Bareikis and R. Katilius (World Scientific, Singapore, 1995) p.8.
2. T. Musha, G. Borbéry and M. Shoji, *Phys. Rev. Lett.* **64**, 2394 (1990).
3. A. J. van Kemenade, P. Hervé and L. K. J. Vandamme, *Electron. Lett.* **30**, 1338 (1994).
4. M. Koch, R. Tetzlaff and D. Wolf, *AIP Conf. Proc.* **285**, 525 (1993).
5. K. Fukamachi, *Europhys.Lett.* **26**, 449 (1994).
6. T. Musha and R. Kobayashi in *Proc. Int. Conf. Noise in Physical Systems and* $1/f$ *Fluctuations*, eds. T. Musha, S. Sato and M. Yamamoto (Ohmsha, Tokyo, 1991) p.559.
7. M. Koch, R. Tetzlaff and D. Wolf, in *Proc. Int. Conf. Noise in Physical Systems and* $1/f$ *Fluctuations*, eds. V. Bareikis and R. Katilius (World Scientific, Singapore, 1995) p.145.
8. H. Kawamura, D. Choi, N. Fuchikami and S. Ishioka, *Jpn. J. Appl. Phys.* **35**, 2387 (1996).
9. J. F. Willemsen, *Phys. Rev. Lett.* **71**, 1172 (1993).
10. Y. Aizawa, *Progr. Theor. Phys.* **81**, 249 (1989).
11. N. Fuchikami, S. Kaneda and S. Ishioka, *J. Phys. Soc. Jpn.* **62**, 3044 (1993).
12. Y. Aizawa, *Bussei Kenkyu (Kyoto)* **61**, 124 (1993).
13. T. Kawakubo and T. Kobayashi, in *Proc. Int. Conf. Noise in Physical Systems and* $1/f$ *Fluctuations*, eds. V. Bareikis and R. Katilius (World Scientific, Singapore, 1995) p.263.
14. K. Anton, R. Tetzlaff and D. Wolf, *AIP Conf. Proc.* **285**, 599 (1993).

THE PECULIAR DYNAMICS OF 1/f NOISE

INVARIANCE OF 1/f SPECTRUM OF GAUSSIAN NOISES AGAINST AMPLITUDE-SATURATION NONLINEARITY

Z. GINGL
Department of Experimental Physics, JATE University, Dom ter 9, Szeged, H-6720 Hungary

L.B. KISS
Uppsala University, Angstrom Laboratory, Box 534, Uppsala, S-75121 Sweden

Gaussian 1/f noise has a strange property: its spectrum is invariant against amplitude saturation nonlinearity. A linear amplifier overloaded by Gaussian 1/f noise produces a 1/f noise at its output even if the output timefunction has become a random square wave due to a heavy overload. The invariance holds for any amplitude truncation level, including assymetric cases like half-wave rectified 1/f noise. Our results are purely empirical: we have no idea how to initiate a theoretical explanation.

1 Introduction

1/f noise is known to be a peculiar kind of noise, especially due to its general occurance in nature and the lack of a general model to explein this feture. It has been discovered at the same time when genaral relativity and quantum mechanics was established, however, even today, the field of 1/f noise is producing unexplained effects and, as a consequence, various controversies.

Here, we would like to report a particularly strange property of Gaussian 1/f noises, which we have first published in 1993 as a side-effect of stochastic resonance under special conditions [1]. Since then, we have thoroughly repeated the experiments and extended the analysis for the case of various $1/f^{\kappa}$ noises, too.

2 Nonlinear amplitude transform of Gaussian $1/f^{\kappa}$ noises

The invariance [1,2] of Gaussian $1/f^{\kappa}$ noises against a special nonlinear transform, the amplitude truncation, has been investigated in the following way. The nonlinear operation is defined by the following expressions:

$$U_2(t) = U_1(t) \quad \text{if} \quad U_{min} \le U_1(t) \le U_{min} \tag{1a}$$

$$U_2(t) = U_{min} \quad \text{if} \quad U_{min} \ge U_1(t) \tag{1b}$$

$$U_2(t) = U_{max} \quad \text{if} \quad U_{max} \le U_1(t) \quad , \tag{1c}$$

where $U_1(t)$ is the original noise, and $U_2(t)$ is the output noise; U_{min} and U_{max} are the lower and higher truncation levels, respectively. By other words, there are two given amplitude levels which limit the amplitude of the input noise. An illustration of the truncation is shown on Fig.1.

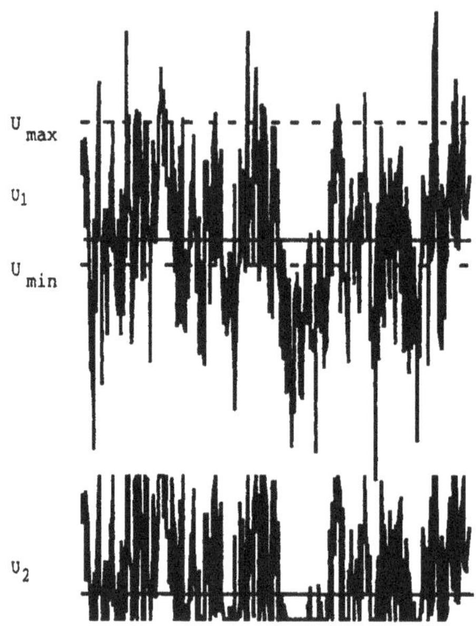

Figure 1: Amplitude truncation of a Gaussian 1/f noise.

3 Methods of investigation

We have carried out not only numerical but also analog simulations for the thorough investigation of the effects. Analog simulations have been very important to prove that the results are not due to aliasing errors which errors can often be found at computers simulation of power density spectra in nowadays' literature.

Using numerical methods, a Gaussian random number sequence having 1/f spectrum was generated. The random number generator was cerafully selected to avoid longrange-correlation artefacts. The spectrum of the measured noise was calculated by FFT by averaging 1000 power spectra obtained by independent runs. The length of one single run was always over 200000.

At the analog simulations the 1/f noise of a transistor was used and a proper analog circuit was used to perform the truncation. The output signal was filtered (anti-aliasing), digitized and the spectrum was calculated using FFT. Here we used lengths of 4096 steps and the averaging of 1000 spectra. The frequency range of the 1/f noise generator was ≈ 0.1Hz-10kHz.

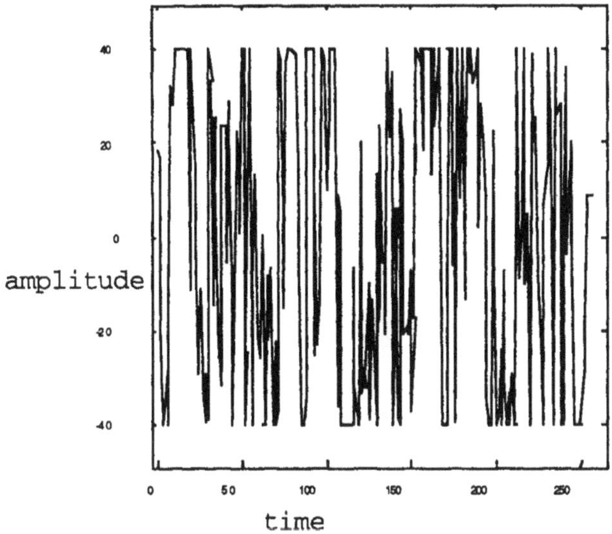

Figure 2a: An example for the amplitude truncation of a Gaussian 1/f noise.

4 Results

Within the experimental errors of our measurement and evaluation, the power spectrum turns out to be invariant against the amlitude truncation of the 1/f noise (see Fig.2). This surprising fact has been found for various assymetrical truncation levels also, including for example U_{min}=0 which corresponds to a half-wave-rectifierd noise. Of course, no one is able to test this phenomenon at each particular amplitude level because of the limitations of the length of simulation and measurement time. As, for the moment, there are no theoretical results explaining this behaviour, our statement is a purely empirical observation, which however, deserves thorough theoretical investigations.

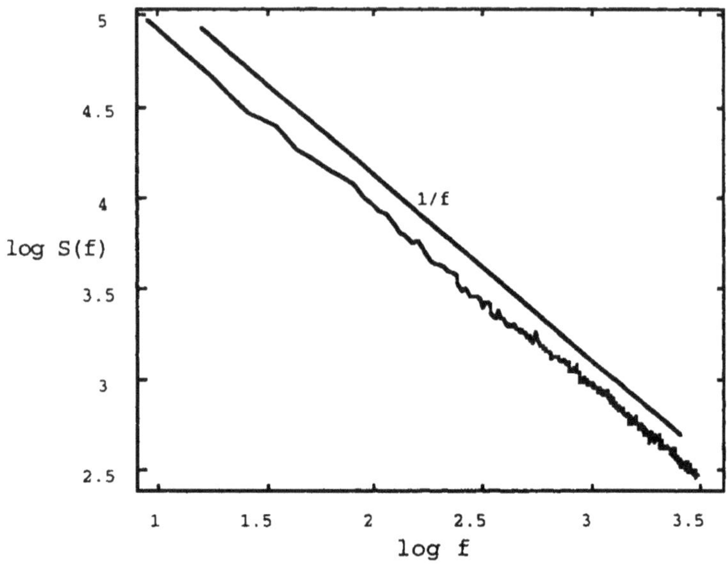

Figure 2b: The power density spectrum belonging to the noise shown at Fig.2.a.

We extended the numerical simulations for the case of $1/f^k$ noises with k not equal to 1. It is obvious from the analogy with the case of diffusion noise that for U_{min} and U_{max} very close to 0 and k=2, the truncated noise should have k=1.5 . For 2>k>1, the k exponent for the truncated noise is lower than 1.5 , it is closer to 1. For k<1 we found a similar invariance as for k=1 noise. The results are summerized on Fig.3.

5 Some of the unsolved problems

i. ***Theory***. Unfortunately there is no theoretical explanation of our results, not even a picture to intuitively explain the effects described here.

ii. ***Frequency range of investigations***. We don't know, whether the invariance of the 1/f shape is valid only up to a limited frequency range. We have carried out investigations with limited frequency band 1/f noises (out of the 1/f range the spectra have tended to zero) and we have found always 1/f spectra in the same frequency band as in the original noise.

iii. ***Relevance of the level crossing dynamics of 1/f noise***. Since the the spectrum seems to be invariant against any truncation of amplitude, we may say, that the level crossing dynamics [4] of the Gaussian 1/f noise, at any choosen level, has a key role in the existence of the 1/f spectrum.

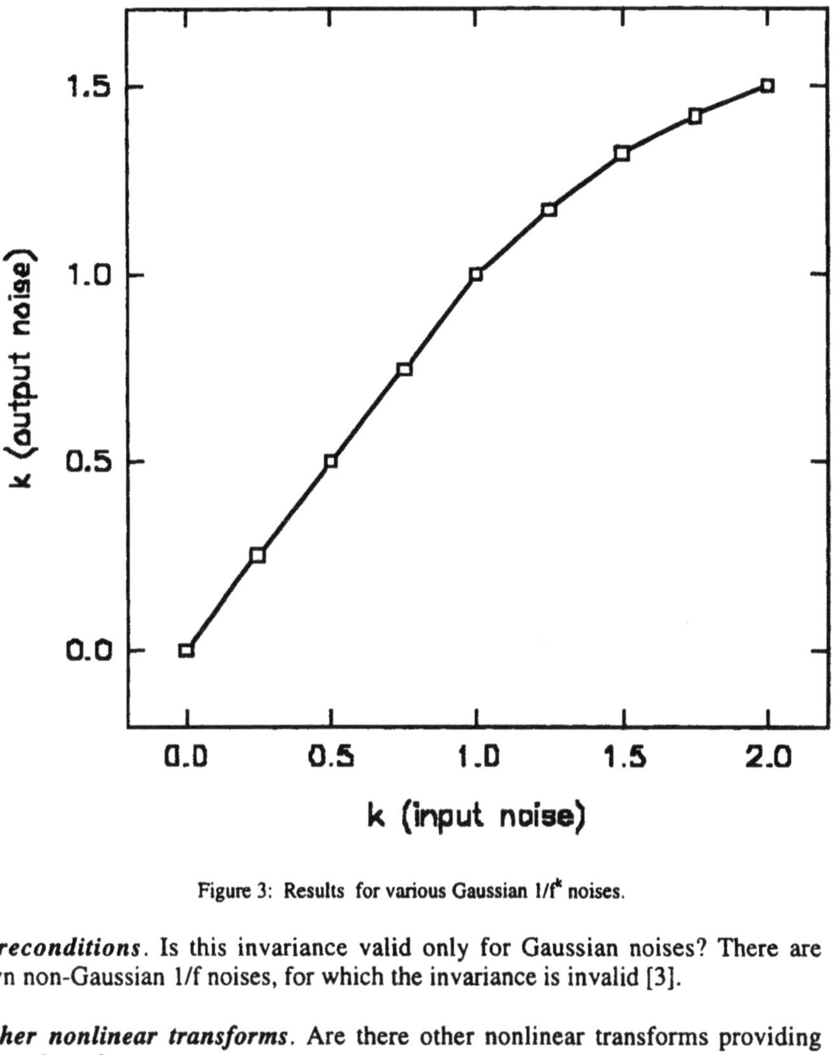

Figure 3: Results for various Gaussian $1/f^k$ noises.

iv. ***Preconditions***. Is this invariance valid only for Gaussian noises? There are known non-Gaussian 1/f noises, for which the invariance is invalid [3].

v. ***Other nonlinear transforms***. Are there other nonlinear transforms providing this invariance?

6. ***Possible convergence from $1/f^2$ to 1/f***. Since the truncation lowers the exponent of the power spectrum for $1/f^k$ noises with $k>1$ (e.g. for $k=2$ it goes to 1.5, etc., see Fig.3), we may find systems which can provide convergence from $1/f^2$ to 1/f, which can also be a method for generating 1/f noise. Of course, such systems have to be rather complex because the non-Gaussian characteristics of the truncated

noise requires to add up a large number of independent truncated noises to provide Gaussian characteristics before the next successive truncation step.

7. ***Usefulness for understanding the generality of 1/f noise***

If we could know the answer for several previous questions, this new property of 1/f noise might be useful to understand the general occurence of 1/f noise. In any case, our result contribute to the knowledge of 1/f fluctuations.

8. *Which real* ***systems can produce this nonlinearity?***

Linear amplifiers obviously do. Else of importance?

Acknowledgements

This work was supported by OTKA (Hungary) and TFR (Sweden). Discussions with Christian Houdre and Alexander Dubkov are appreciated.

References

1. L.B. Kiss, Z. Gingl, Zs. Marton, J. Kertesz, F. Moss, G. Schmera and A. Bulsara, *J.Stat.Phys.* **70**, 451 (1993)
2. Z. Gingl and L.B. Kiss, in *Proc. 7th Vilnius Conference on Fuctuation Phenomena in Phys. Systems*, ed. V. Palenskis (Vilnius Univ. Press, 1994) p. 91
3. A. Ambrózy and L.B. Kiss, in *Proc. 8th Intern. Conf. on Noise in Physical Systems*, eds. A. D'Amico and P. Mazzetti, (North-Holland 1986) p.445
4. B. Kedem, "Time Series Analysis by Higher Order Crossings," (IEEE Press, Piscataway, NJ, 1994); Ch.R. Doering, this proceedings, p. 9; T. Munakata, this proceedings, p. 196

AUTHOR INDEX

Author Index

www.ingramcontent.com/pod-product-compliance
Lightning Source LLC
LaVergne TN
LVHW051052210926
848486LV00002B/88